Differential Geometry
of Curves and Surfaces

Differential Geometry of Curves and Surfaces

Thomas Banchoff

Stephen Lovett

CRC Press
Taylor & Francis Group
Boca Raton London New York

CRC Press is an imprint of the
Taylor & Francis Group, an **informa** business

AN A K PETERS BOOK

A K Peters/CRC Press
Taylor & Francis Group
6000 Broken Sound Parkway NW, Suite 300
Boca Raton, FL 33487-2742

© 2010 by Taylor and Francis Group, LLC
A K Peters/CRC Press is an imprint of Taylor & Francis Group, an Informa business

No claim to original U.S. Government works

Printed in the United States of America on acid-free paper
10 9 8 7 6 5 4 3 2 1

International Standard Book Number: 978-1-56881-456-8 (Hardback)

Visit the Taylor & Francis Web site at
http://www.taylorandfrancis.com

and the A K Peters Web site at
http://www.akpeters.com

Contents

Preface

What is Differential Geometry?

Differential geometry studies the properties of curves, surfaces, and higher-dimensional curved spaces using tools from calculus and linear algebra. Just as the introduction of calculus expands the descriptive and predictive abilities of nearly every field of scientific study, the use of calculus in geometry brings about avenues of inquiry that extend far beyond classical geometry.

Before the advent of calculus, much of geometry consisted of proving consequences of Euclid's postulates. Even conics, which came into vogue in the physical sciences after Kepler observed that planets travel around the sun in ellipses, arise as the intersection of a double cone and a plane, two shapes that fit comfortably within the paradigm of Euclidean geometry. One cannot underestimate the impact of geometry on science, philosophy, and civilization as a whole. Not only did the geometry books in Euclid's *Elements* serve as the model for mathematical proof for over two thousand years in the Western tradition of a liberal arts education, but geometry also produced an unending flow of applications in surveying, architecture, ballistics, astronomy, astrology, and natural philosophy more generally.

The objects of study in Euclidean geometry (points, lines, planes, circles, spheres, cones, and conics) are limited in what they can describe. A boundless variety of curves and surfaces and manifolds arise naturally in areas of inquiry that employ geometry. To address these new classes of objects, various branches of mathematics brought their tools to bear on the expanding horizons of geometry, each with a different bent and set of fruitful results. Techniques from calculus and analysis led to differential geometry, pure set theoretic methods led to topology, and modern algebra contributed the field of algebraic geometry.

The types of questions one typically asks in differential geometry extend far beyond what one can ask in classical geometry and yet the former do not entirely subsume the latter. Differential geometry questions often fall into two categories: local properties, by which one means properties of a curve or surface defined in the neighborhood of a point, or global properties, which refer to properties of the curve or surface taken as a whole. As a comparison to functions of one variable, the derivative of a function f at a point a is a local property, since one only needs information about f near a, whereas the integral of f between a and b is a global property. Some of the most interesting theorems in differential geometry relate local properties to global ones. As a case in point, the celebrated Gauss-Bonnet Theorem single-handedly encapsulates many global results of curves in the plane at the same time as it proves results about spherical and hyperbolic geometry.

Using This Textbook

This book is the first in a pair of books that together are intended to bring the reader through classical differential geometry into the modern formulation of the differential geometry of manifolds. The second book in the pair, by Lovett, is entitled *Differential Geometry of Manifolds with Applications to Physics* [22]. Neither book directly relies on the other but knowledge of the content of this book is quite beneficial for [22].

On its own, the present book is intended as a textbook for a single semester undergraduate course in the differential geometry of curves and surfaces, with only vector calculus and linear algebra as prerequisites. The interactive computer graphics applets that are provided for this book can be used for computer labs, in-class illustrations, exploratory exercises, or simply as intuitive aides for the reader. Each section concludes with a collection of exercises that range from perfunctory to challenging and are suitable for daily or weekly problem sets.

However, the self-contained text, the careful introduction of concepts, the many exercises, and the interactive computer graphics also make this text well-suited for self-study. Such a reader should feel free to primarily follow the textbook and use the software as supporting material; to primarily follow the presentation in the software package and consult the textbook for definitions, theorems, and proofs; or to try to follow both with equal weight. Either way, the authors hope that the dual nature of the software applets and the classic textbook structure will offer the reader both a rigorous and intuitive introduction to the field of differential geometry.

Computer Applets

An integral part of this book is the access to online computer graphics applets that illustrate many concepts and theorems introduced in the text. Though one can explore the computer demos independently of the text, the two are intended as complementary modes of studying the same material: a visual/intuitive approach and an analytical/theoretical approach. Though the text does its best to explain the reason for various definitions and why one might be interested in such and such topic, the graphical applets can often provide motivation for certain definitions, allow the reader to explore examples further, and give a visual explanation for complicated theorems. The ability to change the choice of the parametric curve or the parametrized surface in an applet or to change other properties allows the reader to explore the concepts far beyond what a static book permits.

Any element in the text (Example, Exercise, Definition, Theorem, etc.) that has an associated applet is indicated by the symbol shown in this margin. Each demo comes with some explanatory text. The authors intended the applets to be intuitive enough so that after using just one or two (and reading the supporting text), any reader can quickly understand their functionality. However, the applets are extensible in that they are designed with considerable flexibility so that the reader can often change whether certain elements are displayed or not. Often, there are additional elements that one can display either by accessing the Controls menu on the Demo window or the Plot/Add Plot menu on any display window.

The authors encourage the reader to consult the tutorial page for the applets. All of the applet materials are available online at http://www.akpeters.com/DiffGeo.

Organization of Topics

Chapters 1 through 4 cover alternatively the local and global theory of plane and space curves. In the local theory, we introduce the fundamental notions of curvature and torsion, construct various associated objects (e.g., the evolute, osculating circle, osculating sphere), and present the fundamental theorem of plane or space curves, which is an analog of the fundamental theorem of calculus. The global theory studies how local properties (especially curvature) relate to global properties such as closedness, concavity, winding numbers, and knottedness. The topics in these chapters are particularly well suited for computer investigation. The authors know from experience in teaching how often students make discoveries on their own by being able to quickly manipulate curves and their associated objects and properties.

Chapter 5 rigorously introduces the notion of a regular surface, the type of surface on which the techniques of differential geometry are well defined. Here one first sees the tangent plane and the concept of orientability.

Chapter 6 introduces the local theory of surfaces in $\mathbb{R}^3$, focusing on the metric tensor and the Gauss map from which one defines the essential notions of principal, Gaussian curvature, and mean curvature. In addition, we introduce the study of surfaces that have Gaussian curvature or mean curvature identically 0. One cannot underestimate the importance of this chapter. Even a reader primarily interested in the advanced topic of differentiable manifolds should be comfortable with the local theory of surfaces in $\mathbb{R}^3$ because it provides many visual and tractable examples of what one generalizes in the theory of manifolds. Here again, as in Chapter 8, the use of the software applets is an invaluable aid for developing a good geometric intuition.

Chapter 7 first introduces the reader to the component notation for tensors. It then establishes the famous Theorema Egregium, the celebrated classical result that the Gaussian curvature depends only on the metric tensor. Finally, it outlines a proof for the fundamental theorem of surface theory.

Another title commonly used for Chapter 8 is *Intrinsic Geometry*. Just as Chapter 1 considers the local theory of plane curves, Chapter 8 starts with the local theory of curves on surfaces. Of particular importance in this chapter are geodesics and geodesic coordinates. The book culminates with the famous Gauss-Bonnet Theorem, both in its local and global forms, and presents applications to problems in spherical and hyperbolic geometry.

A Comment on Prerequisites

The mathematics or physics student often first encounters differential geometry at the graduate level. Furthermore, at that point, one is typically immediately exposed to the formalism of manifolds, thereby skipping the intuitive and visual foundation that informs the deeper theory. Indeed, the advent of computer graphics has added a new dimension to and renewed the interest in classical differential geometry. The authors wish to provide a book that introduces the undergraduate student to an interesting and visually stimulating mathematical subject that is accessible with only the typical calculus sequence and linear algebra as prerequisites.

In calculus courses, one typically does not study all the analysis that underlies the theorems one uses. Similarly, in keeping with the stated requirements, this textbook does not always provide all the topological and analytical background for some theorems. The reader who is interested in all the supporting material is encouraged to consult [22].

A few key results presented in this textbook rely on theorems from the theory of differential equations, but either the calculations are all spelled out or a reference to the appropriate theorem has been provided. Therefore, experience with differential equations is occasionally helpful though not at all necessary. In a few cases, the authors choose not to supply the full proofs of certain results but instead refer the reader to the more complete text [22].

A few exercises also require some skills beyond the stated pre-requisites but these are clearly marked. We have marked exercises that require ordinary differential equations with (ODE). Problems marked with (*) indicate difficulty that may be related to technical ability, insight, or length.

Notation

A quick perusal of the literature on differential geometry shows that mathematicians and physicists usually present topics in this field in very different ways. In addition, the classical and modern for-mulations of many differential geometric concepts vary significantly. Whenever different notations or modes of presentation exist for a topic (e.g., differentials, metric tensor, tensor fields), this book at-tempts to provide an explicit coordination between the notation vari-ances.

As a comment on vector notation, this book and [22] consis-tently use the following conventions. A vector or vector function in a Euclidean vector space is denoted by $\vec{v}$, $\vec{X}(t)$, or $\vec{X}(u, v)$. Of-ten γ indicates a curve parametrized by $\vec{X}(t)$ while writing $\vec{X}(t) = \vec{X}(u(t), v(t))$ indicates a curve on a surface. The unit tangent and the binormal vectors of a curve in space are written in the stan-dard notation $\vec{T}(t)$ and $\vec{B}(t)$, respectively, but the principal normal is written $\vec{P}(t)$, reserving $\vec{N}(t)$ to refer to the unit normal vector to a curve on a surface. For a plane curve, $\vec{U}(t)$ is the vector obtained by rotating $\vec{T}(t)$ by a positive quarter turn. Furthermore, we denote by $\kappa_g(t)$ the curvature of a plane curve as one identifies this curvature as the geodesic curvature in the theory of curves on surfaces.

In this book, we often work with matrices of functions. The func-tions themselves are denoted, for example, by a_{ij}, and we denote the matrix by (a_{ij}). Furthermore, it is essential to distinguish between a linear transformation between vector spaces $T : V \rightarrow W$ and its matrix with respect to given bases in V and W. Following notation that is common in current linear algebra texts, if $\mathcal{B}$ is a basis in V and $\mathcal{B}'$ is a basis in W, then we denote by $[T]_{\mathcal{B}}^{\mathcal{B}'}$ the matrix of T with respect to these bases. If the bases are understood by context, we simply write $[T]$ for the matrix associated to T.

Occasionally, there arise irreconcilable discrepancies in definitions or notations (e.g., the definition of a critical point for a function $\mathbb{R}^n \to \mathbb{R}^m$, how one defines θ and ϕ in spherical coordinates, what units to use in electromagnetism). In these instances the authors made a choice that best suits their purposes and indicated commonly used alternatives.

Acknowledgements

Thomas Banchoff

Our work at Brown University on computer visualizations in differential geometry goes back more than forty years, and I acknowledge my collaborator, computer scientist Charles Strauss, for the first fifteen years of our projects. Since 1982, an impressive collection of students have been involved in the creation and development of the software used to produce the applets for this book. The site for my 65th birthday conference, http://www.math.brown.edu/ TFBCON2003, lists dozens of them, together with descriptions of their contributions. Particular thanks for the applets connected with this book belong to David Eigen, Mark Howison, Greg Baltazar, Michael Schwarz, and Michael Morris. For his work as a student, an assistant, and now as a co-author, I am extremely grateful to Steve Lovett. Special thanks for help in the Brown University mathematics department go to Doreen Pappas, Natalie Johnson, Audrey Aguiar, and Carol Oliveira, and to Larry Larrivee for his invaluable computer assistance. Finally, I thank my wife Kathleen for all her support and encouragement.

Stephen Lovett

I would first like to thank Thomas Banchoff, my teacher, mentor, and friend. After one class, he invited me to join his team of students developing electronic books for differential geometry and multivariable calculus. Despite ultimately specializing in algebra, the exciting projects he led and his inspiring course in differential geometry instilled in me a passion for differential geometry. His ability to introduce differential geometry as a visually stimulating and mathematically interesting topic served as one of my personal motivations for writing this book.

I am grateful to the students and former colleagues at Eastern Nazarene College. In particular, I would like to acknowledge the undergraduate students who served as a sounding board for the first few drafts of this manuscript: Luke Cochran, David Constantine, Joseph Cox, Stephen Mapes, and Christopher Young. Special thanks are due to my colleagues Karl Giberson, Lee Hammerstrom, and John Free. In addition, I am indebted to Ellie Waal who helped with editing and index creation.

The continued support from my colleagues at Wheaton College made writing this book a gratifying project. In particular, I must thank Terry Perciante, Chair of the Department of Mathematics and Computer Science, for his enthusiasm and his interest. I am indebted to Dorothy Chapell, Dean of the Natural and Social Sciences, and to Stanton Jones, Provost of the College, for their encouragement and for a grant that freed up my time to finish writing. I am also grateful to Thomas VanDrunen and Darren Craig.

Finally, I cannot adequately express in just a few words how much I am grateful to my wife, Carla Favreau Lovett, and my daughter, Anne. While I was absorbed in this project, they provided a loving home, they braved the significant time commitment, and they encouraged me at every step. They also kindly put up with my occasional geometry comments such as how to see the Gaussian curvature in the reflection of "the Bean" in Chicago.

CHAPTER 1

Plane Curves: Local Properties

Just as one studies real functions of one variable before tackling multivariable calculus, so it makes sense to study curves before studying surfaces and higher-dimensional objects. This first chapter presents local properties of plane curves, where by a *local property* one means properties that are defined in a neighborhood of a point on the curve. For the sake of comparison with calculus, the derivative $f'(a)$ of a function f at a point a is a local property of the function since one only needs knowledge of $f(x)$ for $x \in (a - \varepsilon, a + \varepsilon)$ to define $f'(a)$. In contrast, the definite integral of a function over an interval is a global property since one needs knowledge of the function over the whole stated interval to calculate the integral. In contrast to this present chapter, Chapter 2 introduces global properties of plane curves.

1.1   Parametrizations

Borrowing from a physical understanding of motion in the plane, one could think about plane curves by specifying at time t the coordinates x and y of the position of a point traveling along the curve. Thus we need two functions $x(t)$ and $y(t)$. Using vector notation to locate a point on the curve, we often write $\vec{x}(t) = (x(t), y(t))$ for the above pair and call $\vec{x}(t)$ a *vector function* into $\mathbb{R}^2$. From a mathematical standpoint, t does not have to refer to time and is simply called the *parameter* of the vector function.

Example 1.1.1 (Lines). In basic analytic geometry one learns that every line in the plane can be uniquely specified by two nonequal points $\vec{p}_1 = (x_1, y_1)$ and $\vec{p}_2 = (x_2, y_2)$. The vector

$$\vec{v} = \vec{p}_2 - \vec{p}_1 = (x_2 - x_1, y_2 - y_1)$$

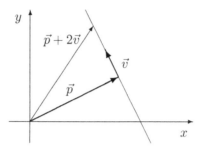

Figure 1.1. A line in the plane.

is called a *direction vector* of the line because it points along the same orientation in the plane as the line does. Any other direction vector of the line is a nonzero multiple of $\vec{v}$. Then every point on the line can be written with a position vector $\vec{p}_1 + t\vec{v}$ for some $t \in \mathbb{R}$. Therefore, we find that a line can also be defined by providing a point and a direction vector.

Using the coordinates of vectors, given a point $\vec{p} = (x_0, y_0)$ and a direction vector $\vec{v} = (v_1, v_2)$, a line through $\vec{p}$ in the direction of $\vec{v}$ is the image of the following vector function:

$$\vec{x}(t) = \vec{p} + t\vec{v} = (x_0 + v_1 t, y_0 + v_2 t).$$

Example 1.1.2 (Circles). The pair of functions

$$\vec{x}(t) = (R \cos t + a, R \sin t + b)$$

trace out a circle of radius R about the point (a, b). To see this, note that by the definition of the $\sin t$ and $\cos t$ functions, $(\cos t, \sin t)$ are the coordinates of the point on the unit circle that is also on the ray out of the origin that makes an angle t with the positive x-axis. Thus,

$$\vec{x}_1(t) = (\cos t, \sin t), \qquad \text{with } t \in [0, 2\pi],$$

traces out the unit circle. Multiplying both coordinate functions by R stretches the circle out by a factor of R away from the origin. Thus, the vector function

$$\vec{x}_2(t) = (R \cos t, R \sin t), \qquad \text{with } t \in [0, 2\pi],$$

has as its image the circle of radius R centered at the origin. Notice also that by writing $\vec{x}_2(t) = (x(t), y(t))$, we deduce that $x(t)^2 + y(t)^2 = R^2$ for all t, which is the algebraic equation of the circle. In order to obtain a vector function that traces out a circle centered at the point (a, b), we must simply translate $\vec{x}_2$ by the vector (a, b). This is simply vector addition, and so we get

$$\vec{x}(t) = (R\cos t + a, R\sin t + b), \qquad \text{with } t \in [0, 2\pi].$$

Two different vector functions can have the same image in $\mathbb{R}^2$. For example,

$$\vec{x}(t) = (\cos \omega t, \sin \omega t), \qquad \text{with } t \in [0, 2\pi],$$

also has the unit circle as its image. However, referring to vocabulary in physics, this latter vector function corresponds to a point moving around the unit circle at an angular velocity of ω.

When trying to establish a suitable mathematical definition of what one usually thinks of as a curve, one does not wish to consider as a curve points that jump around or pieces of segments. We would like to think of a curve as unbroken in some sense. In calculus, one introduces the notion of continuity to describe functions without "jumps" or holes, but one must exercise a little care in carrying over the notion of continuity to vector functions. More generally, we need to define the notion of a limit of a vector function as the parameter t approaches a fixed value. First, however, we remind the reader of the Euclidean distance formula.

Definition 1.1.3. Let $\vec{v}$ be a vector in $\mathbb{R}^n$ with coordinates $\vec{v} = (v_1, v_2, \ldots, v_n)$ in the standard basis. The (Euclidean) length of $\vec{v}$ is given by

$$\|\vec{v}\| = \sqrt{\vec{v} \cdot \vec{v}} = \sqrt{v_1^2 + v_2^2 + \cdots + v_n^2}.$$

If p and q are two points in $\mathbb{R}^n$ with coordinates given by vectors $\vec{v}$ and $\vec{w}$, then the Euclidean distance between p and q is $\|\vec{w} - \vec{v}\|$.

Definition 1.1.4. Let $\vec{x}$ be a vector function from a subset of $\mathbb{R}$ into $\mathbb{R}^n$. We say that the limit of $\vec{x}(t)$ as t approaches a is a vector $\vec{w}$, and we write

$$\lim_{t \to a} \vec{x}(t) = \vec{w},$$

if for all $\varepsilon > 0$ there exists a $\delta > 0$ such that $0 < |t - a| < \delta$ implies $\|\vec{x}(t) - \vec{w}\| < \varepsilon$.

Definition 1.1.5. Let I be an open interval of $\mathbb{R}$, let $a \in I$, and let $\vec{x} : I \to \mathbb{R}^2$ be a vector function. We say that $\vec{x}(t)$ is continuous at a if the limit as t approaches a of $\vec{x}(t)$ exists and

$$\lim_{t \to a} \vec{x}(t) = \vec{x}(a).$$

The above definitions mirror the usual definition of a limit of a real function but must use the length of a vector difference to discuss the proximity between $\vec{x}(t)$ and a fixed vector $\vec{w}$. Though at the outset this definition appears more complicated than the usual definition of a limit of a real function, the following proposition shows that it is not.

Proposition 1.1.6. *Let $\vec{x}$ be a vector function from a subset of $\mathbb{R}$ into $\mathbb{R}^n$ that is defined over an interval containing a, though perhaps not at a itself. Suppose in coordinates we have $\vec{x}(t) = (x(t), y(t))$ wherever $\vec{x}$ is defined. If $\vec{w} = (w_1, w_2)$, then $\lim_{t \to a} \vec{x}(t) = \vec{w}$ if and only if $\lim_{t \to a} x(t) = w_1$ and $\lim_{t \to a} y(t) = w_2$.*

Proof: Suppose first that $\lim_{t \to a} \vec{x}(t) = \vec{w}$. Let $\varepsilon > 0$ be arbitrary and let $\delta > 0$ satisfy the definition of the limit of the vector function. Note that $|x(t) - w_1| < \|\vec{x}(t) - \vec{w}\|$ and that $|y(t) - w_2| < \|\vec{x}(t) - \vec{w}\|$. Hence, $0 < |t - a| < \delta$ implies $|x(t) - w_1| < \varepsilon$ and $|y(t) - w_2| < \varepsilon$. Thus, $\lim_{t \to a} x(t) = w_1$ and $\lim_{t \to a} y(t) = w_2$.

Conversely, suppose that $\lim_{t \to a} x(t) = w_1$ and $\lim_{t \to a} y(t) = w_2$. Let $\varepsilon > 0$ be an arbitrary positive real number. By definition, there exist δ_1 and δ_2 such that $0 < |t - a| < \delta_1$ implies $|x(t) - w_1| < \varepsilon/\sqrt{2}$ and $0 < |t - a| < \delta_2$ implies $|y(t) - w_2| < \varepsilon/\sqrt{2}$. Taking $\delta = \min(\delta_1, \delta_2)$ we see that $0 < |t - a| < \delta$ implies that

$$\|\vec{x}(t) - \vec{w}\| = \sqrt{|x(t) - w_1|^2 + |y(t) - w_2|^2} < \sqrt{\frac{\varepsilon^2}{2} + \frac{\varepsilon^2}{2}} = \varepsilon.$$

This finishes the proof of the proposition. $\qquad\square$

Corollary 1.1.7. *Let I be an open interval of $\mathbb{R}$, let $a \in I$, and consider a vector function $\vec{x} : I \to \mathbb{R}^2$ with $\vec{x}(t) = (x(t), y(t))$. Then $\vec{x}(t)$ is continuous at $t = a$ if and only if $x(t)$ and $y(t)$ are continuous at $t = a$.*

Definition 1.1.5 and Corollary 1.1.7 provide the mathematical framework for what one usually thinks of as a curve in physical intuition. This motivates the following definition.

Definition 1.1.8. Let I be an interval of $\mathbb{R}$. A *parametrized curve* (or *parametric curve*) in the plane is a continuous function $\vec{x} : I \to \mathbb{R}^2$. If we write $\vec{x}(t) = (x(t), y(t))$, then the functions $x : I \to \mathbb{R}$ and $y : I \to \mathbb{R}$ are called the *coordinate functions* or *parametric equations* of the parametrized curve. We call the *locus* of $\vec{x}(t)$ the image of $\vec{x}(t)$ as a subset of $\mathbb{R}^2$.

The following examples begin to provide a library of parametric curves and illustrate how to construct parametric curves to describe a particular shape or trajectory.

Example 1.1.9 (Graphs of Functions). The graph of a continuous function $f : [a, b] \to \mathbb{R}$ over an interval $[a, b]$ is a parametric curve. In order to view the graph of a continuous function as a parametrized curve, we use the coordinate functions $\vec{x}(t) = (t, f(t))$, with $t \in [a, b]$.

Example 1.1.10 (Circles Revisited). Another parametrization for the unit circle (sometimes used in number theory) is

$$\vec{x} = (x(t), y(t)) = \left(\frac{1 - t^2}{1 + t^2}, \frac{2t}{1 + t^2} \right) \qquad \text{for } t \in \mathbb{R}. \qquad (1.1)$$

It is easy to see that for all $t \in \mathbb{R}$, $x(t)^2 + y(t)^2 = 1$, which means that the locus of $\vec{x}$ is on the unit circle. However, this parametrization does not trace out the entire circle as it misses the point $(-1, 0)$. We leave it as an exercise to determine a geometric interpretation of the parameter t and to show that

$$\lim_{t \to \infty} \vec{x} = \lim_{t \to -\infty} \vec{x} = (-1, 0).$$

Example 1.1.11 (Ellipses). Without repeating all the reasoning of the previous exercise, it is not hard to see that

$$\vec{x}(t) = (a \cos t, b \sin t)$$

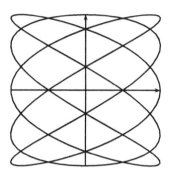

Figure 1.2. A Lissajous figure.

provides a parametrization for the ellipse centered at the origin with axes along the x- and y-axes, with respective half-axes of length a and b. Note that these coordinate functions do indeed satisfy

$$\frac{x(t)^2}{a^2} + \frac{y(t)^2}{b^2} = 1 \qquad \text{for all } t.$$

Example 1.1.12 (Lissajous Figures). It is sometimes amusing to see how $\cos t$ and $\sin t$ relate to each other if we change their respective periods. Lissajous figures, which arise in the context of electronics, are curves parametrized by

$$\vec{x}(t) = (\cos mt, \sin nt),$$

where m and n are positive integers. See Figure 1.2 for an example of a Lissajous figure with $m = 5$ and $n = 3$.

Example 1.1.13 (Cycloids). We can think of a regular cycloid as the locus traced out by a point of light affixed to a bicycle tire as the bicycle rolls forward. We can establish a parametrization of such a curve as follows.

Assume the wheel of radius a begins with its center at $(0, a)$ so that the part of the wheel touching the x-axis is at the origin. We view the wheel as rolling forward on the positive x-axis. As the wheel rolls, the position of the center of the wheel is $\vec{f}(t) = (at, a)$, where t is the angle measuring how much (many times) the wheel has turned since it started. At the same time, the light—at a distance a from

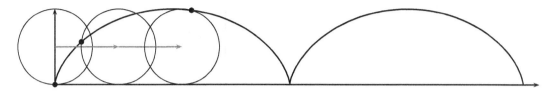

Figure 1.3. Cycloid.

the center of the wheel and first positioned straight down from the
center of the wheel—rotates in a clockwise motion around the center
of the wheel. The motion of the light with respect to the center of
the wheel is

$$\vec{g}(t) = \left(a \cos \left(-t - \frac{\pi}{2} \right), a \sin \left(-t - \frac{\pi}{2} \right) \right) = (-a \sin t, -a \cos t).$$

The locus of the cycloid is the vector function that is the sum of $\vec{f}(t)$
and $\vec{g}(t)$. Thus, a parametrization for the cycloid is

$$\vec{x}(t) = (at - a \sin t, a - a \cos t).$$

One can point out that reflectors on bicycle wheels are usually
not attached directly on the tire but on a spoke of the wheel. We can
easily modify the above discussion to obtain the relevant parametric
equations for when the point of light is located at a distance b from
the center of the rolling wheel. One obtains

$$\vec{x}(t) = (at - b \sin t, a - b \cos t).$$

If $0 < b < a$, one obtains the curve of a realistic bicycle tire reflector,
and this locus is called a *curtate cycloid*. In contrast, the locus
obtained by letting $b > a$ is called a *prolate cycloid*.

Example 1.1.14 (Heart Curve). Arguably the most popular curve around
Valentine's Day is the heart. Here are some parametric equations
that trace out such a curve:

$$\vec{x}(t) = ((1 - \cos^2 t) \sin t, (1 - \cos^3 t) \cos t).$$

We encourage the reader to visit this example in the accompanying
software and to explore ways of modifying these equations to suit
their purposes.

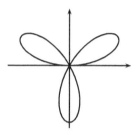

Figure 1.4. Three-leaf flower.

Example 1.1.15 (Polar Functions). Functions in polar coordinates are usually given in terms of the radius r as a function of the angle θ by $r = f(\theta)$. The graphs of such functions can be written as parametrized curves. Recall the coordinate transformation

$$\begin{cases} x = r \cos \theta, \\ y = r \sin \theta. \end{cases} \tag{1.2}$$

Then take θ as the parameter t, and the parametric equations for the graph of $r = f(\theta)$ are

$$\vec{x}(t) = (f(t) \cos t, f(t) \sin t).$$

As an example, the polar function $r = \sin 3\theta$ traces out a curve that resembles a three-leaf flower. As a parametric curve, it is given by

$$\vec{x}(t) = (\sin 3t \cos t, \sin 3t \sin t).$$

Example 1.1.16 (Cardioid). Another common polar function is the cardioid, which is the locus of $r = 1 - \cos \theta$. In parametric equations, we have

$$\vec{x}(t) = ((1 - \cos t) \cos t, (1 - \cos t) \sin t).$$

As mentioned in Example 1.1.2, the set of points $\mathcal{C} = \{\vec{x}(t) \mid t \in I\}$ as a subset of $\mathbb{R}^2$ *does not* depend uniquely on the functions $x(t)$ and $y(t)$. In fact, throughout this text, we make a careful distinction between the notion of a parametrized curve as defined above and the notion of a curve, eventually defined as a one-dimensional manifold. (See [22, Chapter 11].) For the purpose of intuition, one may equate the term "curve" with the locus or, equivalently, the image of a parametrized curve.

Definition 1.1.17. Given a parametrized curve $\vec{x} : I \to \mathbb{R}^2$ and any continuous functions g from an interval J onto the interval I, we can produce a new vector function $\vec{\xi} : J \to \mathbb{R}^2$ defined by $\vec{\xi} = \vec{x} \circ g$. The image of $\vec{\xi}$ is again the set $\mathcal{C}$, and $\vec{\xi} = \vec{x} \circ g$ is called a *reparametrization* of $\vec{x}$.

If g is not onto I, then the image of $\vec{\xi}$ may be a proper subset of $\mathcal{C}$. In this case, we usually do not call $\vec{\xi}$ a reparametrization as it does not trace out the same locus of $\vec{x}$.

Problems

1.1.1. An epicycloid is defined as the locus of a point on the edge of a circle of radius b as this circle rolls on the outside of a fixed circle of radius a. Supposing that at $t = 0$, the moving point is located at $(a, 0)$. Prove that the following are parametric equations for an epicycloid:

$$\vec{x}(t) = \left((a+b)\cos t - b \cos\left(\frac{a+b}{b} t \right), (a+b)\sin t - b\sin\left(\frac{a+b}{b} t \right) \right).$$

1.1.2. A hypocycloid is defined as the locus of a point on the edge of a circle of radius b as this circle rolls on the inside of a fixed circle of radius a. Assuming that $a > b$ and that at $t = 0$, the moving point is located at $(a, 0)$, find parametric equations for a hypocycloid.

1.1.3. Prove that the shortest (orthogonal) distance between a point (x_0, y_0) and a line with equation $ax + by + c = 0$ is

$$d = \frac{|ax_0 + by_0 + c|}{\sqrt{a^2 + b^2}}.$$

[Hint: Minimize the distance function between (x_0, y_0) and a generic point $\vec{x}(t)$ on the line.]

1.1.4. Let $\vec{p}$ be a fixed point, and let $\vec{l}(t) = \vec{a}t + \vec{b}$ be the parametric equations of a line. Prove that the distance between $\vec{p}$ and the line is

$$\left((\vec{p} - \vec{b}) \cdot \vec{a} \right)^2 \left(1 - \frac{2}{\|\vec{a}\|} \right)^2 + \|\vec{p} - \vec{b}\|^2 = \|\vec{a} \times (\vec{p} - \vec{b})\|^2,$$

where for vectors $\vec{v} = (v_1, v_2)$ and $\vec{w} = (w_1, w_2)$ in the plane, we call $\vec{v} \times \vec{w} = (0, 0, v_1 w_2 - v_2 w_1)$, which is the cross product between $\vec{v}$ and $\vec{w}$ when viewed as vectors in $\mathbb{R}^3$.

1.1.5. Equation (1.1) gave the following parametric equations for a circle:

$$\vec{x} = (x(t), y(t)) = \left(\frac{1 - t^2}{1 + t^2}, \frac{2t}{1 + t^2} \right), \qquad \text{for } t \in \mathbb{R}.$$

Prove that the parameter t is equal to $\tan \theta$, where θ is the angle shown in the following picture of the unit circle.

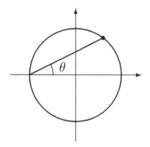

1.1.6. Find the closest point to $(2, 2)$ on the curve $\vec{x}(t) = (t, t^2)$.

1.2   Position, Velocity, and Acceleration

In physics applications, one interprets the vector function $\vec{x}(t) = (x(t), y(t))$ as providing the location along a curve at time t in reference to some fixed frame, where by *frame* we mean a (usually orthonormal) basis attached to a fixed origin. The point $O = (0, 0)$ along with the basis $\{\vec{i}, \vec{j}\}$, where

$$\vec{i} = (1, 0) \qquad \text{and} \quad \vec{j} = (0, 1),$$

form the standard reference frame. One calls this vector function $\vec{x}(t)$ the *position vector*. When one uses the standard reference frame, it is not uncommon to write

$$\vec{x}(t) = x(t)\vec{i} + y(t)\vec{j}.$$

Directly imitating Newton's approach to finding the slope of a curve at a certain point, we wish to ask the question, "What is the direction and rate of change of a curve $\vec{x}$ at point t_0?" If we look at two points on the curve, say $\vec{x}(t_0)$ and $\vec{x}(t_1)$, the direction of motion is described by the vector $\vec{x}(t_1) - \vec{x}(t_0)$, while to specify the rate of change in position, we would need to scale this vector by a factor of $t_1 - t_0$, so the vector rate of change between those two points is

$$\frac{1}{t_1 - t_0}\left(\vec{x}(t_1) - \vec{x}(t_0)\right),$$

To get the instantaneous rate of change of the curve at the point t_0, we need to calculate (if it exists) the following limit:

$$\vec{x}'(t_0) = \lim_{h \to 0} \frac{1}{h}\left(\vec{x}(t_0 + h) - \vec{x}(t_0)\right). \tag{1.3}$$

According to Proposition 1.1.6, if $\vec{x}(t) = (x(t), y(t))$, the limit in Equation (1.3) exists if and only if $x(t)$ and $y(t)$ are both differentiable at t_0. Furthermore, if this limit exists, then $\vec{x}'(t_0) = (x'(t_0), y'(t_0))$. This leads us to the following definition.

Definition 1.2.1. Let $\vec{x} : I \to \mathbb{R}^2$ be a vector function with coordinates $\vec{x}(t) = (x(t), y(t))$. We say that $\vec{x}$ is differentiable at $t = t_0$ if $x(t)$ and $y(t)$ are both differentiable at t_0. If J is the common domain to $x'(t)$ and $y'(t)$, then we define the derivative of the vector function $\vec{x}$ as the new vector function $\vec{x}' : J \to \mathbb{R}^2$ defined by $\vec{x}'(t) = (x'(t), y'(t))$.

If $x(t)$ and $y(t)$ are both differentiable functions on their domain, the derivative vector function $\vec{x}'(t) = (x'(t), y'(t))$ is called the *velocity vector*. Mathematically, this is just another vector function and traces out another curve when placed in the standard reference frame. However, because it illustrates the direction of motion along the curve, one often visualizes the velocity vector corresponding to $t = t_0$ as based at the point $\vec{x}(t_0)$ on the curve.

Following physics language, we call the second derivative of the vector function $\vec{x}''(t) = (x''(t), y''(t))$ the *acceleration vector* related to $\vec{x}(t)$.

In general, our calculations often require that our vector functions be differentiated at least once and sometimes more. Consequently, when establishing theorems, we like to succinctly describe the largest class of functions for which a particular result holds.

Definition 1.2.2. Let I be an interval of $\mathbb{R}$, and let $\vec{x} : I \to \mathbb{R}^2$ be a vector function. We say that $\vec{x}$ is of class C^r on I, or write $\vec{x} \in C^r(I)$, if the rth derivative of $\vec{x}$ exists and is continuous on I. We denote by C^∞ the class of functions that have derivatives of all orders on I. In addition, we also denote by C^ω the subclass of C^∞ functions that are *real analytic* on I, i.e., functions $\vec{x}$ such that, at each point $a \in I$, $\vec{x}$ is equal to its Taylor series centered at a over some open interval $(a - \varepsilon, a + \varepsilon)$.

To say that a function is of class C^0 over the interval I means that it is continuous. By basic theorems in calculus, the condition that a function be of a certain class is an increasingly restrictive condition. In other words, as sets of functions defined over the same interval I, the classes are nested according to

$$C^0 \supset C^1 \supset C^2 \supset \cdots \supset C^\infty \supset C^\omega.$$

Proposition 1.2.3. *Let $\vec{v}(t)$ and $\vec{w}(t)$ be vector functions defined and differentiable over an interval $I \subset \mathbb{R}$. Then the following hold:*

1. *If $\vec{x}(t) = c\vec{v}(t)$, where $c \in \mathbb{R}$, then $\vec{x}'(t) = c\vec{v}'(t)$.*

2. *If $\vec{x}(t) = c(t)\vec{v}(t)$, where $c : I \to \mathbb{R}$ is a real function, then*

$$\vec{x}'(t) = c'(t)\vec{v}(t) + c(t)\vec{v}'(t).$$

3. *If $\vec{x}(t) = \vec{v}(t) + \vec{w}(t)$, then*

$$\vec{x}'(t) = \vec{v}'(t) + \vec{w}'(t).$$

4. *If $\vec{x}(t) = \vec{v}(f(t))$ is a vector function and $f : J \to I$ is a real function into I, then*

$$\vec{x}'(t) = f'(t)\vec{v}'(f(t)).$$

5. *If $f(t) = \vec{v}(t) \cdot \vec{w}(t)$ is the dot product between $\vec{v}(t)$ and $\vec{w}(t)$, then*

$$f'(t) = \vec{v}'(t) \cdot \vec{w}(t) + \vec{v}(t) \cdot \vec{w}'(t).$$

Example 1.2.4. Consider the spiral defined by $\vec{x}(t) = (t \cos t, t \sin t)$
for $t \geq 0$. The velocity and acceleration vectors are

$$\vec{x}'(t) = (\cos t - t \sin t, \sin t + t \cos t),$$
$$\vec{x}''(t) = (-2 \sin t - t \cos t, 2 \cos t - t \sin t).$$

If we wish to calculate the angle between $\vec{x}$ and $\vec{x}'$ as a function of
time, we use the dot product method. In this example, we calculate

$$\|\vec{x}(t)\| = \sqrt{t^2 \sin^2 t + t^2 \cos^2 t} = |t|,$$
$$\|\vec{x}'(t)\| = \sqrt{(\cos t - t \sin t)^2 + (\sin t + t \cos t)^2} = \sqrt{1 + t^2},$$
$$\vec{x}(t) \cdot \vec{x}'(t) = t \cos^2 t - t^2 \cos t \sin t + t \sin^2 t + t^2 \cos t \sin t = t.$$

Thus, the angle $\theta(t)$ between $\vec{x}(t)$ and $\vec{x}'(t)$ is defined for all $t \neq 0$
and is equal to

$$\theta(t) = \cos^{-1} \left(\frac{\vec{x}(t) \cdot \vec{x}'(t)}{\|\vec{x}(t)\| \, \|\vec{x}'(t)\|} \right) = \cos^{-1} \left(\frac{t}{|t|\sqrt{1 + t^2}} \right).$$

Let $\mathcal{C}$ be the arc traced out by a vector function $\vec{x}(t)$ for t ranging
between a and b. Then one approximates the length of the arc $l(\mathcal{C})$
with the Riemann sum

$$l(\mathcal{C}) \approx \sum_{i=1}^{n} \|\vec{x}(t_i) - \vec{x}(t_{i-1})\| \approx \sum_{i=1}^{n} \|\vec{x}'(t_i)\| \Delta t.$$

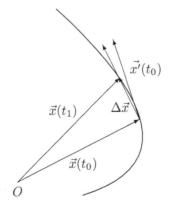

Figure 1.5. Arc length segment.

(See Figure 1.5 as an illustration for the approximation $\|\vec{x}(t_i) - \vec{x}(t_{i-1})\| \approx \|\vec{x}'(t_i)\|\Delta t$.) Taking the limit of this Riemann sum, we obtain the following formula for the arc length of $\mathcal{C}$:

$$l = \int_a^b \sqrt{(x'(t))^2 + (y'(t))^2}\, dt. \tag{1.4}$$

In light of Equation (1.4), we define the arc length function $s : [a, b] \to \mathbb{R}$ related to $\vec{x}(t)$ as the arc length along $\mathcal{C}$ over the interval $[a, t]$. Thus,

$$s(t) = \int_a^t \|\vec{x}'(u)\|\, du = \int_a^t \sqrt{(x'(u))^2 + (y'(u))^2}\, du. \tag{1.5}$$

By the Fundamental Theorem of Calculus, we then have

$$s'(t) = \|\vec{x}'(t)\| = \sqrt{(x'(t))^2 + (y'(t))^2},$$

which, still following the vocabulary from the trajectory of a moving particle, we call the *speed function* of $\vec{x}(t)$.

Example 1.2.5. Consider the parametrized curve defined by $\vec{x}(t) = (t^2, t^3)$. The velocity vector is $\vec{x}'(t) = (2t, 3t^2)$, and thus the arc length function from $t = 0$ is

$$s(t) = \int_0^t \|\vec{x}'(u)\|\, du = \int_0^t \sqrt{4u^2 + 9u^4}\, du = \int_0^t 3|u|\sqrt{\frac{4}{9} + u^2}\, du$$
$$= \operatorname{sign}(t)\left(\frac{4}{9} + t^2\right)^{3/2},$$

where $\operatorname{sign}(t) = 1$ if $t > 0$ and -1 if $t < 0$. The length of $\vec{x}(t)$ between $t = 0$ and $t = 1$ is

$$s(1) - s(0) = \frac{1}{27}\left(13^{3/2} - 8\right).$$

The speed of a vector function $\vec{x}(t)$ at a point is not a geometrical quantity since it can change under any reparametrization, which does not change the locus of the curve. If $t_0 = f(u_0)$ where f is a differentiable function at u_0, then $\vec{x}(f(u_0)) = \vec{x}(t_0)$ and

$$\left\|\frac{d}{du}\vec{x}(f(u_0))\right\| = \|f'(u_0)\vec{x}'(f(u_0))\| = |f'(u_0)|\,\|\vec{x}'(t_0)\| = |f'(u_0)|s'(t_0).$$

However, up to a change in sign, the direction of $\vec{x}'(t)$, by which we mean the unit vector associated with $\vec{x}'(t)$, does not change under reparametrization. More precisely, if $\vec{\xi} = \vec{x} \circ f$, at any parameter value t where $f'(t) \neq 0$ and $\vec{x}'(f(t)) \neq \vec{0}$, we have

$$\frac{\vec{\xi}'(t)}{\|\vec{\xi}'(t)\|} = \frac{f'(t)}{|f'(t)|} \frac{\vec{x}'(f(t))}{\|\vec{x}'(f(t))\|}, \tag{1.6}$$

where $f'(t)/|f'(t)|$ is $+1$ or -1. It bears repeating that in order for the right-hand side to be well defined, one needs $f'(t) \neq 0$. Thus, for a parametrized curve $\vec{x} : I \to \mathbb{R}^2$ and a real function f from an interval J onto I, we call $\vec{\xi} = \vec{x} \circ f$ a *regular reparametrization* if for all $t \in J$, $f'(t)$ is well defined and never 0.

Equation (1.6) shows that the unit vector $\vec{x}'(t)/\|\vec{x}'(t)\|$ is invariant under a regular reparametrization up to a possible change in sign. This leads us to the following important definition.

Definition 1.2.6. Let $\vec{x} : I \to \mathbb{R}^2$ be a plane parametric curve. A point $t_0 \in I$ is called a *critical point* of $\vec{x}(t)$ if $\vec{x}(t)$ is not differentiable at t_0 or if $\vec{x}'(t_0) = \vec{0}$. If t_0 is a critical point, then $\vec{x}(t_0)$ is called a *critical value*. A point $t = t_0$ that is not critical is called a *regular point*. A parametrized curve $\vec{x} : I \to \mathbb{R}^2$ is called *regular* if it is of class C^1 (i.e., continuously differentiable) and $\vec{x}'(t) \neq \vec{0}$ for all $t \in I$. Finally, if t is a regular point of $\vec{x}$, we define the *unit tangent vector* $\vec{T}(t)$ as

$$\vec{T}(t) = \frac{\vec{x}'(t)}{\|\vec{x}'(t)\|}.$$

At any point of a curve where $\vec{T}(t)$ is defined (i.e., at any regular point), the velocity vector can be written as

$$\vec{x}'(t) = s'(t)\vec{T}(t). \tag{1.7}$$

This expresses the velocity vector as the product of its magnitude (speed) and direction (unit tangent vector). In most differential geometry texts, authors simplify their formulas by *reparametrizing by arc length*. From the perspective of coordinates, this means picking an "origin" O on the curve C occurring at some fixed $t = t_0$ and using the arc length along C between O and any other point P as the parameter to locate P on C.

The habit of reparametrizing by arc length has benefits and drawbacks. First, along a curve parametrized by arc length, the speed function s' is identically 1 and the velocity vector is exactly the tangent vector:

$$\vec{x}'(s) = \vec{T}(s).$$

This formulation often simplifies proofs and difficult calculations. However, in practice, most curves do not admit a simple formula for their arc length function, let alone a formula that can be written using elementary functions. Therefore, reparametrizing by arc length is often rather intractable when doing calculations on specific examples.

The notion of "regular" in many ways mirrors geometric properties of continuously differentiable single-variable functions. In particular, if a parametrized curve $\vec{x}(t)$ is regular at t_0, then locally the curve looks linear. As we will see again in Chapter 3, with the formalism of vector functions it becomes particularly easy to express the equation of the tangent line to a curve at a point in any number of dimensions. If $\vec{x}(t)$ is a curve and t_0 is not a critical point for the curve, then the equation $\vec{L}_{t_0}(t)$ of the tangent line at t_0 is

$$\vec{L}_{t_0}(t) = \vec{x}(t_0) + t\vec{x}'(t_0), \qquad \text{with } t \in \mathbb{R},$$

or alternatively

$$\vec{L}_{t_0}(u) = \vec{x}(t_0) + u\vec{T}(t_0), \qquad \text{with } u \in \mathbb{R}. \qquad (1.8)$$

On the other hand, if $t_0 \in I$ is a critical point for the curve $\vec{x}(t)$, the curve may or may not have a tangent line at $t = t_0$. If the following one-sided limits exist, we can define two unit tangent vectors at $\vec{x}(t_0)$:

$$\vec{T}(t_0^+) = \lim_{t \to t_0^+} \vec{T}(t) \qquad \text{and} \qquad \vec{T}(t_0^-) = \lim_{t \to t_0^-} \vec{T}(t).$$

Consequently, we can determine the angle the curve makes with itself at the corner t_0 as

$$\alpha_0 = \cos^{-1}\left(\vec{T}(t_0^+) \cdot \vec{T}(t_0^-)\right).$$

To be more precise, the angle from $\vec{T}(t_0^-)$ to $\vec{T}(t_0^+)$ is the exterior angle of the curve at the corner $t = t_0$. One can only define a tangent line to $\vec{x}(t)$ at $t = t_0$ if $\vec{T}(t_0^-) = \vec{T}(t_0^+)$.

Example 1.2.7. Let $\vec{x}(t) = (t^2, t^3)$. We calculate that $\vec{x}'(t) = (2t, 3t^2)$, and therefore $t = 0$ is a critical point because $\vec{x}'(0) = \vec{0}$. However, if $t \neq 0$, the unit tangent vector is

$$\vec{T}(t) = \frac{1}{\sqrt{4t^2 + 9t^4}}(2t, 3t^2) = \frac{t}{|t|\sqrt{4 + 9t^2}}(2, 3t).$$

Then the right-hand and left-hand side unit tangent vectors are

$$\vec{T}(0^-) = (-1, 0) \qquad \text{and} \qquad \vec{T}(0^+) = (1, 0).$$

These calculations indicate that as t approaches 0, $\vec{x}(t)$ lies in the fourth quadrant but approaches $(0, 0)$ in the horizontal direction $(-1, 0)$. From an intuitive standpoint, we could say that $\vec{x}$ stops at $t = 0$, changes direction, and moves away from the origin in the direction $(1, 0)$ and remaining in the first quadrant.

Example 1.2.8. Let $\vec{x}(t) = (t, |\tan t|)$. We calculate that $\vec{x}'(t) = (1, \text{sign}(t) \sec^2 t)$ and deduce that $t = 0$ is a critical point because $\text{sign}(t)$ is not defined at 0, and hence $\vec{x}$ is not differentiable there. However, we also find that

$$\vec{T}(0^-) = \frac{1}{\sqrt{2}}(1, -1) \qquad \text{and} \qquad \vec{T}(0^+) = \frac{1}{\sqrt{2}}(1, 1).$$

Thus,

$$\vec{T}(0^-) \cdot \vec{T}(0^+) = 0,$$

which shows that $\vec{x}$ makes a right angle with itself at $t = 0$. However, one can tell from the explicit values for $\vec{T}(t_0^-)$ and $\vec{T}(t_0^+)$ that the exterior angle of the curve at $t = 0$ is $\frac{\pi}{2}$.

If t_0 is a critical point, it may still happen that

$$\lim_{t \to t_0} \vec{T}(t)$$

exists, namely when $\vec{T}(t_0^+) = \vec{T}(t_0^-)$, which also means $\alpha_0 = 0$. Then through an abuse of language, we can still talk about the unit tangent vector at that point. As an example of this possibility, consider the curve $\vec{x}(t) = (t^3, t^4)$. We can quickly calculate that

$$\lim_{t \to 0} \vec{T}(t) = (1, 0) = \vec{i}.$$

When this limit exists, even though t_0 is a critical point, one can use Equation (1.8) as parametric equations for the tangent line at t_0, replacing $\vec{T}(t_0)$ in Equation (1.8) with $\lim_{t \to t_0} \vec{T}(t)$.

In geometry, physics and other applications, one must sometimes integrate a function along a curve. To this end, one uses line integrals of scalar functions or vector functions, depending on the particular problem. Since we will make use of them, we remind the reader of the notation for such integrals. Let $\mathcal{C}$ be a curve parametrized by $\vec{x}(t)$ over the interval $[a, b]$. Let $f : \mathbb{R}^2 \to \mathbb{R}$ be a real (scalar) function and $\vec{F} : \mathbb{R}^2 \to \mathbb{R}^2$ a vector field in the plane. Then the line integral of f and of $\vec{F}$ over the curve $\mathcal{C}$ are respectively

$$\int_{\mathcal{C}} f \, ds \qquad \text{means} \qquad \int_a^b f(x(t), y(t)) \sqrt{(x'(t))^2 + (y'(t))^2} dt,$$

$$\int_{\mathcal{C}} \vec{F} \cdot d\vec{s} \qquad \text{means} \qquad \int_a^b \vec{F}(x(t), y(t)) \cdot \vec{x}'(t) \, dt.$$

Problems

1.2.1. Calculate the velocity, the acceleration, the speed, and, where defined, the unit tangent vector function of the following parametric curves:

  (a) The circle $\vec{x}(t) = (R \cos \omega t, R \sin \omega t)$.

  (b) The circle as parametrized in Equation (1.1) in Example 1.1.10.

  (c) The epicycloids defined in Problem 1.1.1.

1.2.2. What can be said about a parametrized curve $\vec{x}(t)$ that has the property that $\vec{x}''(t)$ is identically 0?

1.2.3. Find the arc length function along the parabola $y = x^2$, using as the origin $s = 0$.

1.2.4. (*) Consider the cycloid introduced in Example 1.1.13 given by $\vec{x}(t) = (t - \sin t, 1 - \cos t)$. Prove that the path taken by a point on the edge of a rolling wheel of radius 1 during one rotation has length 8.

1.2.5. Calculate the arc length function of the curve $\vec{x}(t) = (t^2, \ln t)$, defined for $t > 0$.

1.2.6. Let $\vec{x} : I \to \mathbb{R}^2$ be a parametrized curve, and let $\vec{p}$ be a fixed point. Suppose that the closest point on the curve $\vec{x}$ to $\vec{p}$ occurs at $t = t_0$, which is neither of the ends of I. Prove that the line between $\vec{p}$ and the point closest to $\vec{p}$, namely $\vec{x}(t_0)$, is perpendicular to the curve $\vec{x}$ at $t = t_0$.

1.2.7. We say that a plane curve C parametrized by $\vec{x} : I \to \mathbb{R}^2$ intersects itself at a point p if there exist $u \neq t$ in I such that $\vec{x}(t) = \vec{x}(u) = p$. Consider the parametric curve $\vec{x}(t) = (t^2, t^3 - t)$ for $t \in \mathbb{R}$. Find the point(s) of self-intersection. Also determine the angle at which the curve intersects itself at those point(s) by finding the angle between the unit tangent vectors corresponding to the distinct parameters.

1.2.8. Prove the differentiation formulas in Proposition 1.2.3.

1.2.9. Prove that if $\vec{x}(t)$ is a curve that satisfies $\vec{x} \cdot \vec{x}' = 0$ for all values of t, then $\vec{x}$ is a circle. [Hint: Use $\|\vec{x}\|^2 = \vec{x} \cdot \vec{x}$ and apply Proposition 1.2.3 to calculate the derivative of $\|\vec{x}\|^2$.]

1.2.10. Consider the ellipse given by $\vec{x}(t) = (a\cos t, b\sin t)$. Find the extremum values of the speed function.

1.2.11. Consider the linear spiral of Example 1.2.4. Let $n \geq 0$ be a non-negative integer. Prove that the length of the nth derivative vector function is given by

$$\|\vec{x}^{(n)}(t)\| = \sqrt{n^2 + t^2}.$$

1.2.12. Consider the exponential spiral $\vec{x}(t) = (ae^{bt}\cos t, ae^{bt}\sin t)$ where a and b are constants. Calculate the arc length $s(t)$ function of $\vec{x}(t)$. Reparametrize the spiral by arc length.

1.2.13. Consider again the exponential spiral $\vec{x}(t) = (ae^{bt}\cos t, ae^{bt}\sin t)$, with $a > 0$ and $b < 0$.

  (a) Prove that as $t \to +\infty$, $\lim \vec{x}(t) = (0,0)$.

  (b) Show that $\vec{x}'(t) \to (0,0)$ as $t \to +\infty$ and that for any t_0,

$$\lim_{t \to \infty} \int_{t_0}^{t} \|\vec{x}'(u)\| \, du < \infty,$$

  i.e., any part of the exponential spiral that "spirals" in toward 0 has finite arc length.

1.2.14. Recall that polar and Cartesian coordinate systems are related as follows:

$$\begin{cases} x = r\cos\theta, \\ y = r\sin\theta, \end{cases} \quad \text{and} \quad \begin{cases} r = \sqrt{x^2 + y^2}, \\ \tan\theta = \frac{y}{x}. \end{cases}$$

Suppose C is a curve in the plane parametrized using polar coordinate functions $r = r(t)$ and $\theta = \theta(t)$ so that one has a parametrization in Cartesian coordinates as

$$\vec{x}(t) = (x(t), y(t)) = (r(t)\cos(\theta(t)), r(t)\sin(\theta(t))).$$

(a) Express x' and y' in terms of r, θ, r', and θ'.

(b) Express r' and θ' in terms of x, y, x', and y'.

(c) Express $||\vec{x}||$ and $||\vec{x}'||$ in terms of polar coordinate functions.

(All coordinate functions are viewed as functions of t and so r', for example, is a shorthand for $r'(t)$, the derivative of r with respect to t.)

1.2.15. (ODE) Suppose that a curve C parametrized by $\vec{x}(t)$ is such that $\vec{x}(t)$ and $\vec{x}'(t)$ always make a constant angle with each other. Find the shape of this curve. [Hint: Use polar coordinates and the results of Problem 1.2.14.]

1.3   Curvature

Let $\vec{x}: I \to \mathbb{R}^2$ be a twice-differentiable parametrization of a curve $\mathcal{C}$. As we saw in the previous section, the decomposition of the velocity vector $\vec{x}' = s'\vec{T}$ into magnitude and unit tangent direction separates the geometric invariant (the unit tangent $\vec{T}$) from the dynamical aspect (the speed $s'(t)$) of the parametrization. Taking one more derivative, we obtain the decomposition

$$\vec{x}'' = s''\vec{T} + s'\vec{T}'. \tag{1.9}$$

The first component describes a tangential acceleration, while the second component describes a rate of change of the tangent direction—in essence, how the curve is "curving." Since $\vec{T}$ is a unit vector, we always have $\vec{T} \cdot \vec{T} = 1$. Therefore,

$$(\vec{T} \cdot \vec{T})' = 0 \Longrightarrow 2\vec{T} \cdot \vec{T}' = 0$$
$$\Longrightarrow \vec{T} \cdot \vec{T}' = 0.$$

Thus, $\vec{T}'$ is perpendicular to $\vec{T}$.

Just as there are two unit tangent vectors at a regular point of the curve, there are two unit normal vectors as well. Given a particular parametrization, there is no directly preferred way to define "the" unit normal vector, so we make a choice.

Definition 1.3.1. Let $\vec{x} : I \to \mathbb{R}^2$ be a regular parametrized curve and $\vec{T} = (T_1, T_2)$ the tangent vector at a regular value $\vec{x}(t)$. We define the unit normal vector $\vec{U}$ by

$$\vec{U}(t) = (-T_2(t), T_1(t)).$$

Equivalently, $\vec{U}$ is the vector function obtained by rotating $\vec{T}$ by $\frac{\pi}{2}$. In still other words, if we view the xy-plane in three-dimensional space with x-, y-, and z-axes oriented in the usual way, and if we call $\vec{k} = (0, 0, 1)$ the unit vector along the positive z-axis, one can write $\vec{U} = \vec{k} \times \vec{T}$.

Since $\vec{T}' \perp \vec{T}$, the vector function $\vec{T}'$ is a multiple of $\vec{U}$ at all t. This leads to the definition of curvature.

Definition 1.3.2. Let $\vec{x} : I \to \mathbb{R}^2$ be a regular twice-differentiable parametric curve. The *curvature* function $\kappa_g(t)$ is the unique real-valued function defined by

$$\vec{T}'(t) = s'(t)\kappa_g(t)\vec{U}(t). \tag{1.10}$$

This definition does not allow one to directly calculate $\kappa_g(t)$ but we can obtain a formula for it as follows. Equation (1.9) becomes

$$\vec{x}'' = s''\vec{T} + (s')^2\kappa_g\vec{U}. \tag{1.11}$$

Viewing the plane as the xy-plane in three-space, we have $\vec{T} \times \vec{U} = \vec{k}$. Thus,

$$\begin{aligned}
\vec{x}' \times \vec{x}'' &= (s'\vec{T}) \times (s''\vec{T} + (s')^2\kappa_g\vec{U}) \\
&= s's''\vec{T} \times \vec{T} + (s')^3\kappa_g\vec{T} \times \vec{U} \\
&= (s')^3\kappa_g\vec{k}.
\end{aligned}$$

But $s'(t) = \|\vec{x}'(t)\|$, which leads to

$$\kappa_g(t) = \frac{(\vec{x}'(t) \times \vec{x}''(t)) \cdot \vec{k}}{\|\vec{x}'(t)\|^3} = \frac{x'(t)y''(t) - x''(t)y'(t)}{(x'(t)^2 + y'(t)^2)^{3/2}}. \tag{1.12}$$

The reader might well wonder why in the definition of $\kappa_g(t)$ we include the factor $s'(t)$. It is not hard to confirm (see Problem 1.3.11) that including the $s'(t)$ factor renders $\kappa_g(t)$ independent of any regular reparametrization (except perhaps up to a change of sign).

One should note at this point that if a curve $\vec{x}(s)$ is parametrized by arc length, then by Equations (1.7) and (1.9), the velocity and acceleration have the simple expressions

$$\vec{x}'(s) = \vec{T}(s),$$
$$\vec{x}''(s) = \kappa_g(s)\vec{U}(s). \qquad (1.13)$$

Example 1.3.3. Consider the vector function that describes a particle moving on a circle of radius R with a nonzero and constant angular velocity ω given by $\vec{x}(t) = (R\cos\omega t, R\sin\omega t)$. In order to calculate the curvature, we need

$$\vec{x}'(t) = (-R\omega\sin\omega t, R\omega\cos\omega t),$$
$$\vec{x}''(t) = (-R\omega^2\cos\omega t, -R\omega^2\sin\omega t).$$

Thus, using the coordinate form (right-most expression) in Equation (1.12), we get

$$\kappa_g(t) = \frac{R^2\omega^3\sin^2 t + R^2\omega^3\cos^2 t}{(R^2\omega^2\sin^2 t + R^2\omega^2\cos^2 t)^{3/2}} = \frac{R^2\omega^3}{R^3\omega^3} = \frac{1}{R}.$$

The curvature of the circle is a constant function that is equal to the reciprocal of the radius for all t, regardless of the angular velocity ω.

The study of trajectories in physics gives particular names for the components of the first and second derivatives of a vector function in the basis $\{\vec{T},\vec{U}\}$. We already saw that the function $s'(t)$ in Equation (1.7) is called the speed. In Equation (1.11), however, the function $s''(t)$ is called the tangential acceleration, while the quantity $s'(t)^2\kappa_g(t)$ is called the centripetal acceleration. Example 1.3.3 connects the curvature function to the reciprocal of a radius, so if $\kappa_g(t) \neq 0$, then we define the *radius of curvature* to $\vec{x}(t)$ at t as the function $R(t) = \frac{1}{\kappa_g(t)}$. Using the common notation $v(t) = s'(t)$ for the speed function, one recovers the common formula for centripetal acceleration of

$$(s')^2\kappa_g = \frac{v^2}{R}.$$

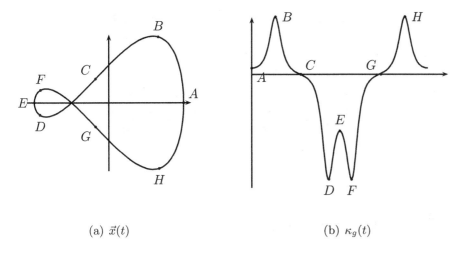

(a) $\vec{x}(t)$ (b) $\kappa_g(t)$

Figure 1.6. Curvature function example.

In introductory physics courses, this formula is presented only in the context of circular motion. With differential geometry at our disposal, we see that centripetal acceleration is always equal to v^2/R at all points on a curve where $\kappa_g \neq 0$, where by R one means the radius of curvature.

In contrast to physics where one tends to refer to the radius of curvature, the curvature function is a more geometrically natural quantity to study. One cannot define a radius of curvature when the curve is a line segment, except to use the awkward expression that "the radius of curvature of a line segment is infinite."

To help build an intuition for the curvature function of a plane curve, consider the parametric curve $\vec{x}(t) = (2\cos t, \sin(2t) + \sin t)$ for $t \in [0, 2\pi]$.

Figure 1.6 shows side by side the locus of the curve and the graph of its curvature function, along with a few labeled points to serve as references. One can see that the curvature function is positive when the curve turns to the left away from the unit tangent vector $\vec{T}$ and negative when the curve turns to the right. The curvature function is 0 when the curve changes from curving to the left of $\vec{T}$ to curving to the right of $\vec{T}$, or vice versa. Also, the curvature function has extrema where it "curves the most" or "least" locally, where by "least" we could mean curving in a negative direction.

Proposition 1.3.4. *A regular parametrized curve* $\vec{x} : I \to \mathbb{R}^2$ *has curvature* $\kappa_g(t) = 0$ *for all* $t \in I$ *if and only if the locus of* $\vec{x}$ *is a line segment or a point.*

Proof: If the locus of $\vec{x}$ traces out a line segment, then (perhaps after a regular reparametrization) we can write $\vec{x}(t) = \vec{a} + \varphi(t)\vec{b}$, where $\varphi(t)$ is a differentiable real function with $\varphi'(t) \neq 0$. Then

$$\vec{x}'(t) = \varphi'(t)\vec{b},$$
$$\vec{x}''(t) = \varphi''(t)\vec{b}.$$

Thus, by Equation (1.12),

$$\kappa_g(t) = \frac{(\varphi'(t)\varphi''(t)\vec{b} \times \vec{b}) \cdot \vec{k}}{(|\varphi'(t)| \, \|\vec{b}\|)^{3/2}} = 0$$

because $\vec{b} \times \vec{b} = \vec{0}$.

Conversely, if $\vec{x}(t) = (x(t), y(t))$ is a curve such that $\kappa_g(t) = 0$, then

$$x'(t)y''(t) - x''(t)y'(t) = 0.$$

We need to find solutions to this differential equation or determine how solutions are related. Since $\vec{x}$ is regular, $\vec{x}' \neq \vec{0}$ for all $t \in I$. Let I_1 be an interval where $y'(t) \neq 0$. Over I_1, we have

$$-\frac{x'y'' - x''y'}{(y')^2} = \frac{d}{dt}\left(\frac{x'}{y'}\right) = 0 \implies \frac{x'}{y'} = C,$$

where C is a constant. Thus, $x' = Cy'$ for all t. Integrating with respect to t we deduce $x = Cy + D$. Similarly, over an interval I_2 where $x'(t) \neq 0$, we deduce that $y = Ax + B$ for some constants A and B. Since I can be covered with intervals where $x'(t) \neq 0$ or $y'(t) \neq 0$, we deduce that the locus of $\vec{x}$ is a piecewise linear curve. However, since $\vec{x}$ is regular, it has no corners, and hence its locus is a line segment. $\qquad\square$

Note that the curvature function arose from calculating $\vec{x}''$ as the derivative of the expression $\vec{x}' = s'\vec{T}$, which involved finding an expression for $\frac{d}{dt}\vec{T}(t)$. Taking the third derivative of $\vec{x}(t)$ and using the decomposition in Equation (1.11), we get

$$\vec{x}''' = s'''\vec{T} + (3s''s'\kappa_g + (s')^2\kappa_g')\vec{U} + (s')^2\kappa_g\vec{U}'.$$

Consequently, we need an expression for the derivative $\vec{U}'(t)$. By definition, for all t, the set $\{\vec{T}, \vec{U}\}$ forms an orthonormal basis and hence

$$\vec{U}' = (\vec{U}' \cdot \vec{T})\vec{T} + (\vec{U}' \cdot \vec{U})\vec{U}.$$

However, just like $\vec{T}(t)$, since $\vec{U}(t)$ is a unit vector for all t, we deduce that $\vec{U} \cdot \vec{U}' = 0$. Furthermore, since $\vec{U} \cdot \vec{T} = 0$ for all t, we also conclude that

$$\vec{U}' \cdot \vec{T} = -\vec{U} \cdot \vec{T}' = -s'\kappa_g.$$

Consequently,

$$\vec{U}'(t) = -s'(t)\kappa_g(t)\vec{T}(t).$$

It is not hard to see that if a curve $\vec{x} : [-a, a] \to \mathbb{R}^2$ is reparametrized by $\vec{x}(-t)$, then the modified curvature function would be $-\kappa_g(t)$. Thus, the sign of the curvature depends on what one might call the "orientation" of the curve, a notion ultimately arbitrary in this case. Excluding this technicality, curvature has a physical interpretation that one can eyeball on particular curves. If a curve is almost a straight line, then the curvature is close to 0, but if a regular curve bends tightly along a certain section, then the curvature is high (in absolute value). Of particular interest are points where the curvature reaches a local extremum.

Definition 1.3.5. Let C be a regular curve parametrized by $\vec{x} : I \to \mathbb{R}^2$ with curvature function $\kappa_g(t)$. A *vertex* of the curve C is a point $P = \vec{x}(t_0)$ where $\kappa_g'(t_0) = 0$.

From the above discussion, one should notice that since $\{\vec{T}, \vec{U}\}$ form an orthonormal basis for all t, every higher derivative of $\vec{x}(t)$, if it exists, can be expressed as a linear combination of $\vec{T}$ and $\vec{U}$. The components of $\vec{x}^{(n)}(t)$ in terms of $\vec{T}$ and $\vec{U}$ involve sums of products of derivatives of $s(t)$ and $\kappa_g(t)$. Furthermore, if $\vec{x}$ is parametrized by arc length, then the coefficients of $\vec{x}^{(n)}(s)$ only involve sums of powers of derivatives of $\kappa_g(s)$.

Problems

1.3.1. Find the curvature function of the following:

    (a) An ellipse $\vec{x}(t) = (a\cos t, b\sin t)$.

    (b) A cycloid $\vec{x}(t) = (t - \sin t, 1 - \cos t)$.

    (c) The curve $\vec{x}(t) = (e^{\cos t}, e^{\sin t})$.

1.3.2. Calculate the curvature function of the flower curve parametrized by

$$\vec{x}(t) = (\sin(nt)\cos t, \sin(nt)\sin t).$$

1.3.3. Consider the curve $\vec{x}(t) = (t^2, t^3 - at)$, where a is a real number. Calculate the curvature function $\kappa_g(t)$ and determine where $\kappa_g(t)$ has extrema.

1.3.4. Consider the graph of a function $y = f(x)$ parametrized as a curve by $\vec{x}(t) = (t, f(t))$. Find a formula for the curvature. Assume that f is a smooth function. Prove that $\kappa_g(t) = 0$ if t is an inflection point. [Hint: Recall that x_0 is the inflection point of a function $y = f(x)$ if $f''(x)$ changes sign at x_0.]

1.3.5. Find an equation in $f(t)$ that describes where the vertices occur for a function graph $\vec{x}(t) = (t, f(t))$. Find an example of a function graph where $\kappa_g'(t) = 0$ but $f'(t) \neq 0$ and vice versa.

1.3.6. Prove that the graph of a polar function $r = f(\theta)$ at angle θ has curvature

$$\kappa_g(\theta) = \frac{2f'(\theta)^2 + f(\theta)^2 - f(\theta)f''(\theta)}{(f'(\theta)^2 + f(\theta)^2)^{3/2}}.$$

1.3.7. Find the vertices of an ellipse with half-axes $a \neq b$ and calculate the curvature at those points.

1.3.8. Calculate the curvature function for all the Lissajous figures: $\vec{x}(t) = (\cos(mt), \sin(nt))$, with $t \in [0, 2\pi]$.

1.3.9. Prove by direct calculation that the following formula holds for the ellipse $\vec{x}(t) = (a\cos t, b\sin t)$:

$$\int_0^{2\pi} \kappa_g \, ds = 2\pi.$$

[Hint: Use a substitution involving $\tan\theta = \frac{a}{b}\tan t$.]

1.3.10. Calculate the curvature function for the cardioid:

$$\vec{x}(t) = ((1 - \cos t)\cos t, (1 - \cos t)\sin t).$$

1.3.11. Let $\vec{x} : I \to \mathbb{R}^2$ be a parametrized curve and $f : J \to I$ a surjective function so that f makes $\vec{\xi} = \vec{x} \circ f$ a regular reparametrization of $\vec{x}$. Call $t_0 = f(u_0)$ so that $\vec{\xi}(u_0) = \vec{x}(t_0)$. Prove that

$$\kappa_{g,\vec{x}}(t_0) = \begin{cases} \kappa_{g,\vec{\xi}}(u_0) & \text{if } f'(u_0) > 0, \\ -\kappa_{g,\vec{\xi}}(u_0) & \text{if } f'(u_0) < 0. \end{cases}$$

1.3.12. We define a *parallel curve* to a parametrized curve $\vec{x} : I \to \mathbb{R}^2$ as a curve that can be parametrized by $\vec{\gamma}_r(t) = \vec{x}(t) + r\vec{U}(t)$, where r is a real number. Suppose that $\vec{x}$ is a regular curve of class C^2 and assume $r \neq 0$. Prove that $\vec{\gamma}_r$ is regular if and only if $\frac{1}{r} \notin [\kappa_m, \kappa_M]$, where

$$\kappa_m = \min_{t \in I}\{\kappa_g(t)\} \qquad \text{and} \qquad \kappa_M = \max_{t \in I}\{\kappa_g(t)\}.$$

1.3.13. Let C be a curve in $\mathbb{R}^2$ defined as the solution to an equation $F(x, y) = 0$. Use implicit differentiation to prove that at any point on this curve, the curvature of C is given by

$$\kappa_g = \frac{F_{xx}F_y^2 - 2F_{xy}F_xF_y + F_{yy}F_x^2}{(F_x^2 + F_y^2)^{3/2}}.$$

1.3.14. Find the vertices of the curve given by the equation $x^4 + y^4 = 1$.

1.3.15. Pedal Curves. Let C be a regular curve and A a fixed point in the plane. The *pedal curve* of C with respect to A is the locus of points of intersection of the tangent lines to C and lines through A perpendicular to these tangents. Given the parametrization $\vec{x}(t)$ of a curve C, provide a parametric formula for the pedal curve to C with respect to A. Find an explicit parametric formula for the pedal curve of the unit circle with respect to $(1, 0)$.

1.4 Osculating Circles, Evolutes, and Involutes

Classical differential geometry introduces the notion of order of contact to measure the degree of intersection between two curves or surfaces at a particular intersection point. One can use various definitions for contact, but we shall use the one provided by Struik in [30, p. 23], rephrased here for use with curves.

Definition 1.4.1. Let two curves C_1 and C_2 have a regular point P in common. Given a point A on C_1 near P, let D_A be the orthogonal projection of A onto C_2, i.e., the point on C_2 closest to A. Then C_2 has *contact of order* n with C_1 at P, when for $A \to P$ along C_1

$$\lim_{A \to P} \frac{AD_A}{AP^k} = \begin{cases} \text{a finite number} \neq 0, & \text{for } k = n+1, \\ 0, & \text{for } k \leq n. \end{cases} \qquad (1.14)$$

This definition is abstract. It applies to curves that are continuous but not differentiable and also to curves in $\mathbb{R}^m$ for any m. The reader should feel free to focus attention on Corollary 1.4.3 in a first reading.

Intuitively, calling two intersecting curves with order of contact n mimics the idea of calling two functions $f(x)$ and $g(x)$ equal of order n at a if $f^{(k)}(a) = g^{(k)}(a)$ for $k \leq n$ and $f^{(n+1)}(a) \neq g^{(n+1)}(a)$. It is not possible to directly adapt this latter definition to curves since one usually cannot simultaneously parametrize two curves by arc length, which would be necessary to provide a well defined criterion. Also, this definition generalizes the order of contact for simple monomial functions in that $y = ax^n$ intersects the x-axis at $(0,0)$, with order of contact $n-1$.

Let us restrict our attention to two parametrized plane curves C_1 and C_2 that intersect at a point P. (We assume that C_1 and C_2 are smooth on neighborhoods of P.) We can parametrize C_1 by its arc length to get a vector function $\vec{\alpha}(s)$ such that P occurs at $s = 0$. Furthermore, near P, we can also parametrize C_2 by $\vec{\beta}(s)$ in such a way that $\vec{\beta}(s)$ is the orthogonal projection of $\vec{\alpha}(s)$ onto C_2. Notice that by construction, there exists a scalar function $f(s)$ such that $\vec{\alpha}(s) = \vec{\beta}(s) + f(s)\vec{U}_\beta(s)$.

Proposition 1.4.2. *With the above setup, C_1 and C_2 have contact of order n at P if and only if the first value of k for which $f^{(k)}(0) \neq 0$ occurs for $k = n+1$.*

Proof: In the definition of order of contact, $AP = s$, where $A = \vec{\alpha}(s)$ and $AD_A = ||\vec{\alpha}(s) - \vec{\beta}(s)||$. Therefore, $AD_A = |f(s)|$. As C_1 and C_2 are smooth near P so is the curve $\vec{\alpha}(s) - \vec{\beta}(s)$. Thus, there exists a neighborhood J of s such that $g(s) = |f(s)|$ is a smooth function.

Then, by l'Hôpital's Rule,

$$\lim_{A \to P} \frac{AD_A}{AP^k} = \lim_{s \to 0} \frac{g(s)}{s^k} = \frac{g^{(k)}(0)}{k!}.$$

Now the result of the proposition follows immediately from the definition of order of contact. $\square$

As an immediate consequence, we deduce the following corollary.

Corollary 1.4.3. *Suppose $C_1 : \vec{\alpha}(t)$ and $C_2 : \vec{\beta}(u)$ are two parametrized curves that intersect at a point P, which corresponds to where $t = t_0$ and $u = u_0$. Reparametrize C_1 by arc length and let s_0 be such that $P = \vec{\alpha}(s_0)$. Let $u(s)$ be the function such that the projection of $\vec{\alpha}(s)$ onto C_2 is located at $\vec{\beta}(u(s))$. Then $\vec{\alpha}$ and $\vec{\beta}$ have contact of order n if and only if*

$$\vec{\alpha}^{(i)}(s_0) = \frac{d^i}{ds^i}\vec{\beta}(u(s))\Big|_{s_0}$$

for all $0 \leq i \leq n$ but that

$$\vec{\alpha}^{(n+1)}(s_0) \neq \frac{d^{n+1}}{ds^{n+1}}\vec{\beta}(u(s))\Big|_{s_0}.$$

The intuitive picture for order of contact indicates that two intersecting curves with contact order $n \geq 1$ have the same tangent line at the intersection point. In particular, we leave it as an exercise to prove the fact that a curve and its tangent line have order of contact 1. In contrast, an intersection point between two curves with contact of order 0 is said to be a *transversal* intersection.

Definition 1.4.4. Let C be a curve parametrized by $\vec{x} : I \to \mathbb{R}^2$, and let t_0 be a regular value of the curve. The *osculating circle* to C at the point t_0 is a circle with contact to C at $\vec{x}(t_0)$ of order 2.

Proposition 1.4.5. *Let C be the locus of a parametrized curve $\vec{x}$ and t_0 a regular value, where $\vec{x}$ is twice differentiable and where $\kappa_g(t_0) \neq 0$. Then,*

1. *There exists a unique osculating circle to C at $\vec{x}(t_0)$;*

2. *It is given by the following vector function:*

$$\vec{\gamma}(t) = \vec{x}(t_0) + \frac{1}{\kappa_g(t_0)}\vec{U}(t_0) + \frac{1}{\kappa_g(t_0)}\left((\sin t)\vec{T}(t_0) - (\cos t)\vec{U}(t_0)\right).$$

Proof: Without loss of generality, we can assume that $\vec{x}$ is parametrized by arc length s and that we are looking for the osculating circle at $s = 0$. Call $\vec{\gamma}(u)$ the parametrization of the proposed osculating circle, which must have the form

$$\vec{\gamma}(u) = (a + R\cos(u), b + R\sin(u)).$$

We assume that $u = u_0$ at the point of contact and that near u_0, the function $u(s)$ gives the projection of $\vec{x}(s)$ onto the curve $\vec{\gamma}$. According to Corollary 1.4.3, in order for there to exist an osculating circle, the parameters a, b, and R and the function $u(s)$ must satisfy

$$\vec{x}(0) = \vec{\gamma}(u(0)), \qquad \vec{x}'(0) = \frac{d}{ds}\vec{\gamma}(u(s))\Big|_0, \quad \vec{x}''(0) = \frac{d^2}{ds^2}\vec{\gamma}(u(s))\Big|_0,$$

which in coordinates is equivalent to

$$a + R\cos(u(0)) = x(0), \qquad (1.15)$$
$$b + R\sin(u(0)) = y(0), \qquad (1.16)$$
$$-Ru'(0)\sin(u(0)) = x'(0), \qquad (1.17)$$
$$Ru'(0)\cos(u(0)) = y'(0), \qquad (1.18)$$
$$-Ru''(0)\sin(u(0)) - Ru'(0)^2\cos(u(0)) = x''(0), \qquad (1.19)$$
$$Ru''(0)\cos(u(0)) - Ru'(0)^2\sin(u(0)) = y''(0), \qquad (1.20)$$

Since $\vec{x}(s)$ is parametrized by arc length, $x'(0)^2 + y'(0)^2 = 1$, from which we conclude that $|u'(0)| = \frac{1}{R}$. From the definition of the function $u(s)$, it is not hard to see that $u'(0) > 0$ so $u'(0) = \frac{1}{R}$. More importantly, however, for all s, $x'(s)^2 + y'(s)^2 = 1$. Taking a derivative of this equation with respect to s, we deduce that $x'(0)x''(0) + y'(0)y''(0) = 0$. This relation, along with Equations (1.17) through (1.20), leads to $u''(0) = 0$.

We obtain the value of R by noting that after calculation

$$\kappa_g(0) = x'(0)y''(0) - x''(0)y'(0) = R^2(u'(0))^3 = \frac{1}{R}.$$

Using Equations (1.17) and (1.18), we find that

$$(\cos(u(0)), \sin(u(0))) = (y'(0), -x'(0)) = -\vec{U},$$

since $\vec{x}$ is parametrized by arc length. Thus the center of the circle $\vec{\gamma}$ is

$$(a, b) = (x(0), y(0)) - R\left(\cos(u(0)), \sin(u(0))\right) = \vec{x}(0) + \frac{1}{\kappa_g(0)}\vec{U}(0).$$

Finally, a quick check that we leave for the reader is to show that given the above determinations for a, b, and R, Equations (1.15) through (1.20) are redundant given the fact that $\vec{x}(s)$ is parametrized by arc length. These results prove part 1) of the proposition.

Part 2) follows easily from part 1). Since $\vec{x}$ is parametrized by arc length, the unit tangent vector is just $\vec{T}(0) = (x'(0), y'(0))$, giving also $\vec{U}(0) = (-y'(0), x'(0))$. In the solutions for Equations (1.15) and (1.16), we see that the center of the osculating circle is at

$$(a, b) = (x(0) - R\cos(\tau(0)), y(0) - R\sin(\tau(0))) = \vec{x}(0) + R\vec{U}(0)$$

Furthermore, for any curve parametrized by arc length, $\kappa_g(s) = x'(s)y''(s) - x''(s)y'(s)$. Thus, $\frac{1}{R} = \kappa_g(0)$. ☐

In light of Proposition 1.4.5 and Example 1.3.3, the curvature function $\kappa_g(t)$ of a plane curve $\vec{x}$ has a nice physical interpretation, namely, the reciprocal of the radius of the osculating circle. A higher order of contact indicates a better geometric approximation, and hence, since there is a unique osculating circle, it is, in a geometric sense, the best approximating circle to a curve at a point. Furthermore, in Problem 1.4.4, we prove that to a curve $\vec{x}$ at $t = t_0$, there does not necessarily exist a touching circle with order of contact greater than 2. Thus, the curvature is the inverse of the radius of the best approximating circle to a curve at a point.

(From a physics point of view, we obtain an additional confirmation of the above interpretation by considering the units of the curvature function. If we view the coordinate functions $x(t)$ and $y(t)$ with the unit of meters and t in any unit, Equation (1.12) gives the unit of 1/meter for $\kappa_g(t)$.)

Definition 1.4.6. Let $\vec{x}$ be a parametrized curve, and let $t = t_0$ be a regular point that satisfies the conditions in Proposition 1.4.5. The center of the osculating circle at $t = t_0$ is called the *center of curvature*.

Definition 1.4.7. The *evolute* of a curve C is the locus of the centers of curvature.

Proposition 1.4.8. *Let $\vec{x} : I \to \mathbb{R}^2$ be a regular parametric plane curve that is of class C^2, i.e., has a continuous second derivative. Let I' be a subinterval of I over which $\kappa_g(t) \neq 0$. Then over the interval I', the evolute of $\vec{x}$ has the following parametrization:*

$$\vec{E}(t) = \vec{x}(t) + \frac{1}{\kappa_g(t)}\vec{U}(t).$$

Example 1.4.9. Consider the parabola $y = x^2$ with parametric equations $\vec{x}(t) = (t, t^2)$. We calculate the curvature with the following steps:

$$\vec{x}'(t) = (1, 2t),$$
$$\vec{x}''(t) = (0, 2),$$
$$s'(t) = \sqrt{1 + 4t^2},$$
$$\kappa_g(t) = \frac{2}{(1 + 4t^2)^{3/2}}.$$

Thus, the parametric equations of the evolute are

$$\vec{E}(t) = (t, t^2) + \frac{1}{2}(1 + 4t^2)^{3/2}\frac{1}{\sqrt{1 + 4t^2}}(-2t, 1) = \left(-4t^3, \frac{1}{2} + 3t^2\right).$$

The Cartesian equation for the evolute of the parabola is then

$$y = \frac{1}{2} + 3\left|\frac{x}{4}\right|^{2/3}$$

(see Figure 1.7).

A closely related curve to the evolute of $\vec{x}(t)$ is the involute, though we leave the exact nature of this relationship to the problems. The involute is defined as follows.

Definition 1.4.10. Let $\vec{x} : I \to \mathbb{R}^2$ be a regular parametrized curve. We call an *involute* to $\vec{x}$ any parametrized curve $\vec{\imath}$ such that for all $t \in I$, $\vec{\imath}(t)$ meets the tangent line to $\vec{x}$ at t at a right angle.

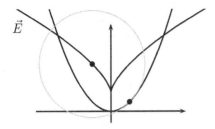

Figure 1.7. Evolute of a parabola.

For all $t \in I$, the point on the involute $\vec{\iota}(t)$ is on the tangent line to $\vec{x}$ at t, so it is natural to write the parametric equations as $\vec{\iota}(t) = \vec{x}(t) + \lambda(t)\vec{T}(t)$. Since we will wish to calculate $\vec{\iota}'(t)$, which involves the derivative $\vec{T}'(t)$, we must assume that the curve $\vec{x}(t)$ is of class C^2. Definition 1.4.10 requires that the vector $\vec{\iota}'(t)$ be in a perpendicular direction to $\vec{T}(t)$, so we have

$$\vec{T} \cdot \vec{\iota}' = 0 \Longrightarrow \vec{T} \cdot \left(s'\vec{T} + \lambda'\vec{T} + \lambda s' \kappa_g \vec{U} \right) = 0$$
$$\Longrightarrow s' + \lambda' = 0$$
$$\Longrightarrow \lambda(t) = C - s(t),$$

where C is some constant of integration. Therefore, if $\vec{x}$ is a regular parametrized curve of class C^2, then the formula for the involute is

$$\vec{\iota}(t) = \vec{x}(t) + (C - s(t))\vec{T}(t).$$

Problems

1.4.1. Prove that the curves of the form $y = ax^n$ have order of contact $n - 1$ with the x-axis at the origin. [Hint for $n \geq 2$: Use the definition of C_1 with the x-axis parametrized by $\vec{\alpha}(t) = (t, 0)$. Call A' the orthogonal projection of D_A onto the x-axis. Prove first that $\lim_{A \to P} A'D_A / AD_A = 1$.]

1.4.2. Prove the claim that the tangent line to a parametrized curve at a regular point has contact of order at least 1.

1.4.3. Let $\vec{x}$ be a regular parametrized curve and let t_0 be an inflection point, i.e. a point where $\kappa_g(t) = 0$. Prove that the tangent line to $\vec{x}(t)$ at $t = t_0$ has contact of order at least 2.

1.4.4. Prove that if the osculating circle to a regular parametrized curve C at a point P has contact of order 3 or more, then P is a vertex of C. Give an example where this does not occur, thereby proving that at a regular point on a parametrized curve, there does not necessarily exist a circle with contact of order more than 2.

1.4.5. Prove that the evolute of the ellipse $\vec{x}(t) = (a\cos t, b\sin t)$ has parametric equations

$$\vec{\gamma}(t) = \left(\frac{a^2 - b^2}{a} \cos^3 t, \frac{b^2 - a^2}{b} \sin^3 t \right).$$

1.4.6. Consider the catenary given by the parametric equation $\vec{x}(t) = (t, \cosh t)$ for $t \in \mathbb{R}$.

   (a) Prove that the curvature of the catenary is $\kappa_g(t) = 1/\cosh^2 t$.

   (b) Prove that the evolute of the catenary is $\vec{\gamma}(t) = (t - \sinh t \cosh t, 2\cosh t)$.

1.4.7. Let $\vec{x}(t) = (t, t^2)$ be the parabola. Define a new curve $\vec{\iota}$ as the involute of $\vec{x}$ such that $\vec{\iota}(0) = \vec{x}(0) = (0,0)$. Calculate parametric equations for $\vec{\iota}(t)$. [Hint: This will involve an integral.]

1.4.8. Continuation of the last problem: Calculate parametric equations for the evolute to $\vec{\iota}$.

1.5   Natural Equations

Recall that for a parametrized curve $\vec{x} : I \to \mathbb{R}^2$, the curvature function is given by

$$\kappa_g(t) = \frac{x'(t)y''(t) - x''(t)y'(t)}{(x'(t)^2 + y'(t)^2)^{3/2}}.$$

Since any rigid motion in the plane (i.e., rotations and translations) does not stretch distances, such a transformation should not distort a plane curve. Therefore, as one might expect, the curvature is preserved under rigid motions, a fact that we now prove.

Theorem 1.5.1. *Let $\vec{x} : I \to \mathbb{R}^2$ be a regular plane curve that is of class C^2. Let $F : \mathbb{R}^2 \to \mathbb{R}^2$ be a rigid motion of the plane given by $F(\vec{v}) = A\vec{v} + \vec{C}$, where A is a rotation matrix and $\vec{C}$ is any vector* *in* $\mathbb{R}^2$. *Consider the regular parametrized curve* $\vec{\xi} = F \circ \vec{x}$. *Then* $\vec{\xi}$ *is a regular parametrized curve that is of class* C^2, *and the curvature function* $\bar{\kappa}_g$ *of* $\vec{\xi}$ *is equal to the curvature function* κ_g *of* $\vec{x}$.

Proof: A rotation matrix in $\mathbb{R}^2$ is of the form

$$A = \begin{pmatrix} a & -b \\ b & a \end{pmatrix},$$

where $a^2 + b^2 = 1$. Let $\vec{C} = (e, f)$. We can write the parametric equation for $\vec{\xi}$ as

$$\vec{\xi}(t) = (ax(t) - by(t) + e, bx(t) + ay(t) + f).$$

We then calculate

$$\vec{\xi}'(t) = (ax'(t) - by'(t), bx'(t) + ay'(t)),$$
$$\vec{\xi}''(t) = (ax''(t) - by''(t), bx''(t) + ay''(t)),$$

and therefore the curvature function of $\vec{\xi}$ is

$$\bar{\kappa}_g(t) = \frac{(ax'(t) - by'(t))(bx''(t) + ay''(t)) - (bx'(t) + ay'(t))(ax''(t) - by''(t))}{((ax'(t) - by'(t))^2 + (bx'(t) + ay'(t))^2)^{3/2}}$$

$$= \frac{abx'x'' + a^2x'y'' - b^2y'x'' - aby'y'' - abx'x'' + b^2x'y'' - a^2y'x'' + aby'y''}{(a^2(x')^2 - 2abx'y' + b^2(y')^2 + b^2(x')^2 + 2abx'y' + a^2(y')^2)^{3/2}}$$

$$= \frac{x'y'' - y'x''}{((x')^2 + (y')^2)^{3/2}} = \kappa_g(t). \qquad \square$$

The curvature function is invariant under any positive isometry, i.e., a composition of rotations and translations. Furthermore, in Problem 1.3.11, we saw that the curvature function is invariant under any regular reparametrization, except up to a sign that depends on "the direction of travel" along the curve. Consequently, $|\kappa_g|$ is a geometric invariant that only depends on the shape of the curve at a particular point and not how the curve is parametrized or where the curve sits in $\mathbb{R}^2$.

It is natural to ask whether a converse relation holds, namely, whether the curvature function is sufficient to determine the parametrized curve up to a positive plane isometry. As posed, the question is not well defined since a curve can have different parametrizations. However, we can prove the following.

Theorem 1.5.2 (Fundamental Theorem of Plane Curves). *Given a function* $\kappa_g(s)$, *there exists a regular curve of class* C^2 *parametrized by arc length* $\vec{x} : I \to \mathbb{R}^2$ *with the curvature function* $\kappa_g(s)$. *Furthermore, the curve is uniquely determined up to a rigid motion in the plane.*

Proof: If a regular curve $\vec{x}(s)$ is parametrized by arc length, then the curvature formula is $\kappa_g(s) = x'(s)y''(s) - y'(s)x''(s)$. The proof of this theorem consists of exhibiting a parametrization that satisfies this differential equation.

For a regular curve of class C^2 that is parametrized by arc length, we have $\vec{x}' = \vec{T}$, and we can view $\vec{T}$ as a function from the interval of definition of $\vec{x}$ into the unit circle. Therefore,

$$\vec{T} = (\cos(\theta(s)), \sin(\theta(s))) \tag{1.21}$$

for some continuous function $\theta(s)$. However, $\vec{T}'(s) = \kappa_g(s)\vec{U}$, and Equation (1.21) shows that

$$\kappa_g(s)\vec{U} = \theta'(s)\left(-\sin(\theta(s)), \cos(\theta(s))\right).$$

Thus, we deduce that $\kappa_g(s) = \theta'(s)$.

The above remarks lead to the following result. Suppose we are given the curvature function $\kappa_g(s)$. Performing two integrations, we see that the only curves $\vec{\alpha}(s)$ with curvature function $\kappa_g(s)$ must be

$$\vec{\alpha}(s) = \left(\int \cos(\theta(s))\, ds + e, \int \sin(\theta(s))\, ds + f\right), \tag{1.22}$$

where

$$\theta(s) = \int \kappa_g(s)\, ds + \theta_0, \tag{1.23}$$

and where θ_0, e, and f are constants of integration. Furthermore, the trigonometric addition formulas show that a nonzero constant θ_0 changes $\vec{\alpha}$ by a rotation of θ_0 and the nonzero constants e and f correspond to a translation along the vector $\vec{C} = (e, f)$. $\square$

If we assume that we are working with plane curves $\vec{x}(s)$ that are real analytic, we can see why Theorem 1.5.2 holds using Taylor series.

Let $\vec{x} : I \to \mathbb{R}^2$ be a real analytic curve parametrized by arc length and assume without loss of generality that $0 \in I$. We can expand the Taylor series of $\vec{x}$ about 0 to get

$$\vec{x}(s) = \vec{x}(0) + s\vec{x}'(0) + \frac{s^2}{2!}\vec{x}''(0) + \frac{s^3}{3!}\vec{x}'''(0) + \cdots .$$

However, since the curve is parametrized by arc length, $\vec{x}'(s) = \vec{T}(s)$, $\vec{x}''(s) = \kappa_g(s)\vec{U}(s)$, $\vec{x}'''(s) = \kappa_g'(s)\vec{U}(s) - (\kappa_g(s))^2\vec{T}(s)$, and so on. The first few terms look like

$$\vec{x}(s) = \vec{x}(0) + \left(s - \frac{1}{6}\kappa_g(0)^2 s^3 + \cdots\right)\vec{T}(0) + \left(\frac{1}{2}s^2 + \frac{1}{6}\kappa_g'(0)s^3 + \cdots\right)\vec{U}(0).$$

$$(1.24)$$

Thus, since the normal vector $\vec{U}(s)$ is just a rotation of $\vec{T}(s)$ by $\frac{\pi}{2}$, given a function $\kappa_g(s)$, once one chooses for initial conditions the point $\vec{x}(0)$ and the direction $\vec{T}(0)$, the Taylor series is uniquely determined. The intersection of the intervals of convergence of the two Taylor series in the $\vec{T}(0)$ and the $\vec{U}(0)$ components is a new interval J that contains $s = 0$. It is possible that J could be trivial, $J = \{0\}$, but if it is not, then the Taylor series uniquely defines $\vec{x}(s)$ over J. Choosing a different $\vec{T}(0)$ amounts to a rotation of the curve in the plane, and choosing a different $\vec{x}(0)$ amounts to a translation. Therefore, we see again that making different choices for the initial conditions corresponds to a rigid motion of the curve in the plane.

For even simple functions for $\kappa_g(s)$, it is often difficult to use the approach in the proof of Theorem 1.5.2 to explicitly solve for $\vec{x}(s)$. However, using a computer algebra system (CAS) with tools for solving differential equations, it is possible to produce a picture of curves that possess a given curvature function $\kappa_g(s)$. One can create a numerical solution using the solution in Equation (1.22) with (1.23). As an equivalent technique, using a CAS, one can solve the system of differential equations

$$\begin{cases} x'(s) &= \cos(\theta(s)), \\ y'(s) &= \sin(\theta(s)), \\ \theta'(s) &= \kappa_g(s), \end{cases}$$

and only plot the solution for the pair $(x(s), y(s))$. A choice of initial conditions determines the position and orientation of the curve in

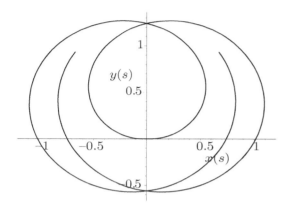

Figure 1.8. Curve with $\kappa_g(s) = 1 + 1/(1 + s^2)$.

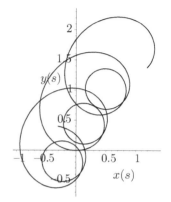

Figure 1.9. Curve with $\kappa_g(s) = 2 + \sin s$.

the plane. Figures 1.8 and 1.9 give a few interesting examples of parametric curves with a given $\kappa_g(s)$.

Problems

1.5.1. Suppose a curve has $\kappa_g(s) = A$. Prove by directly expanding the Taylor series that such a curve is a circle.

1.5.2. Suppose a curve has $\kappa_g(s) = 2As$. Prove by directly expanding the Taylor series that if such a curve has $\vec{x}(0) = \vec{0}$ and $\vec{x}'(s) = (1,0)$, then such a curve is

$$\vec{x}(s) = \left(\int_0^s \cos(As^2)\, ds, \int_0^s \sin(As^2)\, ds \right).$$

1.5.3. Find parametric equations for a curve satisfying $\kappa_g(s) = \frac{1}{1+s^2}$ by direct integration, following the method in the proof of Theorem 1.5.2.

CHAPTER 2

Plane Curves: Global Properties

Most of the properties of curves we have studied so far are called local properties. By definition, a local property of a curve (or surface) is a property that is related to a point on the curve based on information contained just in a neighborhood of that point. In contrast, global properties concern attributes about the curve taken as a whole. In Chapter 1, the arc length of a curve was the only notion introduced that one might consider a global property.

Intuitively speaking, local properties of a curve at a point involve derivatives of the parametric equations while global properties deal with integration along the curve and topological properties and geometric properties of how the curve lies in $\mathbb{R}^2$. Some of the proofs of global properties rely on theorems from topology. These are supplied concisely in the appendix on topology in [22]. (The interested reader is encouraged to consult some references on basic topology, such as Gemignani [15] or Armstrong [1].)

2.1   Basic Properties

Definition 2.1.1. A parametrized curve C is called *closed* if there exists a parametrization $\vec{x} : [a, b] \to \mathbb{R}^2$ of C such that $\vec{x}(a) = \vec{x}(b)$. A regular curve is called closed if, in addition, all the (one-sided) derivatives of $\vec{x}$ at a and at b are equal, i.e., $\vec{x}(a) = \vec{x}(b)$, $\vec{x}'(a) = \vec{x}'(b)$, $\vec{x}''(a) = \vec{x}''(b), \ldots$. A curve C is called *simple* if it has no self-intersections; a closed curve is called simple if it has no self-intersections except at the endpoints.

The conditions on the derivatives of $\vec{x}$ in the above definition seem awkward but attempt to establish the fact that the vector function $\vec{x}$ behaves identically at a and at b. A more topological approach involves using a circle $\mathbb{S}^1$ rather than an interval as the domain for the map $\vec{x}$. We define a topological circle $\mathbb{S}^1$ as the set $[0,1]$ with the points 0 and 1 identified. The topology of $\mathbb{S}^1$ is the identification topology, which in this case means that an open neighborhood U of p contains, for some ε, the subsets

$$\begin{cases} (p - \varepsilon, p + \varepsilon) & \text{with } \varepsilon < \min\{p, 1 - p\}, & \text{if } p \neq 0, \\ [0, \varepsilon) \cup (1 - \varepsilon, 1], & & \text{if } p = 0 \, (= 1). \end{cases}$$

Then we say that a curve C is closed if there exists a continuous surjective function $\varphi : \mathbb{S}^1 \to C$. Furthermore, in this language, we say that a closed curve C is simple if there exists a bijection $\varphi : \mathbb{S}^1 \to C$ that is continuous and such that φ^{-1} is also continuous.

Proposition 2.1.2. *Let C be a closed curve with parametrization $\vec{x} : [a, b] \to \mathbb{R}^2$. C is bounded as a subset of $\mathbb{R}^2$.*

Proof: Let $\vec{x} : [a, b] \to \mathbb{R}^2$ be a parametrization of C with coordinate functions $x(t)$ and $y(t)$. Since $x(t)$ and $y(t)$ are continuous functions with compact domains, then their images J_x and J_y are compact. Note that $J_x \times J_y$ is the smallest horizontal rectangle containing C. Call $J_x = [x_{\min}, x_{\max}]$ and $J_y = [y_{\min}, y_{\max}]$, and let

$$M_x = \max\{|x_{\min}|, |x_{\max}|\} \quad \text{and} \quad M_y = \max\{|y_{\min}|, |y_{\max}|\}.$$

Then for all $t \in [a, b]$,

$$x(t)^2 \leq M_x^2 \quad \text{and} \quad y(t)^2 \leq M_y^2.$$

Thus,

$$\|\vec{x}(t)\| \leq \sqrt{M_x^2 + M_y^2},$$

and so the curve C is contained in a disk of finite radius, and hence C is bounded. $\qquad\qquad\qquad\qquad\qquad\qquad\qquad\qquad\qquad\qquad\qquad\quad\square$

One of the most fundamental properties of global geometry of plane curves is that a simple, closed plane curve C separates the plane into two open components, each with the common boundary of C. In common language, we call these two components the *interior* and *exterior*. However, this intuitive fact, called the Jordan Curve

Theorem, turns out to be rather difficult to prove. We refer the reader to Munkres [24] for a detailed discussion. In [22, Section A.4], we have included a more readable proof of the weaker statement, which assumes that the curve is regular.

We remind the reader of the following theorem from multivariable calculus, which one may view as a global property of plane curves.

Theorem 2.1.3 (Green's Theorem). *Let C be a simple, closed, regular, plane curve with interior region $\mathcal{R}$, and let $\vec{F} = (P(x,y), Q(x,y))$ be a differentiable vector field. Then,*

$$\int_C P \, dx + Q \, dy = \iint_{\mathcal{R}} \left(\frac{\partial Q}{\partial x} - \frac{\partial P}{\partial y} \right) dx \, dy.$$

We remind the reader that if $\vec{x} : [a, b] \to \mathbb{R}^2$ is a regular parametrization for C, with $\vec{x}(t) = (x(t), y(t))$, then the integral on the left means

$$\int_C P \, dx + Q \, dy = \int_C \vec{F} \cdot d\vec{x} = \int_a^b P(x(t), y(t)) x'(t) + Q(x(t), y(t)) y'(t) \, dt.$$

Corollary 2.1.4. *Let C be a simple closed regular plane curve with interior region $\mathcal{R}$. Then the area A of $\mathcal{R}$ is*

$$A = \int_C x \, dy = -\int_C y \, dx = \frac{1}{2} \int_C -y \, dx + x \, dy. \qquad (2.1)$$

Proof: In each of these integrals, one simply chooses a vector field $\vec{F} = (P, Q)$ such that

$$\frac{\partial Q}{\partial x} - \frac{\partial P}{\partial y} = 1$$

For the three integrals, these are, respectively, $\vec{F} = (0, x)$, $\vec{F} = (-y, 0)$ and $\vec{F} = \frac{1}{2}(-y, x)$. Then apply Green's Theorem. □

Example 2.1.5. Consider the ellipse parametrized by $\vec{x}(t) = (a \cos t, b \sin t)$. The techniques from introductory calculus used to calculate area would lead us to evaluate the integral

$$A = 4 \int_0^a b \sqrt{1 - \frac{x^2}{a^2}} \, dx.$$

Though one can calculate this by hand, it is not simple. Much easier would be to use Green's Theorem, which gives

$$A = \int_C x \, dy = \int_0^{2\pi} a \cos t \, b \cos t \, dt$$

$$= ab \int_0^{2\pi} \cos^2 t \, dt = ab \int_0^{2\pi} \frac{1}{2} + \frac{1}{2} \cos(2t) \, dt$$

$$= ab \left[\frac{1}{2} t + \frac{1}{4} \sin(2t) \right]_0^{2\pi}$$

$$= \pi ab.$$

Problems

2.1.1. Calculate the area of the region enclosed by the cardioid $\vec{x}(t) = ((1 - \cos t) \cos t, (1 - \cos t) \sin t)$.

2.1.2. Consider the function graph of a curve in polar coordinates $r = f(\theta)$ parametrized by $\vec{x}(t) = (f(t) \cos t, f(t) \sin t)$. Suppose that for $0 \leq t \leq 2\pi$, the curve is closed and encloses a region $\mathcal{R}$. Use Equation (2.1) and Green's Theorem to prove the following area formula in polar coordinates

$$\text{Area}(\mathcal{R}) = \iint_{\mathcal{R}} r \, dr \, d\theta.$$

2.1.3. The *diameter* of a curve C is defined as the maximum distance between two points, i.e.,

$$\text{diam}(C) = \max\{d(p_1, p_2) \mid p_1, p_2 \in C\}.$$

We call a diameter any chord of C whose length is $\text{diam}(C)$. Let $\vec{x}(t)$ be the parametrization of a closed differentiable curve C. Let $f(t, u) = \|\vec{x}(t) - \vec{x}(u)\|$ be the distance function between two points on the curve.

(a) Prove that if a chord $[p_1, p_2]$ of C is a diameter of C, then the line (p_1, p_2) is simultaneously perpendicular to the tangent line to C at p_1 and the tangent line to C at p_2.

(b) Under what other situations is the line (p_1, p_2) simultaneously perpendicular to the tangent line to C at p_1 and the tangent line to C at p_2?

2.1.4. Prove that the diameter of the cardioid $\vec{x}(t) = ((1-\cos t)\cos t, (1-\cos t)\sin t)$ is $3\sqrt{3}/2$. [Hint: First show that the maximum distance between $\vec{x}(t)$ and $\vec{x}(u)$ occurs when $u = 2\pi - t$.]

2.1.5. (*) This problem studies the relation between the curvature $\kappa_g(s)$ of a curve C and whether or not it is closed.

    (a) Prove that if C is closed, then its curvature $\kappa_g(s)$, as a function of arc length, is periodic.

    (b) Assume that $\kappa_g(s)$ is not constant and is periodic with smallest period p. Use the Fundamental Theorem of Plane Curves (Theorem 1.5.2) to prove that if

$$\frac{1}{2\pi} \int_0^p \kappa(s)\,ds$$

    is an element of $\mathbb{Q} - \mathbb{Z}$, i.e., a rational that is not an integer, then C is a closed curve. [This problem is discussed in [3], though the authors' use of complex numbers is not necessary for this problem.]

2.2   Rotation Index

As a leading example of what we shall term the rotation index of a curve, consider the circle C parametrized by $\vec{x}(t) = (R\cos t, R\sin t)$ with the defining interval of $I = [0, 2\pi]$. A simple calculation shows that for all $t \in [0, 2\pi]$, we have a curvature of $\kappa_g(t) = \frac{1}{R}$. Then it is easy to see that

$$\int_C \kappa_g\,ds = \int_0^{2\pi} \kappa_g(t)s'(t)\,dt = \int_0^{2\pi} \frac{1}{R} \cdot R\,dt = 2\pi.$$

On the other hand, suppose that we use the defining interval $[0, 2\pi n]$ with the same parametrization. In that case, were we to evaluate the same integral, we would obtain

$$\frac{1}{2\pi} \int_C \kappa_g\,ds = n.$$

Now consider any regular, closed, plane curve $\vec{x} : I \to \mathbb{R}^2$. To this curve, we associate the unit tangent vector $\vec{T}(t)$. Placing the base of this vector at the origin, we see that $\vec{T}$ itself draws out a curve $\vec{T} : I \to \mathbb{R}^2$. It is called the *tangential indicatrix* of the plane curve $\vec{x}$.

The tangential indicatrix of a curve lies entirely on the unit circle in the plane but with possibly a complicated parametrization. Depending on the shape of $\vec{x}$, the tangential indicatrix may change speed, stop, and double back along a portion of the circle. Nonetheless, we can use the theory of usual plane curves to study the tangential indicatrix. We are in a position to state the main proposition for this section.

Proposition 2.2.1. *Let C be a closed, regular, plane curve parametrized by $\vec{x} : I \to \mathbb{R}^2$. Then the quantity*

$$\frac{1}{2\pi} \int_C \kappa_g \, ds$$

is an integer. This integer is called the rotation index *of the curve.*

Proof: Rewrite the above integral as follows:

$$\int_C \kappa_g \, ds = \int_I \kappa_g(t) s'(t) \, dt = \int_I \kappa_g(t) s'(t) \|\vec{U}(t)\| \, dt.$$

The curvature function $\kappa_g(t)$ of plane curves is not always positive, but since $\vec{x}(t)$ is smooth, $\kappa_g(t)$ is at least continuous, and the intermediate value theorem applies. Therefore, define two closed subsets (unions of closed subintervals) of the interval I as follows:

$$I^+ = \{t \in I : \kappa_g(t) \geq 0\} \quad \text{and} \quad I^- = \{t \in I : \kappa_g(t) \leq 0\}.$$

Using the definition of curvature from Equation (1.10), we split the above integral into the following two parts:

$$\int_C \kappa_g \, ds = \int_{I^+} \|\vec{T}'(t)\| \, dt - \int_{I^-} \|\vec{T}'(t)\| \, dt.$$

However, the two integrals on the right-hand side of the above equation are integrals of the arc length of the tangential indicatrix traveling in a counterclockwise (respectively clockwise) direction. Since $\vec{x}$ is a closed curve, $\vec{T}$ is as well, and these two integrals represent 2π times the number of times $\vec{T}$ travels around the circle, with a sign indicating the direction. $\qquad\square$

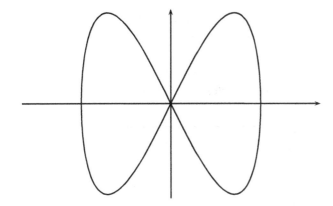

Figure 2.1. The curve $\vec{x}(t) = (\cos t, \sin(2t))$.

Example 2.2.2. Consider the curve C parametrized by $\vec{x}(t) = (\cos t,$ $\sin(2t))$. (See Figure 2.1.) The curvature function is given by

$$\kappa_g(t)s'(t) = \frac{4\sin t\sin(2t) + 2\cos t\cos(2t)}{\sin^2 t + 4\cos^2(2t)} = \frac{6\cos t + 2\cos(3t)}{5 - \cos(2t) + 4\cos(4t)}.$$

Thus, the rotation index n of $\vec{x}$ is

$$n = \frac{1}{2\pi}\int_C \kappa_g(t)s'(t)dt = \frac{1}{2\pi}\int_0^{2\pi} \frac{6\cos t + 2\cos(3t)}{5 - \cos(2t) + 4\cos(4t)}\,dt.$$

Since the integrand is periodic of period 2π, using the substitution $u = t + \frac{\pi}{2}$ and recognizing that we can integrate over any interval of length 2π, we have

$$n = \frac{1}{2\pi}\int_{-\pi}^{\pi} \frac{6\sin u - 2\sin(3u)}{5 + \cos(2u) + 4\cos(4u)}\,du.$$

However, the integrand is now an odd function, so the rotation index of $\vec{x}$ is $n = 0$.

In the proof for Proposition 2.2.1, we analyzed the integral $\int_C \kappa_g ds$ as the signed arc length of the tangential indicatrix $\vec{T}(t)$, which one can view as a map from an interval I to the unit circle $\mathbb{S}^1$. However, the result of Proposition 2.2.1 follows also from a more general result that we wish to explain in detail here. Some of the concepts below come from topology and illustrate the difficulty of analyzing functions from an interval to a circle. The reader should feel free to either skip the technical details of the proofs in the rest of this section or refer to [22, Appendix A] for background material.

Any path $f : I \to \mathbb{S}^1$ on the unit circle may be described by the angle function $\varphi(t)$, so that

$$f(t) = (\cos(\varphi(t)), \sin(\varphi(t))) \,.$$

However, the angle $\varphi(t)$ is defined only up to a multiple of 2π, and hence it need not be a well-defined function, let alone a continuous function. However, since f is continuous, for all $t \in I$ there exists an interval $(t - \varepsilon, t + \varepsilon)$ and a continuous function $\tilde{\varphi}(t)$ such that for all $u \in (t - \varepsilon, t + \varepsilon)$, $\tilde{\varphi}(u)$ differs from $\varphi(u)$ by a multiple of 2π. If $\tilde{\varphi}'(t)$ exists, it is a well-defined function, regardless of any choice made in selecting $\varphi(t)$. Finally, if $I = [a, b]$, we define the *total angle function* related to f as the function

$$\tilde{\varphi}(t) = \int_a^t \tilde{\varphi}'(u) \, du + \varphi(a). \tag{2.2}$$

By construction, $\tilde{\varphi}(t)$ is continuous, it satisfies

$$f(t) = (\cos(\tilde{\varphi}(t)), \sin(\tilde{\varphi}(t))) \,,$$

and it keeps track of how many times and in what direction the path f travels around $\mathbb{S}^1$. The quantity $\tilde{\varphi}(b) - \tilde{\varphi}(a)$ is called the *total angle* swept out by f.

If $\varphi(a)$ is chosen so that $0 \le \varphi(a) < 2\pi$, then it is common to also denote the total angle function $\tilde{\varphi}(t)$ simply by $\tilde{f}(t)$, indicating that this function depends uniquely on the original function f.

For any continuous function between circles $f : \mathbb{S}^1 \to \mathbb{S}^1$, viewing $\mathbb{S}^1$ as an interval $[0, 2\pi]$ with the endpoints identified, we also view f as a continuous function $f : [0, 2\pi] \to \mathbb{S}^1$. Since $f(0) = f(2\pi)$ are points on $\mathbb{S}^1$, then $\tilde{f}(2\pi) - \tilde{f}(0)$ is a multiple of 2π. This leads to the following definition.

Definition 2.2.3. Let $f : \mathbb{S}^1 \to \mathbb{S}^1$ be a continuous map between circles. Let $\tilde{f} : [0, 2\pi] \to \mathbb{R}$ be the lifting of f. Then the *degree* of f is the integer

$$\deg f = \frac{1}{2\pi} \left(\tilde{f}(2\pi) - \tilde{f}(0) \right). \tag{2.3}$$

Intuitively speaking, the degree of a function $f : \mathbb{S}^1 \to \mathbb{S}^1$ between unit circles is how many times around f winds its domain $\mathbb{S}^1$ onto its target $\mathbb{S}^1$.

Returning to the example of the tangential indicatrix $\vec{T}$ of a regular closed curve C, since C is closed, $\vec{T}$ can be viewed as a function $\mathbb{S}^1 \to \mathbb{S}^1$.

Proposition 2.2.4. *Let C be a regular closed curve parametrized by $\vec{x} : I \to \mathbb{R}^2$. The rotation index of C is equal to the degree of $\vec{T}$.*

Proof: Since $\vec{T}$ is a map $\vec{T} : I \to \mathbb{S}^1$, using Equation (2.2), we can write

$$\vec{T} = (\cos(\tilde{\varphi}(t)), \sin(\tilde{\varphi}(t))),$$

so

$$\vec{T}' = \left(-\tilde{\varphi}'(t)\sin(\tilde{\varphi}(t)), \tilde{\varphi}'(t)\cos(\tilde{\varphi}(t)) \right) = \tilde{\varphi}'(t)\vec{U}(t).$$

But $\vec{T}'(t) = \kappa_g(t)s'(t)\vec{U}(t)$, so

$$\kappa_g(t)s'(t) = \tilde{\varphi}'(t). \tag{2.4}$$

Thus, if we call $I = [a, b]$, we have

$$\int_C \kappa_g \, ds = \int_a^b \tilde{\varphi}'(t) \, dt = \tilde{\varphi}(b) - \tilde{\varphi}(a),$$

which concludes the proof. $\qquad\qquad\qquad\qquad\qquad\qquad\square$

The notion of degree of a continuous function $\mathbb{S}^1 \to \mathbb{S}^1$ can be applied in a wider context than can the rotation index of a curve since the latter requires a curve to be regular, while the former concept, as presented here, only requires $\tilde{\varphi}'(t)$ to be integrable. Using the notion of degree, we can make sense out of the question of how often a curve turns around a point.

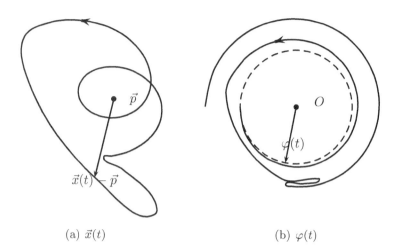

(a) $\vec{x}(t)$ (b) $\varphi(t)$

Figure 2.2. Winding number of 2.

Definition 2.2.5. Let C be a closed, regular, plane curve parametrized by $\vec{x} : I \to \mathbb{R}^2$, and let $\vec{p}$ be a point in the plane not on C. Since C is a closed curve, we may view $\vec{x}$ as a function on $\mathbb{S}^1$. We define the *winding number* $w(p)$ of C around $\vec{p}$ as the degree of the function $\psi : \mathbb{S}^1 \to \mathbb{S}^1$ defined by

$$\psi(t) = \frac{\vec{x}(t) - \vec{p}}{||\vec{x}(t) - \vec{p}||}.$$

In practice, it is often hard to explicitly calculate the winding number of a curve around a point given parametric equations. On the other hand, it is usually easy to see what the degree is by plotting out the function $\psi(t)$. (See Figure 2.2, where in Figure 2.2(b) we've allowed $\psi(t)$ to come off the circle in order to see its graph more clearly.) As one might expect, the winding number of a curve around a point p depends only on what connected component of the curve's complement p lies in.

Proposition 2.2.6. *Let C be a closed, regular, plane curve parametrized by $\vec{x} : I \to \mathbb{R}^2$, and let p_0 and p_1 be two points in the same connected component of $\mathbb{R}^2 - C$. The winding number of C around p_0 is equal to the winding number of C around p_1.*

Proof: Let $\beta : [0,1] \to \mathbb{R}^2$ be a path such that $\beta(0) = p_0$, $\beta(1) = p_1$, and $\beta(u) \neq \vec{x}(t)$ for any $u \in [0,1]$ and $t \in I$. Define the two-variable function $H : I \times [0,1] \to \mathbb{S}^1$ by

$$H(t, u) = \frac{\vec{x}(t) - \beta(u)}{\|\vec{x}(t) - \beta(u)\|}.$$

The function H is continuous since $\vec{x}(t) - \beta(u)$ is continuous and $\|\vec{x}(t) - \beta(u)\|$ is also continuous and never 0. For all $u \in [0,1]$, we consider $H(t, u)$ to be a function of t from I to $\mathbb{S}^1$ and define $\tilde{H}(t, u)$ as its lifting, which is a continuous function $\tilde{H} : I \times [0,1] \to \mathbb{R}$. But then the function of u given by

$$\frac{1}{2\pi} \left(\tilde{H}(2\pi, u) - \tilde{H}(0, u) \right)$$

is continuous and discrete. Thus, it is constant, and hence

$$\frac{1}{2\pi} \left(\tilde{H}(2\pi, 1) - \tilde{H}(0, 1) \right) = \frac{1}{2\pi} \left(\tilde{H}(2\pi, 0) - \tilde{H}(0, 0) \right),$$

which means that the winding numbers of C around p_0 and around p_1 are equal. $\qquad\square$

We are now in a position to discuss some of the details in the proof of the Jordan Curve Theorem.

Theorem 2.2.7 (Regular Jordan Curve Theorem). *Consider a simple, regular, closed, plane curve C parametrized by arc length $\vec{x} : [0, l] \to \mathbb{R}^2$. Then the open set $\mathbb{R}^2 - C$ has exactly two connected components with common boundary $C = \vec{x}([0, l])$. Only one of the connected components is bounded, and we call this component of $\mathbb{R}^2 - C$ the interior.*

In the proof of the Regular Jordan Curve Theorem, one must first establish that $\mathbb{R}^2 - C$ has at least two different components. The strategy is to show that near a point p on the curve C, points on one side of C have a winding number of 0, while on the other side they have a winding number of ± 1. The details for this part and the proof that there are only two components to $\mathbb{R}^2 - C$ are quite long and not particularly informative, so we have relegated it to [22, Section A.4]. However, the proof establishes a fact that is useful to point out for its own right.

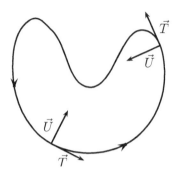

Figure 2.3. A positively oriented curve.

Proposition 2.2.8. *Let* $\vec{\alpha} : [0, l] \to \mathbb{R}^2$ *be a simple, closed, regular, plane curve. The rotation index of* $\vec{\alpha}$ *is* ± 1.

Many properties of regular, simple, closed, plane curves exist in reference to the interior of the curve, whose existence follows from the Jordan Curve Theorem. As a first application, remark that to parametrize a closed curve by arc length one has freedom to choose the point that corresponds to $s = 0$ and to choose the direction. Given a curve parametrized by arc length, we say the curve has a *positive orientation* if at all points $\vec{\alpha}(s)$, the unit normal vector $\vec{U}(s)$ points toward the interior of the curve (see Figure 2.3). If one imagines oneself traveling along the curve, a positive orientation means the interior is on one's left.

Problems

2.2.1. In Example 2.2.2, we calculated directly that the rotation index of a figure eight curve is 0. In contrast, show that the rotation index of the curve $\vec{x}(t) = (\cos t, \sin(3t))$, with $0 \leq t \leq 2\pi$, is 1.

2.2.2. Show that the lemniscate given by

$$\vec{x}(t) = \left(\frac{a \cos t}{1 + \sin^2 t}, \frac{a \cos t \sin t}{1 + \sin^2 t} \right)$$

has a rotation index of 0.

2.2.3. Calculate the rotation index of the limaçon of Pascal given by

$$\vec{x}(t) = ((1 - 2\cos t) \cos t, (1 - 2\cos t) \sin t).$$

2.2.4. Let $\vec{x}(t) = (\cos 3t \cos t, \cos 3t \sin t)$ be the trefoil curve.. Show that $I = [0, \pi]$ is enough of a domain to make $\vec{x}$ into a closed curve. Prove that the rotation index of $\vec{x}$ is 2.

2.3 Isoperimetric Inequality

With the Jordan Curve Theorem at our disposal, we can use Green's Theorem to prove an inequality between the length of a simple closed curve and the area contained in the interior. This inequality, called the isoperimetric inequality, is another example of a global theorem since it relates quantities that take into account the entire curve at once.

Theorem 2.3.1. *Let C be a simple, closed, plane curve with length l, and let A be the area bounded by C. Then*

$$l^2 \geq 4\pi A, \tag{2.5}$$

and equality holds if and only if C is a circle.

Proof: Let L_1 and L_2 be parallel lines that are both tangents to C and such that C is contained in the strip between them. Let $\mathbb{S}^1$ be a circle that is also tangent to both L_1 and L_2 such that $\mathbb{S}^1$ does not intersect C. We call r the radius of this circle, and we set up the coordinate axes in the plane so that the origin is at the center of $\mathbb{S}^1$, the x-axis is perpendicular to L_1 (and L_2), and the y-axis is parallel to L_1 and L_2.

Assume that C is parametrized by arc length $\vec{\alpha}(s) = (x(s), y(s))$ and that the parametrization is positively oriented at the points of tangency with L_1 and L_2. Call $s = 0$ the parameter location for the tangency point $C \cap L_1$ and $s = s_0$ the parameter for $C \cap L_2$. We can also assume that $\mathbb{S}^1$ is parametrized by $\vec{\gamma}(s)$, where $\vec{\gamma}(s)$ is the intersection of

- the upper half of $\mathbb{S}^1$ with the vertical line through $\vec{\alpha}(s)$ if $0 \leq s \leq s_0$;

- the lower half of $\mathbb{S}^1$ with the vertical line through $\vec{\alpha}(s)$ if $s_0 \leq s \leq l$.

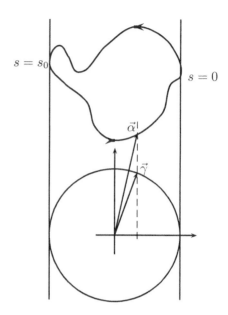

Figure 2.4. Isoperimetric inequality.

Notice that in this parametrization for the circle, writing $\vec{\gamma}(s) = (\bar{x}(s), \bar{y}(s))$, we have $x(s) = \bar{x}(s)$.

Let us call A the area enclosed in the curve $\mathcal{C}$, and let $\bar{A}$ be the area of $\mathbb{S}^1$. Using Green's Theorem, we have the following formulas for area:

$$A = \int_0^l xy' \, ds \quad \text{and} \quad \bar{A} = \pi r^2 = -\int_0^l x'\bar{y} \, ds.$$

Adding the two areas, we get

$$A + \pi r^2 = \int_0^l xy' - x'\bar{y} \, ds \leq \int_0^l \sqrt{(xy' - x'\bar{y})^2} \, ds$$

$$\leq \int_0^l \sqrt{(x^2 + \bar{y}^2)((x')^2 + (y')^2)} \, ds = \int_0^l \sqrt{\bar{x}^2 + \bar{y}^2} \, ds = lr. \tag{2.6}$$

We now use the fact that the geometric mean of two positive real numbers is less than the arithmetic mean. Thus,

$$\sqrt{A \cdot \pi r^2} \leq \frac{1}{2}(A + \pi r^2) \leq \frac{1}{2}lr. \tag{2.7}$$

Squaring both sides and dividing by r^2, we get $4\pi A \leq l^2$. This proves the first part of Theorem 2.3.1.

In order for equality to hold in Equation (2.5), we need equalities to hold in both the inequalities in Equation (2.6). From Equation (2.7), we conclude that $A = \pi r^2$ and $l = 2\pi r$. Furthermore, since A does not change, regardless of the direction of L_1 and L_2, neither does the distance r. We also must have

$$(xy' - x'\bar{y})^2 = (x^2 + \bar{y}^2)((x')^2 + (y')^2) = r^2,$$

which, after expanding both sides, leads to $(xx' + \bar{y}y')^2 = 0$. However, differentiating the relationship $x^2 + \bar{y}^2 = r^2$ for all parameter values s, we find that $xx' + \bar{y}\bar{y}' = 0$ for all s. Thus, for all s, we have $y'(s) = \bar{y}'(s)$, and thus $y(s) = \bar{y}(s) + D$ for some constant D. Since by construction $x(s) = \bar{x}(s)$, then $\mathcal{C}$ is a circle—a translate of $\mathbb{S}^1$ in the direction $(0, D)$.   □

Since nothing in the above proof uses second derivatives, we only need the curve $\vec{\alpha}$ to be C^1, i.e., that it have a continuous first derivative. Furthermore, one can generalize the proof to apply to curves that are only piecewise C^1.

2.4 Curvature, Convexity, and the Four-Vertex Theorem

Definition 2.4.1. A subset of S of $\mathbb{R}^n$ is called *convex* if for all p and q in S, the line segment $[p, q]$ is a subset of S, i.e., lies entirely inside S. If $\mathcal{C}$ is not convex, it is called *concave*. (See Figure 2.5.)

Figure 2.5. A concave curve.

With the Jordan Curve Theorem, one may discuss convexity
properties of a plane curve if one uses the following definition.

 Definition 2.4.2. A closed, regular, simple, parametrized curve C is
called convex if the union of C and the interior of C form a convex
subset of $\mathbb{R}^2$.

In the context of regular curves, it is possible to provide various
characterizations for when a curve is convex based on a curve's local
properties.

Proposition 2.4.3. *A regular, closed, plane curve C is convex if and
only if it is simple and its curvature κ_g does not change sign.*

Proof: Let $\vec{\alpha} : [0, l] \to \mathbb{R}^2$ be a parametrization by arc length with
positive orientation for the curve C. Let $\vec{T} : [0, l] \to \mathbb{S}^1$ be the
tangential indicatrix, and let $\theta : [0, l] \to \mathbb{R}$ be the total angle function
associated to the rotation index. By Equation (2.4), $\kappa_g(s) = \theta'(s)$,
so the condition that $\kappa_g(s)$ does not change sign is equivalent to $\theta(s)$
being monotonic.

We first prove that convexity implies that C is simple. This follows from the definition of convexity, which assumes that C possesses
an interior and, therefore, that $\mathbb{R}^2 - C$ has only two components. By
the Jordan Curve Theorem, since we already assume C is regular and
closed, for it to possess an interior, it must be simple.

Assume from now on that C is simple. Suppose that $\kappa_g(s)$ changes
sign on $[0, l]$. At a point $s = a$, define the height function for all
$s \in [0, l]$ by

$$h_a(s) = (\vec{\alpha}(s) - \vec{\alpha}(a)) \cdot \vec{U}(a).$$

This measures the height of the point $\vec{\alpha}(s)$ from the tangent line L_a
to C at $\vec{\alpha}(a)$, with $\vec{U}(a)$ being considered the positive direction. The
derivatives of this height function are

$$h'(s) = \vec{\alpha}'(s) \cdot \vec{U}(a) = \vec{T}(s) \cdot \vec{U}(a),$$
$$h''(s) = \kappa_g(s)\vec{U}(s) \cdot \vec{U}(a).$$

Consequently, at $s = a$, the height function satisfies $h'(a) = 0$ and
$h''(a) = \kappa_g(a)$.

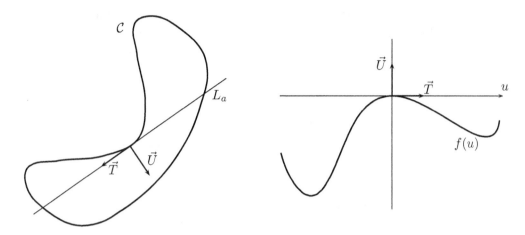

(a) The curve with tangent line L_a (b) The height function

Figure 2.6. Height Function $f(u)$.

From the definition of differentiation, one deduces that in a neighborhood of a, say over the interval $(a - \delta, a + \delta)$, the projection $p_a : C \to L_a$ of C onto L_a is a bicontinuous bijection (i.e., both the function and its inverse are continuous). Let

$$p_a(a - \delta) = \vec{\alpha}(a) - \varepsilon_1 \vec{T}(a) \qquad \text{and} \qquad p_a(a + \delta) = \vec{\alpha}(a) + \varepsilon_2 \vec{T}(a),$$

where $\varepsilon_1, \varepsilon_2 > 0$. Define the function $g : (-\varepsilon_1, \varepsilon_2) \to [0, l]$ by

$$g(u) = \vec{\alpha}^{-1}(p_a^{-1}(\vec{\alpha}(a) + u\vec{T}(a))).$$

Finally, define $f : (-\varepsilon_1, \varepsilon_2) \to \mathbb{R}$ by $f = h_a \circ g$. The graph of the function f placed in the frame $(\vec{T}(a), \vec{U}(a))$ with origin $\vec{\alpha}(a)$ traces out the curve C in a neighborhood of $s = a$ (Figure 2.6).

The derivatives of f are

$$f'(u) = h'(g(u))g'(u),$$
$$f''(u) = h''(g(u))(g'(u))^2 + h'(g(u))g''(u).$$

Since $g(0) = a$ and $h'(a) = 0$, without knowing $g'(0)$, we deduce that $f(0) = f'(0) = 0$ and that $f''(0)$ has the same sign as $\kappa_g(a)$.

If $\kappa_g(a) < 0$, then from basic calculus we deduce that f is concave down over $(-\varepsilon_1, \varepsilon_2)$, which implies that every line segment between points on the graph of f forms a chord that is below the graph. As an example, use the segment between $(u_1, f(u_1))$ and $(u_2, f(u_2))$. Furthermore, since the curve $\mathcal{C}$ has a positive orientation, the interior of $\mathcal{C}$ is in the direction of $\vec{U}(a)$, which corresponds to being above the graph of f. Thus, the line segment between the points $\vec{\alpha}(g(u_1))$ and $\vec{\alpha}(g(u_2))$ is in the exterior of the curve. Consequently, this proves that κ_g changes sign if and only if $\mathcal{C}$ is not convex and the proposition follows. $\square$

This proposition and a portion of the proof lead to another more geometric characterization of convex curves.

Proposition 2.4.4. *A regular, simple, closed curve $\mathcal{C}$ is convex if and only if it lies on one side of every tangent line to $\mathcal{C}$.*

Proof: Suppose we parametrize $\mathcal{C}$ by arc length with a positive orientation. From the proof of Proposition 2.4.3, one observes that at any point p of the curve $\mathcal{C}$ where the curvature is negative (given the positive orientation), in a neighborhood of p, the tangent line L to $\mathcal{C}$ at p is in the interior of $\mathcal{C}$. Consequently, since $\mathcal{C}$ is bounded and L is not, L must intersect the curve $\mathcal{C}$ at another point. By Proposition 2.4.3, this proves that $\mathcal{C}$ is concave if and only if there exists a tangent line that has points in the interior of $\mathcal{C}$ and the exterior of $\mathcal{C}$. $\square$

In Section 1.3, we gave the term vertex for a point on the curve where the curvature reaches an extremum. More precisely, for any regular parametrized curve $\vec{x} : I \to \mathbb{R}^2$, we defined a vertex to be a point $\vec{x}(t_0)$ where $\kappa_g'(t_0) = 0$. Note that since the curvature of a curve is independent of the parametrization up to a possible change of sign, vertices are independent of any parametrization of $\mathcal{C}$.

The concept of a vertex is obviously a local property of the curve, but if one were to experiment with a variety of closed curves, one would soon guess that there must be a restriction on the number of vertices. In Problem 1.3.7, the reader calculated that the noncircular ellipse has four vertices. Figure 2.7(b) shows another simple closed curve with six vertices. (In each figure, the dots indicate the vertices of the curve.)

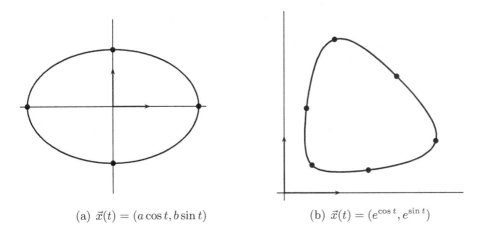

(a) $\vec{x}(t) = (a\cos t, b\sin t)$ (b) $\vec{x}(t) = (e^{\cos t}, e^{\sin t})$

Figure 2.7. Two closed curves and their vertices.

Theorem 2.4.5 (Four-Vertex Theorem). *Every simple, closed, regular, convex, plane curve $\mathcal{C}$ has at least four vertices.*

Proof: Let $\vec{\alpha} : [0, l] \to \mathbb{R}^2$ be a parametrization by arc length for the curve $\mathcal{C}$. Since $[0, l]$ is compact and $\kappa_g : [0, l] \to \mathbb{R}$ is continuous, then by the Extreme Value Theorem, κ_g attains both a maximum and a minimum over $[0, l]$. These values already assure that $\mathcal{C}$ has two vertices. Call these points p and q and call L the line between them. Let β be the arc along $\mathcal{C}$ from p to q, and let γ be the arc along $\mathcal{C}$ from q to p.

We claim that β and γ are contained in opposite half-planes defined by the line L. Assume one of the arcs is not contained in one of the half-planes. Then $\mathcal{C}$ meets L at another point r. By convexity, in order for the line segments $[p, q]$, $[p, r]$, and $[q, r]$ to all lie inside $\mathcal{C}$, all three points would need to have L as its tangent line to $\mathcal{C}$. By Problem 2.4.1, this is a contradiction. Assume now that the arcs β and γ are contained in half-planes but in the same half-plane. Then again by convexity, the only possibility is that one of the arcs is a line segment along L. Then along that line segment, the curvature $\kappa_g(s)$ is identically 0. However, this implies that the curvature at p and q is 0, but since p and q are extrema of κ_g, this would force $\mathcal{C}$ to be a line. This is a contradiction since $\mathcal{C}$, being closed, is bounded.

If there are no other vertices on the curve, then $\kappa'_g(s)$ does not change sign on β or on γ. If L has equation $Ax + By + C = 0$, then

$$\int_0^l (Ax + By + C)\frac{d\kappa_g}{ds}\, ds \tag{2.8}$$

is positive. However, this leads to a contradiction because for all real constants A, B, and C, the integral in Equation(2.8) is always 0 (Problem 2.4.2).

This proves that there is at least one other vertex. But then, since $\kappa'_g(s)$ changes sign at least three times, it must change signs a fourth time as well. Hence, $\kappa'_g(s)$ is equal to 0 at least four times.□

Problems

2.4.1. A *bitangent* line L to a regular curve $\mathcal{C}$ is a line that is tangent to $\mathcal{C}$ at a minimum of two points p and q such that between p and q on $\mathcal{C}$, there are points that do not have L as a tangent line. Prove that a simple, regular curve is not convex if and only if it has a bitangent line.

2.4.2. Let $\vec{x} : [0, l] \to \mathbb{R}^2$ be a regular, closed, plane curve parametrized by arc length and write $\vec{x}(s) = (x(s), y(s))$.

(a) Show that $x'' = -\kappa y'$ and $y'' = \kappa x'$.

(b) Prove that

$$\int_0^l x(s)\kappa'_g(s)\, ds = -\int_0^l \kappa_g(s)x'(s)\, ds,$$

and do the same for $y(s)$ instead of $x(s)$.

(c) Use the above to show that for any constants A, B, and C, we have

$$\int_0^l (Ax + By + C)\frac{d\kappa_g}{ds}\, ds = 0.$$

2.4.3. If a closed curve $\mathcal{C}$ is contained inside a disk of radius r, prove that there exists a point $P \in \mathcal{C}$ such that the curvature $\kappa_g(P)$ of $\mathcal{C}$ at P satisfies

$$|\kappa_g(P)| \geq \frac{1}{r}.$$

2.4.4. Let $\vec{\alpha} : I \to \mathbb{R}^2$ be a regular, closed, simple curve with positive orientation. Define the parallel curve at a distance r as

$$\vec{\beta}(t) = \vec{\alpha}(t) - r\vec{U}(t).$$

Call $l(\vec{\alpha})$ the length of the curve $\vec{\alpha}$ and $A(\vec{\alpha})$ the area of the interior of the curve. Prove the following:

(a) $\vec{\beta}$ is a regular curve if and only if $-\frac{1}{r}$ is not between the extremal values of $\kappa_g(t)$.

(b) If $\vec{\beta}$ is regular, then $l(\vec{\beta}) = l(\vec{\alpha}) + 2\pi r$.

(c) If $\vec{\beta}$ is regular, then $A(\vec{\beta}) = A(\vec{\alpha}) + rl(\vec{\alpha}) + \pi r^2$.

(d) If we call $\kappa_{g,r}$ the curvature function of $\vec{\beta}$, then

$$\kappa_{g,r}(t) = \frac{\kappa_g(t)}{1 + r\kappa_g(t)}.$$

2.4.5. Show that the limaçon $\vec{x}(t) = ((1 - 2\cos t)\cos t, (1 - 2\cos t)\sin t)$ for $t \in [0, 2\pi]$ has exactly two vertices. Explain why it does not contradict the Four-Vertex Theorem.

2.4.6. (*) Consider the curve in Figure 2.7(b). Determine the coordinates of the vertices of $\vec{x}(t)$.

2.4.7. Let $[A, B]$ be a line segment in the plane, and let $l > AB$. Show that the curve $\mathcal{C}$ that joins A and B and has length l such that, together with the segment $[A, B]$, bounds the largest possible area is an arc of a circle passing through A and B.

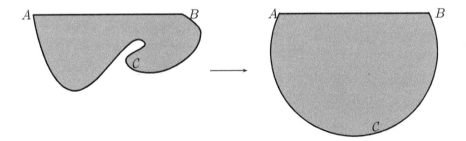

CHAPTER 3

Curves in Space: Local Properties

The local theory of space curves is similar to the theory for plane curves, but differences arise in the richer variety of configurations available for loci of curves in $\mathbb{R}^3$. In the study of plane curves, we introduced the curvature function, a function of fundamental importance that measures how much a curve deviates from being a straight line. As we shall see, in the study of space curves, we again introduce a curvature function that measures how much a space curve deviates from being linear, and we also introduce a torsion function that measures how much the curve twists away from being planar.

3.1   Definitions, Examples, and Differentiation

As in the study of plane curves, one must take some care in defining what one means by a space curve. If I is an interval of $\mathbb{R}$, to call a space curve any function $\vec{x} : I \to \mathbb{R}^3$ (or the image thereof) would allow for separate pieces or even a set of scattered points. By a curve, one typically thinks of a connected set of points, and, just as with plane curves, the desired property is continuity.

Instead of repeating the definitions provided in Section 1.1, we point out that the definition for the limit (Definition 1.1.4) and for continuity (Definition 1.1.5) of a vector function, Proposition 1.1.6, and Corollary 1.1.7 continue to hold for vector functions into $\mathbb{R}^n$.

Definition 3.1.1. Let I be an interval of $\mathbb{R}$. A *parametrized curve* in $\mathbb{R}^n$ is a continuous function $\vec{x} : I \to \mathbb{R}^n$. If $n = 3$, we call $\vec{x}$ a *space curve*. If for all $t \in I$ we have

$$\vec{x}(t) = (x_1(t), x_2(t), \ldots, x_n(t)),$$

then the functions $x_i : I \to \mathbb{R}$ for $1 \le i \le n$ are called the *coordinate functions* or *parametric equations* of the parametrized curve. The *locus* is the image $\vec{x}(I)$ of the parametrized curve.

In order to develop an intuition for space curves, we present a number of examples. These illustrate only some of the great variability in the shape of parametric curves, but they also provide a short library for examples we revisit later.

Example 3.1.2 (Lines). In space or in $\mathbb{R}^n$, a line is defined by a point $\vec{p}$ on the line and a direction vector $\vec{v}$. Parametric equations for a line are then $\vec{x} : \mathbb{R} \to \mathbb{R}^n$ with $\vec{x}(t) = \vec{v}t + \vec{p}$.

Example 3.1.3 (Planar Curves). We define a planar curve as any space curve whose image lies in a plane. The notion of torsion, which we will define in Section 3.2, provides a characterization for when a curve is planar. However, if one wishes to concoct parametric equations for a particular planar curve in space, one needs three data: a point in the plane and two noncollinear vectors in the given plane. Then one uses this point as an origin and these two vectors as a basis of the plane and then provides parametric equations in this reference frame.

As an example, one can obtain the equations for a circle of radius 7 in the plane $3x - 2y + 2z = 6$ with center $\vec{p} = (2, 1, 1)$ as follows. A normal vector to the plane is $\vec{n} = (3, -2, 2)$. Two vectors that are not collinear and in this plane are $\vec{a} = (2, 3, 0)$ and $\vec{b} = (2, 0, -3)$. We use the Gram-Schmidt orthonormalization process to find an orthonormal basis of $\mathrm{Span}(\vec{a}, \vec{b})$. The first vector in the orthonomal basis is

$$\vec{u_1} = \frac{\vec{a}}{\|\vec{a}\|} = \left(\frac{2}{\sqrt{13}}, \frac{3}{\sqrt{13}}, 0 \right).$$

Then calculate

$$\vec{v}_2 = \vec{b} - \mathrm{proj}_{\vec{u}_1} \vec{b} = \vec{b} - (\vec{u}_1 \cdot \vec{b})\vec{u}_1 = \left(\frac{18}{13}, -\frac{12}{13}, -3 \right)$$

and get a second basis vector

$$\vec{u_2} = \frac{\vec{v}_2}{\|\vec{v}_2\|} = \left(\frac{6}{\sqrt{221}}, -\frac{4}{\sqrt{221}}, -\frac{13}{\sqrt{221}} \right).$$

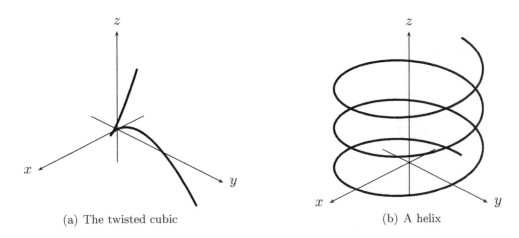

(a) The twisted cubic   (b) A helix

Figure 3.1. A few space curves.

Finally, a parametrization for the circle we are looking for is given by

$$\vec{x}(t) = \vec{p} + (7\cos t)\vec{u_1} + (7\sin t)\vec{u_2}.$$

Example 3.1.4 (Twisted Cubic). One of the simplest non-planar curves is called the twisted cubic, which has the parametrization $\vec{x}(t) = (t, t^2, t^3)$ (see Figure 3.1(a)).

One might wonder why this should be considered the simplest nonplanar curve. After all, could one not create a nonplanar curve with just quadratic polynomials? In fact, the answer is no. A curve $\vec{x}(t)$ with quadratic polynomials for coordinate functions can be written as

$$\vec{x}(t) = \vec{a}t^2 + \vec{b}t + \vec{c},$$

where $\vec{a}$, $\vec{b}$, and $\vec{c}$ are constant vectors. If $\vec{a}$ and $\vec{b}$ are linearly independent, then the image of $\vec{x}(t)$ lies in the plane through $\vec{c}$ with direction vectors $\vec{a}$ and $\vec{b}$, in other words, the plane through $\vec{c}$ with normal vector $\vec{a} \times \vec{b}$.

Example 3.1.5 (Circular Helix). A circular helix is a space curve that wraps around a vertical circular cylinder, climbing in altitude at a constant rate. We have for equations

$$\vec{x}(t) = (a\cos t, a\sin t, bt).$$

See Figure 3.1(b) for an example with $\vec{x}(t) = (2\cos t, 2\sin t, 0.2t)$.

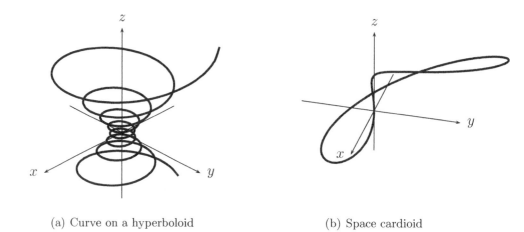

(a) Curve on a hyperboloid   (b) Space cardioid

Figure 3.2. Examples of parametric curves.

Example 3.1.6.   The coordinate functions of the parametric curve $\vec{x}(t) = (at\cos t, at\sin t, bt)$, for a and b not zero, satisfy the algebraic equation

$$\frac{x^2}{a^2} + \frac{y^2}{a^2} - \frac{z^2}{b^2} = 0$$

for all t. Thus, the image of $\vec{x}$ lies on the circular cone described by this equation.

Example 3.1.7. Consider the parametric curve $\vec{x}(t) = (a\cosh t\cos t, a\cosh t\sin t, b\sinh t)$, for a and b not zero. It is not hard to see that the coordinate functions of $\vec{x}$ satisfy

$$\frac{x^2}{a^2} + \frac{y^2}{a^2} - \frac{z^2}{b^2} = 1$$

so that the image of $\vec{x}$ lies on the hyperboloid of one sheet. See Figure 3.2(a) for an example with $\vec{x}(t) = (\cosh t\cos(10t), \cosh t\sin(10t), \sinh t)$.

Example 3.1.8 (Space Cardioid). The parametric curve with equation

$$\vec{x}(t) = ((1 - \cos t)\cos t, (1 - \cos t)\sin t, \sin t)$$

is called the *space cardioid* (see Figure 3.2(b)). One interesting property of this curve is that it is a closed curvein $\mathbb{R}^3$ with no critical points. However, projected into the xy-plane, it gives the cardioid. In an intuitive sense, we have stretched the cardioid out of the plane and removed its critical point (or cusp).

Using the same vocabulary as in Section 1.1, we call a *reparametrization* of a parametrized curve $\vec{x} : I \to \mathbb{R}^n$ any other continuous function $\vec{\xi} : J \to \mathbb{R}^n$ defined by

$$\vec{\xi} = \vec{x} \circ g$$

for some surjective function $g : J \to I$. Then the image $\mathcal{C} \subseteq \mathbb{R}^n$ of $\vec{\xi}$ is the same as the image of $\vec{x}$. When g is not surjective, the image of $\vec{\xi}$ could be a proper subset of $\mathcal{C}$, and we do not call $\vec{\xi}$ a reparametrization.

The definition in Equation (1.3) of the derivative for a vector function was purposefully presented irrespective of the dimension. We restate the definition in an alternate form that is sometimes better suited for proofs.

Definition 3.1.9. Let I be an interval in $\mathbb{R}$, and let $t_0 \in I$. If $\vec{f} : I \to \mathbb{R}^n$ is a continuous vector function, we say that $\vec{f}$ is *differentiable* at t_0 with derivative $\vec{f}'(t_0)$ if there exists a vector function $\vec{\varepsilon}$ such that

$$\vec{f}(t_0 + h) = \vec{f}(t_0) + \vec{f}'(t_0)h + h\vec{\varepsilon}(h) \qquad \text{and} \qquad \lim_{h \to 0} \|\vec{\varepsilon}(h)\| = 0.$$

It follows as a consequence of Proposition 1.1.6 (modified to vector functions in $\mathbb{R}^n$) that a vector function $\vec{x} : I \to \mathbb{R}^n$ is differentiable at a given point if and only if all its coordinate functions are differentiable at that point. Furthermore, if $\vec{x} : I \to \mathbb{R}^n$ is a continuous vector function that is differentiable at $t = t_0$ and if the coordinate functions of $\vec{x}$ are

$$\vec{x}(t) = (x_1(t), x_2(t), \ldots, x_n(t)),$$

then the derivative $\vec{x}'(t_0)$ is

$$\vec{x}'(t_0) = \left(x_1'(t_0), x_2'(t_0), \ldots, x_n'(t_0)\right).$$

As in $\mathbb{R}^2$, for any vector function $\vec{x} : I \to \mathbb{R}^n$, borrowing from the language of trajectories in mechanics, we often call $\vec{x}$ the *position function*, $\vec{x}'$ the *velocity function*, and $\vec{x}''$ the *acceleration function*. Furthermore, we define the *speed* function associated to $\vec{x}$ as the function $s' : I \rightarrow \mathbb{R}$ defined by

$$s'(t) = \|\vec{x}'(t)\|.$$

It is also a simple matter to define equations for the tangent line to a parametrized curve $\vec{x}(t)$ at $t = t_0$ as long as $\vec{x}'(t_0) \neq \vec{0}$. The parametric equations for the tangent line are

$$\vec{l}(u) = \vec{x}(t_0) + u\,\vec{x}'(t_0).$$

Proposition 3.1.10. *Proposition 1.2.3 holds for differentiable vector functions $\vec{v}$ and $\vec{w}$ from $I \subset \mathbb{R}$ into $\mathbb{R}^n$. Furthermore, suppose that $\vec{v}(t)$ and $\vec{w}(t)$ are vector functions into $\mathbb{R}^3$ that are defined on and differentiable over an interval $I \subset \mathbb{R}$. If $\vec{x}(t) = \vec{v}(t) \times \vec{w}(t)$, then $\vec{x}$ is differentiable over I and*

$$\vec{x}'(t) = \vec{v}'(t) \times \vec{w}(t) + \vec{v}(t) \times \vec{w}'(t).$$

Proof: (Left as an exercise for the reader. See Problem 3.1.10.) $\square$

As in Section 1.1, in this introductory section for space curves, the problems focus on properties of vectors and vector functions in $\mathbb{R}^3$.

Problems

3.1.1. Calculate the velocity, acceleration, and speed of the following space curves:

   (a) The space cardioid $\vec{x}(t) = ((1-\cos t)\cos t, (1-\cos t)\sin t, \sin t)$.
   (b) The twisted cubic $\vec{x}(t) = (t, t^2, t^3)$.
   (c) $\vec{x}(t) = (\tan^{-1}(t), \sin t, \cos 2t)$.

3.1.2. Let $\vec{a}$, $\vec{b}$, and $\vec{c}$ be three vectors in $\mathbb{R}^3$. Prove that

$$(\vec{a} \times \vec{b}) \cdot \vec{c} = (\vec{b} \times \vec{c}) \cdot \vec{a} = (\vec{c} \times \vec{a}) \cdot \vec{b}.$$

3.1.3. Let $\vec{a}$, $\vec{b}$, $\vec{c}$, and $\vec{d}$ be four vectors in $\mathbb{R}^3$. Prove that

$$(\vec{a} \times \vec{b}) \cdot (\vec{c} \times \vec{d}) = \begin{vmatrix} (\vec{a} \cdot \vec{c}) & (\vec{b} \cdot \vec{c}) \\ (\vec{a} \cdot \vec{d}) & (\vec{b} \cdot \vec{d}) \end{vmatrix}.$$

3.1.4. Let $\vec{\alpha}(t)$ be a regular, parametrized curve in the xy-plane, viewed as a subset of $\mathbb{R}^3$. Let $\vec{p}$ be a fixed point in the plane not on the curve, and let $\vec{u}$ be a fixed vector. Let $\theta(t)$ be the angle $\vec{\alpha}(t) - \vec{p}$ makes with the direction $\vec{u}$. Prove that (up to a change in sign)

$$\theta'(t) = \frac{\|\vec{\alpha}'(t) \times (\vec{\alpha}(t) - \vec{p})\|}{\|\vec{\alpha}(t) - \vec{p}\|^2}.$$

Conclude that the angle function $\theta(t)$ of $\vec{\alpha}(t) - \vec{p}$ with respect to the direction $\vec{u}$ has a local extremum at a point t_0 if and only if $\vec{\alpha}'(t_0)$ is parallel to $\vec{\alpha}(t_0) - \vec{p}$.

3.1.5. Consider the space curve $\vec{x}(t) = (t\cos t, t\sin t, t)$. Find where the extremal values of the angle between $\vec{x}'$ and $\vec{x}''$ occur and determine whether they are maxima or minima.

3.1.6. Consider the plane in $\mathbb{R}^3$ given by the equation $ax + by + cz + f = 0$ and a point $P = (x_0, y_0, z_0)$. Prove that the distance between the plane and the point P is

$$d = \frac{|ax_0 + by_0 + cz_0 + f|}{\sqrt{a^2 + b^2 + c^2}}.$$

3.1.7. Determine the angle of intersection between the lines $\vec{u}(t) = (2t-1, 3t+2, -2t+3)$ and $\vec{v}(t) = (3t-1, 5t+2, 3t+3)$.

3.1.8. Consider the two lines given by $\vec{u}(t) = (2t-1, 3t+2, -2t+3)$ and $\vec{v}(t) = (3t+1, -5t-3, 3t-1)$. Find the shortest distance between these two lines. [Hint: Consider the function $f(s,t) = \|\vec{u}(s) - \vec{v}(t)\|$.]

3.1.9. Consider two nonparallel lines given by the equations

$$\vec{l}_1(s) = \vec{a} + s\vec{u} \qquad \text{and} \qquad \vec{l}_2(t) = \vec{b} + t\vec{v}.$$

Prove that the distance d between these two lines is

$$d = \frac{\|(\vec{u} \times \vec{v}) \cdot (\vec{b} - \vec{a})\|}{\|\vec{u} \times \vec{v}\|}.$$

3.1.10. Let $\vec{\alpha}(t)$ and $\vec{\beta}(t)$ be two differentiable parametrized curves $\mathbb{R} \to \mathbb{R}^n$. Prove the following:

(a) If $f(t) = \vec{\alpha}(t) \cdot \vec{\beta}(t)$, then $f'(t) = \vec{\alpha}'(t) \cdot \vec{\beta}(t) + \vec{\alpha}(t) \cdot \vec{\beta}'(t)$.

(b) Suppose that $n = 3$. If $\vec{\gamma}(t) = \vec{\alpha}(t) \times \vec{\beta}(t)$, then $\vec{\gamma}'(t) = \vec{\alpha}'(t) \times \vec{\beta}(t) + \vec{\alpha}(t) \times \vec{\beta}'(t)$.

3.1.11. Let $\vec{x}(t)$ be any parametrized curve that is of class C^3. Prove that

$$\frac{d}{dt}\left(\vec{x}(t) \cdot (\vec{x}'(t) \times \vec{x}''(t))\right) = \vec{x}(t) \cdot (\vec{x}'(t) \times \vec{x}'''(t)).$$

[Hint: Use Problem 3.1.10.]

3.2   Curvature, Torsion, and the Frenet Frame

Let I be an interval of $\mathbb{R}$, and let $\vec{x} : I \to \mathbb{R}^3$ be a differentiable space curve. Following the setup with plane curves, we can talk about the unit tangent vector $\vec{T}(t)$ defined by

$$\vec{T}(t) = \frac{\vec{x}'(t)}{\|\vec{x}'(t)\|}. \tag{3.1}$$

Obviously, the unit tangent vector is not defined at a value $t = t_0$ if $\vec{x}$ is not differentiable at t_0 or if $\vec{x}'(t_0) = 0$. This leads to the following definition (which has been generalized to curves in $\mathbb{R}^n$).

Definition 3.2.1. Let $\vec{x} : I \to \mathbb{R}^n$ be a parametrized curve. We call any point $t = t_0$ a *critical point* if $\vec{x}'(t_0)$ is not defined or is equal to $\vec{0}$. A point $t = t_0$ that is not critical is called a *regular point*. A parametrized curve $\vec{x} : I \to \mathbb{R}^n$ is called *regular* if it is of class C^1 (i.e., continuously differentiable) and if $\vec{x}'(t) \neq \vec{0}$ for all $t \in I$.

In practice, if $\vec{x}'(t_0) = \vec{0}$ but $\vec{x}'(t) \neq \vec{0}$ for all t in some interval J around t_0, it is possible for

$$\lim_{t \to t_0} \frac{\vec{x}'(t)}{\|\vec{x}'(t)\|} \tag{3.2}$$

to exist. In such cases, it is common to think of $\vec{T}$ as completed by continuity by calling $\vec{T}(t_0)$ the limit in Equation (3.2).

Since $\vec{T}(t)$ is a unit vector for all $t \in I$, we have $\vec{T} \cdot \vec{T} = 1$ for all $t \in I$. Therefore, if $\vec{T}$ is itself differentiable,

$$\frac{d}{dt}\left(\vec{T}(t) \cdot \vec{T}(t)\right) = 2\vec{T}'(t) \cdot \vec{T}(t) = 0.$$

Thus, the derivative of the unit tangent vector $\vec{T}'$ is perpendicular to $\vec{T}$. At any point t on the curve, where $\vec{T}'(t) \neq \vec{0}$, we define the *principal normal vector* $\vec{P}(t)$ to the curve at t to be the unit vector

$$\vec{P} = \frac{\vec{T}'}{\|\vec{T}'\|}.$$

Again, it is still possible sometimes to complete $\vec{P}$ by continuity even at points $t = t_0$, where $\vec{T}'(t_0) = \vec{0}$. In such cases, it is not uncommon to assume that $\vec{P}(t)$ is completed by continuity wherever possible. Nonetheless, by Equation (3.1), the requirement that $\vec{T}$ be differentiable is tantamount to $\vec{x}$ being twice differentiable.

Finally, for a space curve defined over any interval J, where $\vec{T}(t)$ and $\vec{P}(t)$ exist (perhaps when completed by continuity), we complete the set $\{\vec{T}, \vec{P}\}$ to an orthonormal frame by adjoining the so-called *binormal vector* $\vec{B}(t)$ given as

$$\vec{B} = \vec{T} \times \vec{P}.$$

Definition 3.2.2. Let $\vec{x} : I \to \mathbb{R}^3$ be a continuous space curve of class C^2 (i.e., has a continuous second derivative). To each point $\vec{x}(t)$ on the curve, we associate the *Frenet frame* as the triple of vectors $(\vec{T}, \vec{P}, \vec{B})$.

Figure 3.3 illustrates the Frenet frame as it moves through $t = 0$ on the space cardioid

$$\vec{x}(t) = ((1 - \cos t)\cos t, (1 - \cos t)\sin t, \sin t).$$

(In this figure, the basis vectors of the Frenet frame were scaled down by a factor of $\frac{1}{2}$ to make the picture clearer.) In this example, it is interesting to see that near $t = 0$, though the unit tangent vector does not change too quickly, the vectors $\vec{P}$ and $\vec{B}$ rotate quickly about the tangent line. The functions that measure how fast $\vec{T}$ (respectively $\vec{B}$) changes are called the curvature (respectively torsion) functions of the space curve.

Similar to the basis $\{\vec{T}(t), \vec{U}(t)\}$ for plane curves, the Frenet frame provides a geometrically natural basis in which to study local properties of space curves. We now analyze the derivatives of a space curve $\vec{x}'(t)$ in reference to the Frenet frame.

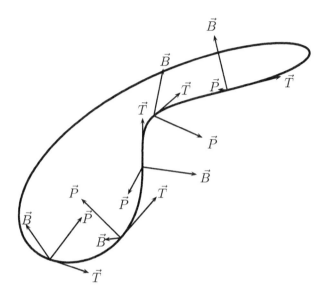

Figure 3.3. Moving Frenet frame on the space cardioid.

Recall that the speed $s'(t)$ is $\|\vec{x}'(t)\|$. By definition of the unit tangent vector, we have

$$\vec{x}'(t) = s'(t)\vec{T}(t).$$

If $\vec{x}$ is twice differentiable at t, then $\vec{T}'(t)$ exists. If, in addition, $\vec{P}$ is defined at t, then by definition of the principal normal vector, $\vec{T}'$ is parallel to $\vec{P}$. This allows us to define the curvature of a space curve as follows.

Definition 3.2.3. Let $\vec{x}$ be a parametrized curve of class C^2. The curvature $\kappa : I \to \mathbb{R}^+$ of a space curve is

$$\kappa(t) = \frac{\|T'(t)\|}{s'(t)}.$$

Note that at any point where $P(t)$ is defined, $\kappa(t)$ is the nonnegative number defined by

$$\vec{T}'(t) = s'(t)\kappa(t)\vec{P}(t).$$

We would like to determine how the other unit vectors of the Frenet frame behave under differentiation. Remember that $\vec{B} = \vec{T} \times \vec{P}$.

Taking a derivative of this cross product, one obtains

$$\vec{B}' = \vec{T}' \times \vec{P} + \vec{T} \times \vec{P}' = \vec{T} \times \vec{P}',$$

since $\vec{T}'$ is parallel to $\vec{P}$. However, just as $\vec{T}' \perp \vec{T}$, we have the same for $\vec{P}' \perp \vec{P}$. Thus, since we are in three dimensions, we can write $\vec{P}'(t) = f(t)\vec{T}(t) + g(t)\vec{B}(t)$ for some continuous functions $f, g : I \rightarrow \mathbb{R}$. Consequently, we deduce that

$$\vec{B}' = \vec{T} \times (f(t)\vec{T} + g(t)\vec{B}) = f(t)\vec{T} \times \vec{T} + g(t)\vec{T} \times \vec{B} = -g(t)\vec{P}.$$

Thus, the derivative of the unit binormal vector is parallel to the principal normal vector.

Definition 3.2.4. Let $\vec{x} : I \rightarrow \mathbb{R}^3$ be a regular space curve of class C^2 for which the Frenet frame is defined everywhere. We define the *torsion function* $\tau : I \rightarrow \mathbb{R}$ as the unique function such that

$$\vec{B}'(t) = -s'(t)\tau(t)\vec{P}(t).$$

We now need to determine $\vec{P}'(t)$, and we already know that it has the form $f(t)\vec{T}(t) + g(t)\vec{B}(t)$. However, we can say much more without performing any specific calculations. Since $(\vec{T}, \vec{P}, \vec{B})$ form an orthonormal frame for all t, we have the following equations:

$$\vec{T} \cdot \vec{P} = 0 \qquad \text{and} \qquad \vec{P} \cdot \vec{B} = 0.$$

Taking derivatives with respect to t, we have

$$\vec{T}' \cdot \vec{P} + \vec{T} \cdot \vec{P}' = 0 \qquad \text{and} \qquad \vec{P}' \cdot \vec{B} + \vec{P} \cdot \vec{B}' = 0,$$

that is,

$$\vec{P}' \cdot \vec{T} = -\vec{T}' \cdot \vec{P} \qquad \text{and} \qquad \vec{P}' \cdot \vec{B} = -\vec{P} \cdot \vec{B}'.$$

Thus, we deduce that

$$\vec{P}'(t) = -s'(t)\kappa(t)\vec{T}(t) + s'(t)\tau(t)\vec{B}(t). \qquad (3.3)$$

If we assume, as one does in linear algebra, that $\vec{T}$, $\vec{P}$, and $\vec{B}$ are column vectors and write $\begin{pmatrix} \vec{T} & \vec{P} & \vec{B} \end{pmatrix}$ as the matrix that has these vectors as columns, we can summarize Definition 3.2.3, Definition 3.2.4 and Equation (3.3) in matrix form by

$$\frac{d}{dt}\begin{pmatrix} \vec{T} & \vec{P} & \vec{B} \end{pmatrix} = \begin{pmatrix} \vec{T} & \vec{P} & \vec{B} \end{pmatrix} \begin{pmatrix} 0 & -s'\kappa & 0 \\ s'\kappa & 0 & -s'\tau \\ 0 & s'\tau & 0 \end{pmatrix}. \tag{3.4}$$

As provided, the definitions for curvature and torsion of a space curve do not particularly lend themselves to direct computations when given a particular curve. We now obtain formulas for $\kappa(t)$ and $\tau(t)$ in terms of $\vec{x}(t)$. In order for our formulas to make sense, we assume for the remainder of this section that the parametrized curve $\vec{x} : I \to \mathbb{R}^3$ is regular and of class C^3.

First, to find the curvature of a curve, we take the following derivatives:

$$\vec{x}'(t) = s'(t)\vec{T}(t), \tag{3.5}$$

$$\vec{x}''(t) = s''(t)\vec{T}(t) + (s'(t))^2\kappa(t)\vec{P}(t). \tag{3.6}$$

Taking the cross product of these two vectors, we now obtain

$$\vec{x}'(t) \times \vec{x}''(t) = (s'(t))^3\kappa(t)\vec{B}(t).$$

However, by definition of curvature for a space curve, $\kappa(t)$ is a nonnegative function. Furthermore, $s'(t) = \|\vec{x}'(t)\|$, so we get

$$\kappa(t) = \frac{\|\vec{x}'(t) \times \vec{x}''(t)\|}{\|\vec{x}'(t)\|^3}. \tag{3.7}$$

Secondly, to obtain the torsion function from $\vec{x}(t)$ directly, we will need to take the third derivative

$$\vec{x}'''(t) = s'''(t)\vec{T} + s''(t)s'(t)\kappa(t)\vec{P} + 2s'(t)s''(t)\kappa(t)\vec{P}$$
$$+ (s'(t))^2\kappa'(t)\vec{P} + (s'(t))^3\kappa(t)(-\kappa(t)\vec{T} + \tau(t)\vec{B}),$$

which, without writing the dependence on t explicitly, reads

$$\vec{x}''' = \left(s''' - (s')^3\kappa^2\right)\vec{T} + \left(3s''s'\kappa + (s')^2\kappa'\right)\vec{P} + (s')^3\kappa\tau\vec{B}. \tag{3.8}$$

Taking the dot product of $\vec{x}' \times \vec{x}''$ with $\vec{x}'''$ eliminates all the terms of $\vec{x}'''$ associated to $\vec{T}$ and $\vec{P}$. One obtains

$$(\vec{x}' \times \vec{x}'') \cdot \vec{x}''' = (s'(t))^6 (\kappa(t))^2 \tau(t),$$

from which one deduces a formula for $\tau(t)$ only in terms of the curve $\vec{x}(t)$ as follows:

$$\tau(t) = \frac{(\vec{x}'(t) \times \vec{x}''(t)) \cdot \vec{x}'''(t)}{\|\vec{x}'(t) \times \vec{x}''(t)\|^2}. \tag{3.9}$$

Example 3.2.5 (Helices). We wish to calculate the curvature and torsion of the circular helix $\vec{x}(t) = (a\cos t, a\sin t, bt)$. We need the following:

$$\vec{x}'(t) = (-a\sin t, a\cos t, b),$$
$$s'(t) = \sqrt{a^2 \sin^2 t + a^2 \cos^2 t + b^2} = \sqrt{a^2 + b^2},$$
$$\vec{x}''(t) = (-a\cos t, -a\sin t, 0),$$
$$\vec{x}'(t) \times \vec{x}''(t) = (ab\sin t, -ab\cos t, a^2),$$
$$\vec{x}'''(t) = (a\sin t, -a\cos t, 0).$$

Note that one can easily parametrize the helix by arc length since $s'(t) = \sqrt{a^2 + b^2}$ is a constant function and, therefore, we have $s(t) = t\sqrt{a^2 + b^2}$. Thus, the parametrization by arc length for the circular helix is

$$\vec{x}(s) = \left(a\cos\left(\frac{s}{c}\right), a\sin\left(\frac{s}{c}\right), \frac{b}{c}s \right), \qquad \text{where } c = \sqrt{a^2 + b^2}.$$

We now calculate the curvature and torsion as

$$\kappa(t) = \frac{\|\vec{x}' \times \vec{x}''\|}{(s')^3} = \frac{a\sqrt{a^2 + b^2}}{(a^2 + b^2)^{3/2}} = \frac{a}{a^2 + b^2},$$
$$\tau(t) = \frac{(\vec{x}' \times \vec{x}'') \cdot \vec{x}'''}{\|\vec{x}' \times \vec{x}''\|^2} = \frac{a^2 b}{a^2(a^2 + b^2)} = \frac{b}{a^2 + b^2}.$$

Consequently, we find that this circular helix has constant curvature and constant torsion.

The circular helix, however, is a particular case of a larger class of curves simply called *helices*. These are defined by requiring that the unit tangent make a constant angle with a fixed line in space. Thus, $\vec{x}(t)$ is a helix if and only if for some unit vector $\vec{u}$,

$$\vec{T} \cdot \vec{u} = \cos \alpha = \text{const.} \tag{3.10}$$

Taking a derivative of Equation (3.10), we obtain

$$\vec{P} \cdot \vec{u} = 0.$$

Hence for all t, $\vec{u}$ is in the plane determined by $\vec{T}$ and $\vec{B}$, so we can write

$$\vec{u} = \cos \alpha \vec{T} + \sin \alpha \vec{B}$$

for all t. Taking the derivative, we obtain

$$0 = s'\kappa \cos \alpha \vec{P} - s'\tau \sin \alpha \vec{P},$$

which implies that

$$\frac{\tau(t)}{\kappa(t)} = \cot \alpha$$

and thus, for any helix, the ratio of curvature to torsion is a constant. This ratio $\frac{\tau}{\kappa}$ is called the *pitch* of the helix.

One can follow the above discussion in reverse and also conclude that the converse is true. Therefore, a curve is a helix if and only if the ratio of curvature to torsion is a constant.

Example 3.2.6 (Space Cardioid). We consider the space cardioid again as a follow-up to Figure 3.3. Figure 3.3 shows that in the vicinity of $t = 0$, the Frenet frame twists quickly about the tangent line, even while the tangent line does not move much. This indicates that near 0, $\kappa(t)$ is not large, while $\tau(t)$ is relatively large.

We leave the precise calculation of the curvature and torsion functions to the space cardioid as an exercise for the reader but plot their graphs in Figure 3.4. The graphs of $\kappa(t)$ and $\tau(t)$ justify the intuition provided by Figure 3.3 concerning how the Frenet frame moves through $t = 0$. In particular, the torsion function has a high peak at $t = 0$, which indicates that the Frenet frame rotates quickly about the tangent line.

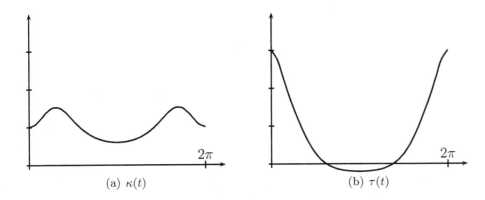

(a) $\kappa(t)$ (b) $\tau(t)$

Figure 3.4. Curvature and torsion of the space cardioid.

In some proofs that we will encounter later, it is often useful to assume that a curve is parametrized by arc length. In this case, in all of the above formulas, one has $s' = 1$ and $s'' = 0$ as functions. The transformation properties of the Frenet frame then read

$$\frac{d}{ds}\begin{pmatrix} \vec{T} & \vec{P} & \vec{B} \end{pmatrix} = \begin{pmatrix} \vec{T} & \vec{P} & \vec{B} \end{pmatrix} \begin{pmatrix} 0 & -\kappa & 0 \\ \kappa & 0 & -\tau \\ 0 & \tau & 0 \end{pmatrix}. \qquad (3.11)$$

If $\vec{x}(s)$ is parametrized by arc length and is of class C^3, then the curvature is given by

$$\kappa(s) = \|\vec{x}'(s) \times \vec{x}''(s)\|. \qquad (3.12)$$

Furthermore, at any point where $\kappa(s) \neq 0$, the torsion function is

$$\tau(s) = \frac{(\vec{x}'(s) \times \vec{x}''(s)) \cdot \vec{x}'''(s)}{\kappa(s)^2}. \qquad (3.13)$$

A key property of the curvature and torsion functions of a space curve is summarized in the following proposition.

Proposition 3.2.7. *Let $\vec{x} : I \to \mathbb{R}^3$ be a regular parametric curve.*

1. *Suppose that $\vec{x}$ is of class C^2. If the curvature $\kappa(t)$ is identically 0, then the locus of $\vec{x}$ is a line segment.*

2. *Suppose that $\vec{x}$ is of class C^3. If the torsion $\tau(t)$ is identically 0, then the locus of $\vec{x}$ lies in a plane.*

Proof: (Left as an exercise for the reader. See Problem 3.2.7.) □

Problems

3.2.1. Calculate the curvature and torsion of the following curves:

(a) The twisted cubic $\vec{x}(t) = (t, t^2, t^3)$.

(b) The space cardioid $\vec{x}(t) = ((1-\cos t)\cos t, (1-\cos t)\sin t, \sin t)$.

(c) $\vec{x}(t) = (t, f(t), g(t))$, where f and g are smooth functions.

(d) $\vec{x} = (a(t - \sin t), a(1 - \cos t), bt)$.

(e) $\vec{x}(t) = (t, \frac{1+t}{t}, \frac{1-t^2}{t})$.

3.2.2. In Chapter 5, we will encounter a surface called a torus. Curves that lie on torus are often knotted. Define a *torus knot* as a curve parametrized by

$$\vec{x}(t) = ((a + b\cos(qt))\cos(pt), (a + b\cos(qt))\sin(pt), b\sin(qt))$$

where a and b are real numbers, with $a > b > 0$, and p and q are positive integers. Let $a = 2$ and $b = 1$. Calculate the curvature functions of the corresponding torus knot.

3.2.3. Calculate the curvature function of the figure-eight knot given with the parametrization

$$\vec{x}(t) = (10\cos t + 10\cos 3t + cos2t + \cos 4t, 6\sin t + 10\sin 3t, 4\sin 3t \sin(5t/2) - 2\sin 6t).$$

3.2.4. Let $\vec{x}$ be a parametrized curve, and let $\vec{\xi} = \vec{x} \circ g$ be a reparametrization of $\vec{x}$.

(a) Prove that the curvature function κ is unchanged under a reparametrization, i.e., that $\kappa_{\vec{x}}(g(u)) = \kappa_{\vec{\xi}}(u)$.

(b) Prove that the absolute value of the torsion function $|\tau|$ is unchanged under reparametrization.

3.2.5. Consider the curve $\vec{x}(t) = (a\sin t \cos t, a\sin^2 t, a\cos t)$.

(a) Prove that the locus of $\vec{x}$ lies on a sphere.

(b) Calculate the curvature and torsion functions of $\vec{x}(t)$.

3.2.6. If $\vec{x}(t)$ is a parametrization for a planar curve, then for some fixed vectors $\vec{a}$, $\vec{b}$, and $\vec{c}$, with $\vec{b}$ and $\vec{c}$ not collinear, and for some real functions $f(t)$ and $g(t)$, we can write

$$\vec{x}(t) = \vec{a} + f(t)\vec{b} + g(t)\vec{c}.$$

Prove that for a planar curve, its torsion function is identically 0.

3.2.7. (ODE) Prove Proposition 3.2.7.

3.2.8. As with plane curves, we define the evolute to a space curve $\vec{x}(t)$ as the curve

$$\vec{\gamma}(t) = \vec{x}(t) + \frac{1}{\kappa(t)}\vec{P}(t).$$

Prove that the evolute to the circular helix $\vec{x}(t) = (a\cos t, a\sin t, bt)$ is another circular helix and find the pitch of this new helix.

3.2.9. Consider the circular cone with equation

$$\frac{x^2}{a^2} + \frac{y^2}{a^2} = \frac{z^2}{b^2}.$$

(a) Find parametric equations for a curve that lies on this cone such that the tangent makes a constant angle with the z-axis.

(b) Prove that they project to logarithmic spirals on the xy-plane.

(c) Find the natural equations for such a helix (i.e., find the curvature and torsion).

3.2.10. Let $\vec{\alpha} : I \to \mathbb{R}^3$ be a parametrized regular curve with $\kappa(t) \neq 0$ and $\tau(t) \neq 0$ for $t \in I$. The curve $\vec{\alpha}$ is called a *Bertrand curve* if there exists another curve $\vec{\beta} : I \to \mathbb{R}^3$ such that the principal normal lines to $\vec{\alpha}$ and $\vec{\beta}$ are equal at all $t \in I$. The curve $\vec{\beta}$ is called a Bertrand mate of $\vec{\alpha}$.

(a) Prove that we can write

$$\vec{\beta}(t) = \vec{\alpha}(t) + r\vec{P}(t)$$

for some constant r.

(b) Prove that $\vec{\alpha}$ is a Bertrand curve if and only if there exists a linear relation

$$a\kappa(t) + b\tau(t) = 1 \qquad \text{for all } t \in I,$$

where a and b are nonzero constants and $\kappa(t)$ and $\tau(t)$ are the curvature and the torsion of $\vec{\alpha}$ respectively.

(c) Prove that a curve $\vec{\alpha}$ has more than one Bertrand mate if and only if $\vec{\alpha}$ is a circular helix or is planar.

3.3   Osculating Plane and Osculating Sphere

As with plane curves, if $\vec{x}(t)$ is a regular space curve of class C^2, one can talk about the osculating circle to $\vec{x}(t)$ at a point $t = t_0$. Recall that an osculating circle to a curve at a point is a circle with contact of order 2 at that point. Following the proof of Proposition 1.4.5, one determines that at any point $t = t_0$, where $\kappa(t_0) \neq 0$, the osculating circle exists and that a parametric formula for it is

$$\vec{\gamma}(u) = \vec{x}(t_0) + \frac{1}{\kappa(t_0)}\vec{P}(t_0) + \frac{1}{\kappa(t_0)}\left((\sin u)\vec{T}(t_0) - (\cos u)\vec{P}(t_0)\right).$$

$$(3.14)$$

Even without reference to osculating circles, given any parametric curve $\vec{x} : I \rightarrow \mathbb{R}^3$, the second-order Taylor approximation to $\vec{x}$ at $t = t_0$ is a planar curve with contact of order 2. Furthermore, if $\vec{T}'(t_0) \neq \vec{0}$, this second-order approximation $\vec{f}$ is

$$\vec{f}(t) = \vec{x}(t_0) + (t - t_0)\vec{x}'(t_0) + \frac{1}{2}(t - t_0)^2\vec{x}''(t_0)$$

$$= \vec{x}(t_0) + \left(s'(t_0)(t - t_0) + \frac{1}{2}s''(t_0)(t - t_0)^2\right)\vec{T}(t_0)$$

$$+ \frac{1}{2}s'(t_0)^2\kappa(t_0)(t - t_0)^2\vec{P}(t_0).$$

The vector function $\vec{f}$ is a planar curve that lies in the plane that goes through the point $\vec{x}(t_0)$ and has $\vec{T}(t_0)$ and $\vec{P}(t_0)$ as direction vectors. (If $\kappa(t_0) = 0$, then $\vec{P}(t_0)$ is not strictly defined and might not even be defined by completing by continuity. In this case, the second-order approximation $\vec{f}$ to $\vec{x}$ at t_0 lies on a line.) This leads to the following definition:

Definition 3.3.1. Let $\vec{x} : I \rightarrow \mathbb{R}^3$ be a parametrized curve, and let $t_0 \in I$. Suppose that $\vec{x}$ is of class C^2 over an open interval containing t_0 and that $\kappa(t_0) \neq 0$. The *osculating plane* to $\vec{x}$ at $t = t_0$ is the plane through $\vec{x}(t_0)$ spanned by $\vec{T}(t_0)$ and $\vec{P}(t_0)$. In other words, the osculating plane is the set of points $\vec{u} \in \mathbb{R}^3$ such that

$$\vec{B}(t_0) \cdot (\vec{u} - \vec{x}(t_0)) = 0.$$

We introduced the notion of order of contact between two curves in Section 1.4, but we can also talk about the order of contact between a curve C and a surface Σ by defining this latter notion as

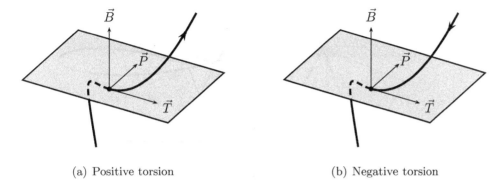

(a) Positive torsion   (b) Negative torsion

Figure 3.5. Torsion and the osculating plane.

the order of contact between C and C_2, where C_2 is the orthogonal projection of C onto Σ.

Proposition 3.3.2. *Let $\vec{x} : I \to \mathbb{R}^3$ be a regular parametrized curve of class C^2. The osculating plane to $\vec{x}$ at $t = t_0$ is the unique plane in $\mathbb{R}^3$ of contact order 2 or greater. Furthermore, assuming $\vec{x}$ is of class C^3, the osculating plane has contact order 3 or greater if and only if $\tau(t_0) = 0$ or $\kappa(t_0) = 0$.*

Proof: Set $A = \vec{x}(t_0)$ and reparametrize $\vec{x}$ by arc length so that $s = 0$ corresponds to the point A. The orthogonal distance $f(s)$ between $\vec{x}(s)$ and the osculating plane $\mathcal{P}$ is

$$f(s) = |\vec{B}(0) \cdot (\vec{x}(s) - \vec{x}(0))|.$$

From Equations (3.5), (3.6), and (3.8), using the Taylor approximation of $\vec{x}(s)$ near $s = 0$ and the fact that $\vec{B} \cdot \vec{T} = \vec{B} \cdot \vec{P} = 0$, we deduce that

$$f(s) = \left| \frac{1}{6}\kappa(0)\tau(0)s^3 + \text{ higher order terms} \right|.$$

Thus,

$$\lim_{s \to 0} \frac{f(s)}{s^3} = \frac{|\kappa(0)\tau(0)|}{6}$$

and the proposition follows. □

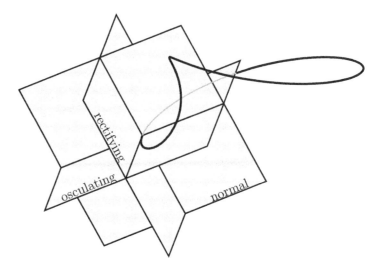

Figure 3.6. Moving trihedron.

One can now interpret the sign of the torsion function $\tau(t)$ of a parametrized curve in terms of the curve's position with respect to its osculating plane at a point. In fact, $\tau(t_0) < 0$ at $\vec{x}(t_0)$ when the curve comes up through the osculating plane (where the binormal $\vec{B}$ defines the up direction) and $\tau(t_0) > 0$ when the curve goes down through the osculating plane (see Figure 3.5).

The osculating plane along with two other planes form what is called the moving trihedron (Figure 3.6), which consists of the co-ordinate planes in the Frenet frame. The plane through $\vec{x}(t_0)$ and spanned by the principal normal and binormal is called the *normal plane* and is the set of points $\vec{u}$ that satisfy

$$\vec{T}(t_0) \cdot (\vec{u} - \vec{x}(t_0)) = 0.$$

The plane through the tangent and binormal is called the *rectifying plane* and is the set of points $\vec{u}$ that satisfy

$$\vec{P}(t_0) \cdot (\vec{u} - \vec{x}(t_0)) = 0.$$

We now apply the theory of order of contact from Section 1.4 to find the *osculating sphere*—a sphere that has order of contact 3 or higher to a curve at a point.

Suppose that a sphere has center $\vec{c}$ and radius r so that its points $\vec{X}$ satisfy the equation

$$\|\vec{X} - \vec{c}\|^2 = r^2.$$

Consider a curve $\mathcal{C}$ parametrized by arc length by $\vec{x} : I \to \mathbb{R}^3$. The distance $f(s)$ between the point $\vec{x}(s)$ on the curve and the sphere is

$$f(s) = \big| \, \|\vec{x}(s) - \vec{c}\| - r \big|.$$

Since derivatives of $G(t) = \sqrt{g(t)}$ are equal to 0 if and only if $g'(t) = 0$, then the derivatives of $f(s)$ are equal to 0 if and only if the derivatives of

$$h(s) = (\vec{x}(s) - \vec{c}) \cdot (\vec{x}(s) - \vec{c})$$

are equal to 0. The first three derivatives of $h(s)$ lead to

$$h'(s) = 0 \iff (\vec{x} - \vec{c}) \cdot \vec{T} = 0,$$
$$h''(s) = 0 \iff (\vec{x} - \vec{c}) \cdot \kappa\vec{P} + 1 = 0,$$
$$h'''(s) = 0 \iff (\vec{x} - \vec{c}) \cdot (\kappa'\vec{P} - \kappa^2\vec{T} + \kappa\tau\vec{B}) = 0.$$

Consequently, at any point $\vec{x}(s_0)$ on the curve such that $\kappa(s_0) \neq 0$ and $\tau(s_0) \neq 0$, the first three derivatives can be equal to 0 if we have

$$(\vec{x} - \vec{c}) \cdot \vec{T} = 0, \qquad (\vec{x} - \vec{c}) \cdot \vec{P} = -\frac{1}{\kappa}, \qquad (\vec{x} - \vec{c}) \cdot \vec{B} = \frac{\kappa'}{\kappa^2\tau}. \quad (3.15)$$

The osculating sphere to a curve $\vec{x}(s)$ at the point $s = s_0$ is then the sphere of center

$$\vec{c} = \vec{x}(s_0) + R(s_0)\vec{P}(s_0) + \frac{R'(s_0)}{\tau(s_0)}\vec{B}(s_0), \qquad (3.16)$$

where $R(s) = \frac{1}{\kappa(s)}$, and with radius

$$r = \sqrt{R(s_0)^2 + \left(\frac{R'(s_0)}{\tau(s_0)}\right)^2}.$$

That the first three derivatives of the distance function $f(s)$ are 0 implies that the curve $\mathcal{C}$ and the osculating sphere have contact of order 3.

If $\tau(s_0) = 0$, then $h'''(s)$ may still be 0 as long as $\kappa'(s_0) = 0$ at the same time. In that case, the curve admits a one-parameter family of osculating spheres at $\vec{x}(s_0)$, where the $\vec{B}$ component of the center $\vec{c}$ can be anything. This discussion leads to the following proposition.

Proposition 3.3.3. *Let $\vec{x} : I \to \mathbb{R}^3$ be a regular parametrized curve of class C^3. Let $t_0 \in I$ be a point on the curve where $\kappa(t_0) \neq 0$. The curve $\vec{x}$ admits an osculating sphere at $t = t_0$ if either (1) $\tau(t_0) \neq 0$ or (2) $\tau(t_0) = 0$ and $\kappa'(t_0) = 0$. Define $R(t) = \frac{1}{\kappa(t)}$. If $\tau(t_0) \neq 0$, then at t_0 the curve admits a unique osculating sphere with center $\vec{c}$ and radius r, where*

$$\vec{c} = \vec{x}(t_0) + R(t_0)\vec{P}(t_0) + \frac{R'(t_0)}{s'(t_0)\tau(t_0)}\vec{B}(t_0)$$

and

$$r = \sqrt{R(t_0)^2 + \left(\frac{R'(t_0)}{s'(t_0)\tau(t_0)}\right)^2}.$$

If $\tau(t_0) = 0$ and $\kappa'(t_0) = 0$, then at t_0 the curve admits as an osculating sphere any sphere with center $\vec{c}$ and radius r where

$$\vec{c} = \vec{x}(t_0) + R(t_0)\vec{P}(t_0) + c_B\vec{B}(t_0) \qquad and \qquad r = \sqrt{R(t_0)^2 + c_B^2},$$

where c_B is any real number.

Proof: The only matter to address beyond the previous discussion is to see how the various quantities in Equation (3.16) change under a reparametrization.

Let $J \subset \mathbb{R}$ be an interval, and let $f : J \to I$ be of class C^3 such that f' doesn't change sign. Then $\vec{\xi} = \vec{x} \circ f$ is a regular reparametrization of $\vec{x}$. It is not hard to check that

$$\vec{T}_\xi = \text{sign}(f')\vec{T},$$
$$\vec{P}_\xi = \vec{P},$$
$$\vec{B}_\xi = \text{sign}(f')\vec{B}.$$

Also, in Problem 3.2.4, the reader showed that

$$\kappa_\xi = \kappa,$$
$$\tau_\xi = \text{sign}(f')\tau.$$

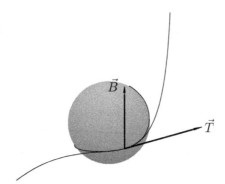

Figure 3.7. Osculating sphere.

Consquently, we have $R_\xi = R$ and also that $\frac{1}{\tau}\vec{B}$ is invariant under any regular reparametrization.

On the other hand, if $t = f(u)$ so that $\vec{\xi}(u) = \vec{x}(t)$, then $R_\xi(u) = R(f(u))$ and therefore, $R'_\xi(u) = R'(f(u))f'(u) = R'(t)f'(u)$. In particular, $\frac{dR}{ds} = \frac{1}{s'(t)}\frac{dR}{dt}$.

The proposition then follows from the prior discussion.   □

Figure 3.7 illustrates an example of a curve and its osculating sphere at a point. The figure also shows the Frenet frame at the point in question, along with the orthogonal projection of the curve onto the sphere, which one can easily obtain by

$$\vec{\gamma}(t) = \frac{\vec{x}(t) - \vec{c}}{\|\vec{x}(t) - \vec{c}\|}r.$$

Problems

3.3.1. Find the osculating plane to the space cardioid $\vec{x}(t) = ((1-\cos t)\cos t, (1 - \cos t)\sin t, \sin t)$ at $t = \pi$.

3.3.2. Let $\vec{x}(s) : I \to \mathbb{R}^3$ be the parametrization by arc length of a curve $\mathcal{C}$. Consider a circle that passes through the three points $\vec{x}(s)$, $\vec{x}(s+h_1)$ and $\vec{x}(s + h_2)$. Prove that as $(h_1, h_2) \to (0,0)$, the limiting position of this circle is precisely the osculating circle to $\mathcal{C}$ at the point $\vec{x}(s)$.

3.3.3. Let $\vec{x}(t) : I \to \mathbb{R}^3$ be a parametrized curve, and let $t \in I$ be a fixed point where $\kappa(t) \neq 0$. Define $\pi : \mathbb{R}^3 \to \mathbb{R}^2$ as the orthogonal projection of $\mathbb{R}^3$ onto the osculating circle to $\vec{x}$ at t. Define $\gamma = \pi \circ \vec{x}$

as the orthogonal projection of the space curve $\vec{x}$ onto the osculating plane. Prove that the curvature $\kappa(t)$ of $\vec{x}$ is equal to the curvature $\kappa_g(t)$ of the plane curve $\vec{\gamma}$.

3.3.4. Calculate the osculating sphere of the twisted cubic $\vec{x}(t) = (t, t^2, t^3)$ at the point $(0, 0, 0)$.

3.3.5. Prove that if all the normal planes to a curve $\mathcal{C}$ pass through a fixed point $\vec{p}$, then $\mathcal{C}$ lies on a sphere of center $\vec{p}$.

3.3.6. Let $\mathcal{C}$ be a curve whose curvature and torsion in terms of arc length are $\kappa(s)$ and $\tau(s)$. Set $R(s) = \frac{1}{\kappa(s)}$. Prove that $\mathcal{C}$ lies on a sphere if and only if

$$R(s)^2 + \left(\frac{R'(s)}{\tau(s)} \right)^2 = \text{const.}$$

3.3.7. Determine a condition where the osculating circle has contact of order 3 or higher.

3.4   Natural Equations

In Section 1.5, we showed that the curvature (in terms of arc length) function uniquely specifies a regular curve up to its location and orientation in $\mathbb{R}^2$. For space curves, one must introduce the torsion function for a measurement of how much the curve twists away from being planar, and Equation (3.8) shows how the torsion function appears as a component of $\vec{x}'''$ in the Frenet frame. Since we know how $\vec{T}$, $\vec{P}$, and $\vec{B}$ change with respect to t, we can express all higher derivatives $\vec{x}^{(n)}(t)$ of $\vec{x}(t)$ in the Frenet frame in terms of $s'(t)$, $\kappa(t)$, and $\tau(t)$ and their derivatives. This leads one to posit that a curve is to some degree determined uniquely by its curvature and torsion functions. The following theorem shows that this is indeed the case.

Theorem 3.4.1 (Fundamental Theorem of Space Curves). *Given functions $\kappa(s) \geq 0$ and $\tau(s)$ continuously differentiable over some interval $J \subseteq \mathbb{R}$ containing 0, there exists an open interval I containing 0 and a regular vector function $\vec{x} : I \to \mathbb{R}^3$ that parametrizes its locus by arc length, with $\kappa(s)$ and $\tau(s)$ as its curvature and torsion functions, respectively. Furthermore, any two curves C_1 and C_2 with curvature function $\kappa(s)$ and torsion function $\tau(s)$ can be mapped onto one another by a rigid motion of $\mathbb{R}^3$.*

Proof: Let $\kappa(s)$ and $\tau(s)$ be functions defined over an interval $J \subset \mathbb{R}$ with $\kappa(s) \geq 0$. Consider the following system of 12 linear first-order differential equations:

$$
\begin{aligned}
x_1'(s) &= t_1(s), & p_1'(s) &= -\kappa(s)t_1(s) + \tau(s)b_1(s), \\
x_2'(s) &= t_2(s), & p_2'(s) &= -\kappa(s)t_2(s) + \tau(s)b_2(s), \\
x_3'(s) &= t_3(s), & p_3'(s) &= -\kappa(s)t_3(s) + \tau(s)b_3(s), \\
t_1'(s) &= \kappa(s)p_1(s), & b_1'(s) &= -\tau(s)p_1(s), \\
t_2'(s) &= \kappa(s)p_2(s), & b_2'(s) &= -\tau(s)p_2(s), \\
t_3'(s) &= \kappa(s)p_3(s), & b_3'(s) &= -\tau(s)p_3(s)
\end{aligned}
\tag{3.17}
$$

where $x_i(s)$, $t_i(s)$, $p_i(s)$, and $b_i(s)$, with $i = 1, 2, 3$, are unknown functions. With the stated conditions on $\kappa(s)$ and $\tau(s)$, according to the existence and uniqueness theorem for first-order systems of differential equations (see [2, Section 31.8]) there exists a solution to the above system defined for s in a neighborhood of 0. Furthermore, there exists a unique solution with specified initial conditions

$$
\begin{aligned}
x_1(0) &= x_{10}, & x_2(0) &= x_{20}, & x_3(0) &= x_{30}, \\
t_1(0) &= t_{10}, & t_2(0) &= t_{20}, & t_3(0) &= t_{30}, \\
p_1(0) &= p_{10}, & p_2(0) &= p_{20}, & p_3(0) &= p_{30}, \\
b_1(0) &= b_{10}, & b_2(0) &= b_{20}, & b_3(0) &= b_{30}.
\end{aligned}
$$

Define the two matrices of functions

$$
A(s) = \begin{pmatrix} 0 & -\kappa(s) & 0 \\ \kappa(s) & 0 & -\tau(s) \\ 0 & \tau(s) & 0 \end{pmatrix}, \quad M(s) = \begin{pmatrix} t_1(s) & p_1(s) & b_1(s) \\ t_2(s) & p_2(s) & b_2(s) \\ t_3(s) & p_3(s) & b_3(s) \end{pmatrix}.
$$

Recall from Equation (3.11) that $M'(s) = M(s)A(s)$. It is possible to show that since $A(s)$ is antisymmetric, the function $f(s) = M(s)^T M(s)$ is constant as a matrix of functions. Therefore, any solution to Equation (3.17) is such that $M(s)^T M(s)$ remains constant. In particular, if we choose initial conditions such that

$$
\begin{pmatrix} t_{10} \\ t_{20} \\ t_{30} \end{pmatrix}, \quad \begin{pmatrix} p_{10} \\ p_{20} \\ p_{30} \end{pmatrix}, \quad \begin{pmatrix} b_{10} \\ b_{20} \\ b_{30} \end{pmatrix}
\tag{3.18}
$$

form an orthonormal basis of $\mathbb{R}^3$, then solutions to Equation (3.17)
are such that for all s in the domain of the solution, the vectors

$$\begin{pmatrix} t_1(s) \\ t_2(s) \\ t_3(s) \end{pmatrix}, \quad \begin{pmatrix} p_1(s) \\ p_2(s) \\ p_3(s) \end{pmatrix}, \quad \begin{pmatrix} b_1(s) \\ b_2(s) \\ b_3(s) \end{pmatrix} \qquad (3.19)$$

form an orthonormal basis.

Therefore, we have shown that with a choice of initial conditions
such that the vectors in Equation (3.18) form an orthonormal basis,
the corresponding solution to Equation (3.17) is such that $\vec{x}(s) =$
$(x_1(s),\ x_2(s),\ x_3(s))$ is a regular curve of class C^3 with curvature
$\kappa(s)$ and torsion $\tau(s)$. In addition, the vector functions in Equation
(3.19) are the $\vec{T}$, $\vec{P}$, and $\vec{B}$ vectors of the Frenet frame associated to
$\vec{x}(s)$. This proves existence.

The existence and uniqueness theorem of systems of differential
equations states that a solution is unique once (the right number
of) initial conditions are specified. However, we have imposed the
additional condition that the vectors in Equation (3.18) form an
orthonormal set. Any different choice for the initial conditions in
Equation (3.18) corresponds to a rotation in $\mathbb{R}^3$. Also, two different
choices of initial conditions x_{10}, x_{20}, x_{30} correspond to a translation
in $\mathbb{R}^3$. Therefore, different allowed initial conditions correspond to a
rigid motion of the locus of $\vec{x}(s)$ in $\mathbb{R}^3$. This proves the theorem. $\square$

Problems

3.4.1. Assume that $\kappa(s)$ and $\tau(s)$ are real analytic over an interval I and
assume that we only consider space curves $\vec{x} : I \to \mathbb{R}^3$ that are
also real analytic. Use the Taylor expansion of $\vec{x}(s)$ to prove Theorem 3.4.1.

CHAPTER 4

Curves in Space: Global Properties

Paralleling our presentation of curves in the plane, we now turn from local properties of space curves to global properties. As before, global properties of curves are properties that involve the curve as a whole as opposed to properties that are defined in the neighborhood of a point on the curve.

The Jordan Curve Theorem does not apply to curves in $\mathbb{R}^3$, so Green's Theorem, the isoperimetric inequality, and theorems connecting curvature and convexity do not have an equivalent for space curves. On the other hand, curves in space exhibit new types of global properties, in particular, knottedness and linking.

4.1 Basic Properties

Definition 4.1.1. A parametrized space curve C is called *closed* if there exists a parametrization $\vec{x} : [a, b] \to \mathbb{R}^3$ of C such that $\vec{x}(a) = \vec{x}(b)$. A regular curve is called closed if, in addition, all the (one-sided) derivatives of $\vec{x}$ at a and at b are equal, i.e., $\vec{x}(a) = \vec{x}(b)$, $\vec{x}'(a) = \vec{x}'(b)$, $\vec{x}''(a) = \vec{x}''(b), \dots$. A curve C is called *simple* if it has no self-intersections, and a closed curve is called simple if it has no self-intersections except at the endpoints.

A closed space curve can be understood as a function $f : \mathbb{S}^1 \to \mathbb{R}^3$ that is continuous as a function between topological spaces (see [22, Appendix A]). In this context, to say that the curve is simple is tantamount to saying that f is injective.

Our first result has an equivalent in the theory of plane curves.

Proposition 4.1.2. *If a regular curve is closed, then it is bounded.*

Proof: (Left as an exercise for the reader.) □

Stokes' Theorem presents a global property of curves in space in that it relates a quantity calculated along the curve with a quantity that depends on any surface that has that curve as a boundary. However, the theorem involves a vector field in $\mathbb{R}^3$, and it does not address "geometric" properties of the curve, by which we mean properties that are independent of the curve's location and orientation in space. Therefore, we simply state Stokes' Theorem as an example of a global theorem and trust the reader has seen it in a previous calculus course.

Theorem 4.1.3 (Stokes' Theorem). *Let $\mathcal{S}$ be an oriented, piecewise smooth surface bounded by a closed, piecewise regular curve $\mathcal{C}$. Let $\vec{F} : \mathbb{R}^3 \to \mathbb{R}^3$ be a vector field over $\mathbb{R}^3$ that is of class C^1. Then*

$$\int_{\mathcal{C}} \vec{F} \cdot d\vec{s} = \iint_{\mathcal{S}} (\vec{\nabla} \times \vec{F}) \cdot d\vec{S}. \tag{4.1}$$

If $\mathcal{S}$ has no boundary curve, then the line integral on the left is defined to be zero.

Note as a reminder that the line element $d\vec{s}$ stands for $\vec{\gamma}'(t)dt$, where $\vec{\gamma}(t)$ is a parametrization of $\mathcal{C}$, and $d\vec{S}$ stands for $\vec{n}dA$ where $\vec{n}$ is the unit normal and dA is the surface element at a point on the surface. If a surface is parametrized by a vector function $\vec{X}(u,v)$ into $\mathbb{R}^3$, then $d\vec{S} = \vec{X}_u \times \vec{X}_v \, du \, dv$.

The following is an interesting corollary to Stokes' Theorem.

Corollary 4.1.4. *Let $\mathcal{C}$ be a simple, regular, closed, space curve parametrized by $\vec{\gamma} : I \to \mathbb{R}^3$. There exists no function $f : \mathbb{R}^3 \to \mathbb{R}$ of class C^2 such that the gradient satisfies $\vec{\nabla} f = \vec{T}(t)$ at all points of the curve.*

Proof: Recall that for all functions $f : \mathbb{R}^3 \to \mathbb{R}$ of class C^2, the curl of the gradient is identically 0, namely, $\vec{\nabla} \times \vec{\nabla} f = \vec{0}$. If $\mathcal{S}$ is any orientable piecewise smooth surface that has $\mathcal{C}$ as a boundary, then

$$\iint_{\mathcal{S}} (\vec{\nabla} \times \vec{\nabla} f) \cdot d\vec{S} = \iint_{\mathcal{S}} \vec{0} \cdot d\vec{S} = 0.$$

If f did satisfy $\vec{\nabla} f = \vec{T}(t)$ at all points of the curve $\mathcal{C}$, then

$$\int_{\mathcal{C}} \vec{\nabla} f \cdot d\vec{s} = \int_I \vec{T}(t) \cdot \vec{T}(t) s'(t) dt = \int_{\mathcal{C}} ds,$$

which is the length of the curve. By Stokes' Theorem, this leads to a contradiction, assuming the curve has length greater than 0.   □

Problems

4.1.1. Let $\vec{\alpha} : I \rightarrow \mathbb{R}^3$ be a regular, closed, space curve, and let $\vec{p}$ be a point. Prove that a point t_0 of maximum distance on the curve away from $\vec{p}$ is such that $\vec{\alpha}(t_0) - \vec{p}$ is in the normal plane to the curve at $\vec{\alpha}(t_0)$.

4.1.2. Recall that the diameter of a curve $\vec{\alpha}$ is the maximum of the function

$$f(t, u) = \|\vec{\alpha}(t) - \vec{\alpha}(u)\|.$$

Use the second derivative test in multivariable calculus to show that if (t_0, u_0) gives a diameter of a space curve, then the following hold:

(a) The vector $\vec{\alpha}(u_0) - \vec{\alpha}(t_0)$ is in the intersection of the normal planes of $\alpha(t_0)$ and $\alpha(u_0)$.

(b) The vector $\vec{\alpha}(u_0) - \vec{\alpha}(t_0)$ is on the side of the rectifying plane of $\vec{\alpha}(t_0)$ that makes an acute angle with $\vec{P}(t_0)$ (and similarly for $\alpha(u_0)$).

(c) The diameter $\|\vec{\alpha}(t_0) - \vec{\alpha}(u_0)\| \geq \max\{1/\kappa(t_0), 1/\kappa(u_0)\}$.

[Hint: One can assume that $\vec{\alpha}$ is parametrized by arclength. Also, the extrema of $f(t, u)$ occur at and have the same properties as the extrema of $g(t, u) = f(t, u)^2$.]

4.1.3. Prove Proposition 4.1.2.

4.2   Indicatrices and Total Curvature

Definition 4.2.1. Given a space curve $\vec{x} : I \rightarrow \mathbb{R}^3$, define the *tangent*, *principal* and *binormal indicatrices* respectively, as the loci of the space curves given by $\vec{T}(t)$, $\vec{P}(t)$, and $\vec{B}(t)$, as defined in Section 3.2.

Since the vectors of the Frenet frame have a length of 1, the indicatrices are curves on the unit sphere $\mathbb{S}^2$ in $\mathbb{R}^3$. In contrast to the tangent indicatrix for plane curves, there are more possibilities for curves on the sphere than for curves on a circle. Therefore, results such as the theorem on the rotation index or results about the winding number do not have an immediate equivalent for curves in space.

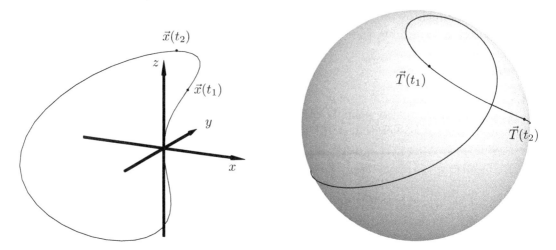

Figure 4.1. The space cardioid and its tangent indicatrix.

Example 4.2.2 (Lines). The tangent indicatrix of a line is a single point $\vec{v}/\|\vec{v}\|$ on the sphere, where the line is parametrized by

$$\vec{x}(t) = \vec{v}t + \vec{p}.$$

Example 4.2.3 (Helices). Consider a circular helix around an axis L through the origin. The tangent indicatrix of this helix is the circle on the unit sphere $\mathbb{S}^2$, given as the intersection of $\mathbb{S}^2$ with the plane perpendicular to L at a distance of $1/\sqrt{1 + (\kappa/\tau)^2}$ from the origin, where $\frac{\kappa}{\tau}$ is the pitch of the helix.

Example 4.2.4 (Space Cardioid). The space cardioid, given by the parametrization

$$\vec{x}(t) = ((1 - \cos t) \cos t, (1 - \cos t) \sin t, \sin t), \qquad t \in [0, 2\pi],$$

is a closed curve. The tangent indicatrix is again a closed curve and is shown in Figure 4.1.

Definition 4.2.5. The *total curvature* of a closed curve C parametrized by $\vec{x} : [a, b] \to \mathbb{R}^3$ is

$$\int_C \kappa \, ds = \int_a^b \kappa(t) s'(t) \, dt.$$

This is a nonnegative real number since $\kappa(t) \geq 0$ for space curves.

Though we cannot define the concept of a rotation index, Fenchel's Theorem gives a lower bound for the total curvature of a space curve. Since $\vec{T}' = s'\kappa\vec{P}$, the speed of the tangent indicatrix is $s'(t)\kappa(t)$. Therefore, the total curvature of $\vec{x}$ is the length of the tangent indicatrix. One can also note that the tangent indicatrix has a critical point at the point where $\kappa(t) = 0$.

Theorem 4.2.6 (Fenchel's Theorem). *The total curvature of a regular closed space curve C is greater than or equal to 2π. It is equal to 2π if and only if C is a convex plane curve.*

Before we prove Fenchel's Theorem, we need to discuss the concept of distance between points on the unit sphere $\mathbb{S}^2$. We can define the distance between two points p and q on $\mathbb{S}^2$ as

$$d(p, q) = \inf\{\text{length}(\Gamma) \mid \Gamma \text{ is a curve connecting } p \text{ and } q\}. \quad (4.2)$$

Let O be the center of the unit sphere. It is not difficult to show that the path connecting p and q of shortest distance is an arc between p and q on the circle defined by the intersection of the sphere and the plane containing O, p, and q. (We will obtain this result in Example 8.2.7 when studying geodesics, but it is possible to prove this claim without the techniques of geodesics.) Then the spherical distance between two points on the unit sphere is

$$d(p, q) = \cos^{-1}(p \cdot q), \quad (4.3)$$

where we view p and q as vectors in $\mathbb{R}^3$.

We will denote by AB the Euclidean distance between two points A and B as elements in $\mathbb{R}^3$. If $A, B \in \mathbb{S}^2$, we use the notation $\overline{AB}$ to denote the distance between A and B on the sphere. Since OAB forms an isosceles triangle, we see that the spherical and Euclidean distance (see Figure 4.2) are related via

$$AB = 2\sin\left(\frac{\overline{AB}}{2}\right) \quad \text{and} \quad \overline{AB} = 2\sin^{-1}\left(\frac{AB}{2}\right). \quad (4.4)$$

Lemma 4.2.7 (Horn's Lemma). *Let Γ be a regular curve on the unit sphere $\mathbb{S}^2$. If Γ has length less than 2π, then there exists a great circle C on the sphere such that Γ does not intersect C.*

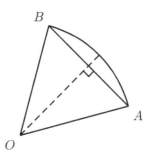

Figure 4.2. Spherical distance.

Proof: The proof we give to this lemma is due to R. A. Horn [18].

Let A and B be two points of Γ that divide the curve into two arcs of equal length $L/2$. The points A and B divide Γ into two curves Γ_1 and Γ_2 of equal length. Since $L \leq 2\pi$, the spherical distance between A and B is less than the length of Γ_1, which is strictly less than π. Consequently, A and B are not antipodal (i.e., the segment $[A, B]$ is not a diameter of the sphere) and therefore there exists a unique point M on the minor arc from A to B midway between A and B. We claim that Γ does not meet the equator $\mathcal{C}$ that has M as the north pole, so the curve lies in the hemisphere centered at M.

To show that Γ_1 does not meet the equator, consider a copy Γ_1' of Γ_1 rotated one half turn about M. The curve Γ_1' is a curve from B to A of length $L/2$. Define the closed curve Γ'' as the curve that follows Γ_1 from A and B and then follows Γ_1' from B back to A. The curve Γ'' has length L. Furthermore, if Γ_1 intersected $\mathcal{C}$, then so would Γ_1'. Hence, Γ'' would contain antipodal points R and R'. But, the spherical distance $\overline{RR'} = \pi$. Therefore, the distance from R to R' along Γ'' would be greater or equal to π and, by the symmetry of Γ'', similarly for the path from R' to R. This contradicts the fact that the length of Γ'' is less than 2π.

Thus any curve with length less than 2π lies in an open hemisphere. □

We are now in a position to prove Fenchel's Theorem.

Proof (of Theorem 4.2.6): Let $\vec{\gamma} : [a, b] \to \mathbb{R}^3$ be a parametrization for the regular closed curve C, and let p be any point on the unit sphere. Consider the function $g(t) = p \cdot \vec{\gamma}(t)$. Since $[a, b]$ is a closed and bounded interval and $g(t)$ is continuous, then it attains a maximum and minimum value in $[a, b]$. This value occurs where

$$g'(t) = p \cdot \vec{\gamma}'(t) = 0.$$

We remark that the set of points $\vec{x} \in \mathbb{S}^2$ such that $p \cdot \vec{x} = 0$ is the great circle on $\mathbb{S}^2$ that is on the plane perpendicular to the line (Op), where O is the center of the sphere. Since $\vec{\gamma}'(t)$ and $\vec{T}(t)$ are collinear and since p was chosen arbitrarily, we conclude that the tangent indicatrix intersects every great circle on $\mathbb{S}^2$. Consequently, by Lemma 4.2.7, the length of the tangent indicatrix is greater than or equal to 2π. Thus, since the length of the tangent indicatrix is the total curvature,

$$\int_a^b \kappa(t) s'(t)\, dt = \int_C \kappa\, ds \geq 2\pi.$$

To prove the second part of the theorem, first note that if $\vec{\gamma}(t)$ traces out a convex plane curve, then $\kappa(t) = |\kappa_g(t)|$ for $\vec{\gamma}$ as a plane curve. Furthermore, by Propositions 2.4.3 and 2.2.8, the total curvature is 2π. Therefore, to finish proving the theorem, we only need to prove the converse.

Suppose that the total curvature of $\vec{\gamma}$ is 2π. The length of the tangent indicatrix is therefore 2π. Furthermore, since by the above reasoning the tangent indicatrix must intersect every great circle, the tangent indicatrix must itself be a great circle. Thus $\vec{T}(t)$, $\vec{T}'(t)$, and $\vec{T}''(t)$ are coplanar and so $\tau(t) = 0$. Thus, by Proposition 3.2.7, $\vec{\gamma}(t)$ is planar. In this case, it is not hard to check that, again, $\kappa(t) = |\kappa_g(t)|$, where $\kappa_g(t)$ is the curvature of $\vec{\gamma}(t)$ as a plane curve. Thus,

$$2\pi = \int_C |\kappa_g|\, ds$$

is the total distance that $\vec{T}(t)$ travels on the unit circle, and this is the length of the unit circle. Consequently, if $t_1, t_2 \in [a, b]$, with $\vec{T}(t_1) = \vec{T}(t_2)$, then $\vec{T}(t)$ is constant over $[t_1, t_2]$ because, otherwise, the total distance $\vec{T}(t)$ travels on the unit circle would exceed 2π by at least

$$\int_{t_1}^{t_2} |\kappa_g(t)|\, ds.$$

We conclude then that $\vec{\gamma}$ has no bitangent lines, and by Problem 2.4.1, we conclude that C is a convex plane curve.   □

As an immediate corollary, we obtain the following result.

Corollary 4.2.8. *If a regular, closed, space curve has a curvature function κ that satisfies*

$$\kappa \leq \frac{1}{R},$$

then C has length greater than or equal to $2\pi R$.

Proof: Suppose that a regular, closed, space curve has a curvature function $\kappa(s)$, given in terms of arclength, that satisfies the given hypothesis. Then if L is the length of the curve, we have

$$L = \int_0^L ds \geq \int_0^L R\kappa(s)\,ds = R\int_0^L \kappa(s)\,ds \geq 2\pi R,$$

where the last inequality follows from Fenchel's Theorem.   □

Though it is outside the scope of our current techniques, we wish to state here Jacobi's Theorem since it is a global theorem on space curves. The proof follows as an application of the Gauss-Bonnet Theorem (Problem 8.4.8).

Theorem 4.2.9 (Jacobi's Theorem). *Let $\vec{\alpha} : I \rightarrow \mathbb{R}^3$ be a closed, regular, parametrized curve whose curvature is never 0. Suppose that the principal normal indicatrix $\vec{P} : I \rightarrow \mathbb{S}^2$ is simple. Then the locus $\vec{P}(I)$ of the principal normal indicatrix separates the sphere into two regions of equal area.*

The reader should note that Jacobi's Theorem is quite profound in the following sense. Given a parametrized curve $\vec{\gamma} : I \rightarrow \mathbb{R}^3$ such that $\|\vec{\gamma}(t)\| = 1$ for all $t \in I$, the problem of calculating the surface area of the sphere lying on one side or the other of the locus of this curve is not a tractable problem. (Green's Theorem does give a simple formula for calculating the area of the interior of a simple, closed, parametrized curve in the plane. However, when considering a curve $\vec{\gamma}$ enclosing a region $\mathcal{R}$ on a sphere, one might attempt to use Stokes' Theorem

$$\int_{\vec{\gamma}(I)} \vec{F} \cdot d\vec{s} = \iint_{\mathcal{R}} \vec{\nabla} \times \vec{F} \cdot d\vec{S},$$

with a vector field $\vec{F}$ such that $\vec{\nabla} \times \vec{F} = \vec{r} = (x, y, z)$. The right-hand side of the formula would give the area of $\mathcal{R}$. However, no such $\vec{F}$ exists since when calculating the divergence, $\vec{\nabla} \cdot \vec{r} = 3$, but $\vec{\nabla} \cdot (\vec{\nabla} \times \vec{F}) = 0$ for all vector fields over $\mathbb{R}^3$ of class C^2.)

We now present Crofton's Theorem, which we will use in the next section to prove a theorem by Fary and Milnor on the total curvature of a knot.

Let O be the center of the unit sphere $\mathbb{S}^2$. Each great circle $\mathcal{C}$ is uniquely defined by a line L through the origin of the sphere as the intersection between $\mathbb{S}^2$ and the plane through O perpendicular to L. Furthermore, L is defined by two "poles," the points of intersection of L with $\mathbb{S}^2$. However, there exists a bijective correspondence between oriented great circles and points on $\mathbb{S}^2$ by associating to $\mathcal{C}$ the pole of L that is in the direction on L that is positive in the sense of the right-hand rule of motion on $\mathcal{C}$.

Consider a set Z of oriented great circles on $\mathbb{S}^2$. We call the *measure* $m(Z)$ of the set Z the area of the region traced out on $\mathbb{S}^2$ by the positive poles of the oriented circles in Z.

Theorem 4.2.10 (Crofton's Theorem). *Let Γ be a curve of class C^1 on $\mathbb{S}^2$. The measure of the great circles of $\mathbb{S}^2$ that meet Γ is equal to four times the length of Γ.*

Proof: Suppose that $\vec{e}_1 : [0, L] \to \mathbb{S}^2$ parametrizes Γ by arc length. Complete $\{\vec{e}_1(s)\}$ to form an orthonormal basis $\{\vec{e}_1(s), \vec{e}_2(s), \vec{e}_3(s)\}$ so that $\vec{e}_i(s)$ for $i = 2, 3$ are of class C^1. Without loss of generality, we can construct $\vec{e}_2$ and $\vec{e}_3$ so that

$$\det(\vec{e}_1, \vec{e}_2, \vec{e}_3) = 1$$

for all $s \in [0, L]$. Using the same reasoning that established Equation (3.4), one can determine that

$$\frac{d}{ds} \begin{pmatrix} \vec{e}_1 & \vec{e}_2 & \vec{e}_3 \end{pmatrix} = \begin{pmatrix} \vec{e}_1 & \vec{e}_2 & \vec{e}_3 \end{pmatrix} \begin{pmatrix} 0 & a_2 & a_3 \\ -a_2 & 0 & a_1 \\ -a_3 & -a_1 & 0 \end{pmatrix} \qquad (4.5)$$

for some continuous functions $a_i : [0, L] \to \mathbb{R}$. Furthermore, since $\vec{e}_1$ is parametrized by arc length we know that it has unit speed, so $a_2(s)^2 + a_3(s)^2 = 1$. Consequently, there exists a function $\alpha : [0, L] \rightarrow \mathbb{R}$ of class C^1 such that

$$\vec{e}_1'(s) = \cos(\alpha(s))\vec{e}_2 + \sin(\alpha(s))\vec{e}_3.$$

The set of oriented great circles that meet Γ at $\vec{e}_1(s)$ is parametrized by its positive poles

$$(\cos\theta)\vec{e}_2(s) + (\sin\theta)\vec{e}_3(s)$$

for $\theta \in [0, 2\pi]$. Therefore, the region traced out by the positive poles of oriented great circles that meet Γ is parametrized by

$$\vec{Y}(s, \theta) = (\cos\theta)\vec{e}_2(s) + (\sin\theta)\vec{e}_3(s) \qquad \text{for } (s, \theta) \in [0, L] \times [0, 2\pi].$$

We wish to determine the area element $|dA|$ for the set of poles traced out by the set of oriented great circles meeting Γ. However,

$$|dA| = \|\vec{Y}_s \times \vec{Y}_\theta\| \, d\theta \, ds,$$

so after some calculation and using Equation (4.5), one obtains

$$|dA| = |a_2(s)\cos\theta + a_3(s)\sin\theta| \, d\theta \, ds = |\cos(\alpha(s) - \theta)| \, d\theta \, ds.$$

Call $\mathcal{C}_{\theta,s}$ the oriented great circle with positive pole $\vec{Y}(\theta, s)$ and denote by $n(\mathcal{C}_{\theta,s})$ the number of points in $\mathcal{C}_{\theta,s} \cap \Gamma$. Then the measure m of oriented great circles in $\mathbb{S}^2$ that meet Γ is

$$m = \iint n(\mathcal{C}_{\theta,s})|dA| = \int_0^L \int_0^{2\pi} |\cos(\alpha(s) - \theta)| \, d\theta \, ds. \qquad (4.6)$$

However, for all fixed α_0, we have

$$\int_0^{2\pi} |\cos(\alpha_0 - \theta)| \, d\theta = 4,$$

so we conclude that $m = 4L$, which establishes the theorem.   □

Problems

4.2.1. Consider the twisted cubic $\vec{x}(t) = (t, t^2, t^3)$ with $t \in \mathbb{R}$.

    (a) Prove that the closure of the tangential indicatrix of the twisted cubic is a closed curve, in particular, that

$$\lim_{t \to \infty} \vec{T}(t) = \lim_{t \to -\infty} \vec{T}(t) = (0, 0, 1).$$

    (b) Prove that the tangential indicatrix does not have a corner at $(0, 0, 1)$, namely, that

$$\lim_{t \to \infty} \vec{T}'(t) = \lim_{t \to -\infty} \vec{T}'(t).$$

    (c) Prove directly that the total curvature of the twisted cubic is bounded.

4.2.2. Consider the helix $\vec{x}(t) = (a\cos t, a\sin t, bt)$ for $t \in \mathbb{R}$. Determine the locus of the tangent indicatrix of $\vec{X}(t)$.

4.2.3. Calculate directly the total curvature of the space cardioid

$$\vec{x}(t) = ((1 - \cos t)\cos t, (1 - \cos t)\sin t, \sin t) \qquad \text{for } t \in [0, 2\pi].$$

4.2.4. Prove Fenchel's Theorem as a corollary to Crofton's Theorem.

4.2.5. (*) Let $\vec{\alpha} : [0, L] \to \mathbb{R}^3$ be a regular closed curve (parametrized by arc length) whose image lies on a sphere. Suppose also that the curvature is nowhere 0. Prove that

$$\int_0^L \tau(s)\, ds = 0.$$

4.3 Knots and Links

The reader should be forewarned that the study of knots and links is a vast and fruitful area that one usually considers as a branch of topology. In this section, we only have space to give a cursory introduction to the concept of a knotted curve in $\mathbb{R}^3$ or two linked curves in $\mathbb{R}^3$. However, the property of being knotted or linked is a global property of a curve or curves (since it depends on the curve as a whole), which motivates us to briefly discuss these topics in this chapter. This section presents two main theorems: the Fary-Milnor Theorem, which shows how the property of knottedness imposes a condition on the total curvature of a curve, and Gauss's formula for the linking number of two curves.

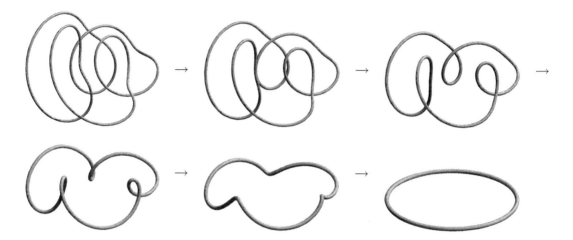

Figure 4.3. An unknotting homotopy in $\mathbb{R}^3$.

4.3.1   Knots

Intuitively speaking, a knot is a simple closed curve in $\mathbb{R}^3$ that cannot be deformed into a circle without breaking the curve and reconnecting it. In other words, a knot cannot be deformed into a circle by a continuous process without passing through a stage where it is not a simple curve.

The following gives a precise definition to the above intuition.

Definition 4.3.1. A simple closed curve Γ in $\mathbb{R}^3$ is called *unknotted* if there exists a continuous function $H : \mathbb{S}^1 \times [0,1] \to \mathbb{R}^3$ such that $H(\mathbb{S}^1 \times \{0\}) = \Gamma$ and $H(\mathbb{S}^1 \times \{1\}) = \mathbb{S}^1$ and such that $\Gamma_t = H(\mathbb{S}^1 \times \{t\})$ is a curve that is homeomorphic to a circle. If there does not exist such a function H, the curve Γ is called *knotted*.

The function H described in the above definition is called a *homotopy* in $\mathbb{R}^3$ between Γ and a circle $\mathbb{S}^1$. Figure 4.3 illustrates four intermediate stages of a homotopy between a space curve and a circle.

Figure 4.4 shows the trefoil knot realized as a curve in space, along with a two-dimensional diagram. As it is somewhat tedious to plot general knotted curves, even with the assistance of a computer algebra system, one often uses a diagram that shows the "crossings," i.e., which part of the curve passes above the other whenever the

 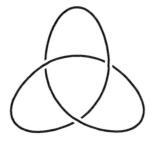

Figure 4.4. A trefoil knot and its diagram.

diagram would intersect in the given perspective. (The interested reader is encouraged to read [21] for an advanced introduction to knot theory.)

The following theorem gives a necessary relationship between a knotted curve and the curvature of a space curve.

Theorem 4.3.2 (Fary-Milnor Theorem). *The total curvature of a knot is greater than or equal to 4π.*

Proof: Let $\vec{x} : [0, L] \to \mathbb{R}^3$ be a closed, regular, space curve parametrized by arc length, and let $\vec{T} : [0, L] \to \mathbb{S}^2$ be the tangent indicatrix. Call C the image of $\vec{x}$ in $\mathbb{R}^3$ and Γ the image of $\vec{T}$ on the sphere.

Recall the notations in the proof of Crofton's Theorem (Theorem 4.2.10) and, in particular, that $n(\mathcal{C}_{\theta_0, s_0})$ is the number of times that the great circle $\mathcal{C}_{\theta_0, s_0}$ intersects the curve Γ (see Figure 4.5). By part of the proof of Crofton's Theorem, $n(\mathcal{C}_{\theta_0, s_0}) > 0$ over the domain $(\theta, s) \in [0, 2\pi] \times [0, L]$, and $n(\mathcal{C}_{\theta_0, s_0})$ is even.

Define the height function $h(s) = \vec{Y}(\theta_0, s_0) \cdot \vec{x}(s)$. Note that $h'(s) = \vec{Y}(\theta_0, s_0) \cdot \vec{T}(s)$ so that the relative maxima and minima of $h(s)$ occur where Γ intersects $\mathcal{C}_{\theta_0, s_0}$, so $n(\mathcal{C}_{\theta_0, s_0})$ is the number of critical points of $h(s)$.

Suppose that the total curvature of the curve $\vec{x}$, which is also the length of Γ, is less than 4π. By Equation (4.6) in the proof of Crofton's Theorem,

$$\iint n(\mathcal{C}_{\theta, s}) |dA| < 16\pi.$$

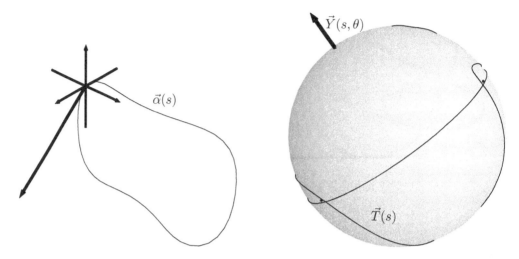

Figure 4.5. Proof of the Fary-Milnor Theorem.

Since the area of the sphere is 4π, there exists (θ_0, s_0) such that $n(\mathcal{C}_{\theta_0,s_0}) = 2$, which means that the height function corresponding to $\vec{Y}(\theta_0, s_0)$ has two critical points, namely, one maximum and one minimum occuring at s_1 and s_2. The points $\vec{x}(s_1)$ and $\vec{x}(s_2)$ partition C into two curves over which the height function $h(s)$ is monotonic, one increasing and the other decreasing. Consequently, any plane perpendicular to $\vec{Y}(\theta_0, s_0)$ (that intersects $\vec{x}$ between the points $\vec{x}(s_1)$ and $\vec{x}(s_2)$ of extremal height function) intersects $\vec{x}$ in exactly two points. This fact allows us to construct a homotopy between C and a circle, which we describe below.

Call v the height parameter for a point on the curve $\vec{x}(s)$. Call $v_{\min}$ and $v_{\max}$ the minimum and maximum values of $h(s)$ over the domain $[0, L]$ of $\vec{x}$. Every plane perpendicular to $\vec{Y}(\theta_0, s_0)$ is determined uniquely by the height parameter v as the unique plane perpendicular to $\vec{Y}(\theta_0, s_0)$ and going through the point $v\vec{Y}(\theta_0, s_0)$. According to the discussion in the previous paragraph, the plane $\mathcal{P}_v$ perpendicular to $\vec{Y}(\theta_0, s_0)$ at height v intersects C in two points. Therefore, C can be expressed as the union of the locus of two continuous curves

$$\vec{\gamma}_1 : [v_{\min}, v_{\max}] \longrightarrow \mathbb{R}^3,$$
$$\vec{\gamma}_2 : [v_{\min}, v_{\max}] \longrightarrow \mathbb{R}^3,$$

such that $\mathcal{P}_v \cap C = \{\vec{\gamma}_1(v), \vec{\gamma}_2(v)\}$. Note that

$$\vec{\gamma}_1(v_{\max}) = \vec{\gamma}_2(v_{\max}) = \vec{x}(s_1), \qquad \text{and}$$
$$\vec{\gamma}_1(v_{\min}) = \vec{\gamma}_2(v_{\min}) = \vec{x}(s_2).$$

Call $\vec{\gamma}(v) = \frac{1}{2}(\vec{\gamma}_1(v) + \vec{\gamma}_2(v))$ the midpoint of the pair $\mathcal{P}_v \cap C = \{\vec{\gamma}_1(v), \vec{\gamma}_2(v)\}$ (which degenerates to a singleton set for $v = v_{\min}$ or $v_{\max}$).

We now proceed to show that the union of all the segments connecting $\vec{\gamma}_1(v)$ and $\vec{\gamma}_2(v)$ can be continuously deformed to a disk. Complete $\{\vec{Y}(\theta_0, s_0)\}$ to a fixed orthonormal basis $\{\vec{Y}(\theta_0, s_0), \vec{e}_2, \vec{e}_3\}$. In this basis, we can write $\vec{\gamma}(v)$ as

$$\vec{\gamma}(v) = (v, a(v)\cos(\alpha(v)), a(v)\sin(\alpha(v)))$$

for some functions $a(v) \geq 0$ and $\alpha(v)$ defined on $[v_{\min}, v_{\max}]$. Furthermore, since $\vec{\gamma}(v)$, $\vec{\gamma}_1(v)$, and $\vec{\gamma}_2(v)$ as position vectors all lie in $\mathcal{P}_v$, there exist functions $b(v)$ and $\theta(v)$ such that we can write

$$\vec{\gamma}_1(v) = \vec{\gamma}(v) + b(v)\left((\cos\theta(v))\vec{e}_2 + (\sin\theta(v))\vec{e}_3\right),$$
$$\vec{\gamma}_2(v) = \vec{\gamma}(v) - b(v)\left((\cos\theta(v))\vec{e}_2 + (\sin\theta(v))\vec{e}_3\right).$$

Call $v_0 = \frac{1}{2}(v_{\min} + v_{\max})$ and $R = \frac{1}{2}(v_{\max} - v_{\min})$ and define the function $v(u) = v_0 + R\cos(u)$. Finally, define the function $H : [0, 2\pi] \times [0, 1]$ by

$$
\begin{aligned}
H(u, t) = & v(u)\vec{Y}(\theta_0, s_0) + (1-t)a(v(u))\cos(\alpha(v(u))) \\
& + \sin u\left((1-t)r(u) + tR\right)\cos\left((1-t)\theta(v(u))\right)\vec{e}_2 \\
& + (1-t)a(v(u))\sin(\alpha(v(u))) \\
& + \sin u\left((1-t)r(u) + tR\right)\sin\left((1-t)\theta(v(u))\right)\vec{e}_3,
\end{aligned}
\tag{4.7}
$$

where $r(u)$ is a nonnegative function over $[0, 2\pi]$ such that $r(u)|\sin u| = b(v(u))$. We now leave it as an exercise to the reader to prove that $H(u, t)$ is a homotopy between C and a circle of radius R such that for all $t_0 \in (0, 1)$, the image of $H(u, t_0)$ is homeomorphic to a circle (Problem 4.3.1).

By Definition 4.3.1, the existence of H as defined above allows us to conclude that C is unknotted. Therefore, if C is a knot, we conclude that the total curvature of C is greater than or equal to 4π. $\square$

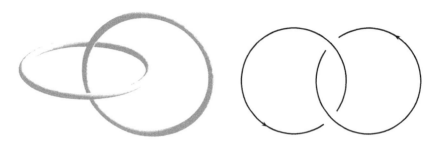

Figure 4.6. Linked curves; linking number 1.

4.3.2 Links

A simple, closed, space curve in itself resembles a circle. Knottedness is a property that concerns how a simple closed curve "sits" in the ambient space. The technical language for this scenario is that any simple closed curve is homeomorphic to a circle, but knottedness concerns how the curve is embedded in $\mathbb{R}^3$. In a similar way, the notion of linking between two simple closed curves is a notion that considers how the curves are embedded in space in relation to each other.

Figure 4.6 illustrates a basic scenario of linked curves. Since it is difficult to accurately sketch curves in $\mathbb{R}^3$ and to see which curve is in front of which, one commonly uses a diagram to depict the curves. In Figure 4.6, the diagram on the right corresponds to the curves on the left. The important part of the diagram is to clearly indicate the crossings, i.e., given the perspective of the diagram, which curve is on top.

The standard definition for the linking number of two curves uses the link diagram along with the sign associated to each crossing, as shown below:

$$\times -1 \qquad \times +1$$

Definition 4.3.3. The linking number $\operatorname{link}(C_1, C_2)$ of two simple closed curves C_1 and C_2 is half the sum of the signs of all the crossings in the diagram of the pair (C_1, C_2).

For example, in Figure 4.6, both crossings have sign $+1$. Thus half the sum of the signs of the crossings is $+1$.

One of the surprising results about linking is Gauss's formula for the linking number between two curves. Though linking and the linking number are obviously global properties of the curves, Gauss's formula calculates the linking number in terms of the parametrizations of the curves, thereby connecting the global property to local properties.

Theorem 4.3.4 (Gauss's Linking Formula). *Let C_1 and C_2 be two closed, regular, space curves parametrized by $\vec{\alpha} : I \rightarrow \mathbb{R}^3$ and $\vec{\beta} : J \rightarrow \mathbb{R}^3$, respectively. The linking number between C_1 and C_2 is*

$$\text{link}(C_1, C_2) = \frac{1}{4\pi} \int_I \int_J \frac{\det(\vec{\alpha}(u) - \vec{\beta}(v), \vec{\alpha}'(u), \vec{\beta}'(v))}{\|\vec{\alpha}(u) - \vec{\beta}(v)\|^3} \, du \, dv. \quad (4.8)$$

Though we cannot give a proof for this formula at this time, we briefly sketch a few of the concepts and techniques that go into it.

In Section 2.2, we introduced the notion of degree for a continuous map between circles, $f : \mathbb{S}^1 \rightarrow \mathbb{S}^1$. Intuitively speaking, the degree of f counts how many times f covers its codomain and with what orientation. The degree of f takes into account when f might double back. For example, if $q \in \mathbb{S}^1$ has six preimages from f and, for four of the preimages, f pass through q with the same orientation as on the domain and, for the other two preimages, f passes through q in an opposite orientation, then the degree of f is $4 - 2 = 2$.

Similarly, in the field of algebraic topology, a topic that is outside the scope of this book, one can define the degree of a map between spheres $f : \mathbb{S}^2 \rightarrow \mathbb{S}^2$ as how often f covers $\mathbb{S}^2$. Again, one should note the difficulty inherent in making this definition precise since one must take into account a form of orientation and doubling back, i.e., whether f folds back over itself over some region of the codomain. In the same way, one can define the degree of a continuous map $F : S \rightarrow \mathbb{S}^2$, where S is a regular surface without boundary. (We give the definition for a regular surface in Chapter 5.) The conditions that S is regular and has no boundary guarantees that F generically covers $\mathbb{S}^2$ by the same amount at all its points, so long as we take into account orientation and assign a negative covering value to F when F covers $\mathbb{S}^2$ negatively.

As discussed in Definition 2.2.5 and in the following, it is more appropriate to view the parametrization of a simple closed curve C

as a continuous function $\vec{x} : \mathbb{S}^1 \to \mathbb{R}^3$. So consider two simple closed curves with parametrizations $\vec{\alpha} : \mathbb{S}^1 \to \mathbb{R}^3$ and $\vec{\beta} : \mathbb{S}^1 \to \mathbb{R}^3$. The function

$$\Psi : \mathbb{S}^1 \times \mathbb{S}^1 \to \mathbb{S}^2 \quad \text{given by} \quad \Psi(t, u) = \frac{\vec{\beta}(u) - \vec{\alpha}(t)}{\|\vec{\beta}(u) - \vec{\alpha}(t)\|}$$

defines a continuous function from the torus $\mathbb{S}^1 \times \mathbb{S}^1$ to the unit sphere $\mathbb{S}^2$. The technical definition of the linking number of two curves is the degree of $\Psi(t, u)$. The proof of Theorem 4.3.4 consists of calculating the degree of $\Psi(t, u)$.

Equation (4.8) is difficult to use in general, and in the exercises, we often content ourselves with using a computer algebra system to calculuate the integrals. Some basic facts about linking are not difficult to show with the appropriate topological background but are difficult using Equation (4.8). For example, if two simple closed curves can be separated by a plane, then $\text{link}(C_1, C_2) = 0$. This fact is simple once one has a few facts about the degrees of maps to $\mathbb{S}^2$, but using Gauss' Formula to show the same result is a difficult problem.

Problems

4.3.1. This exercise finishes the proof of the Fary-Milnor Theorem. Consider the function $H(u, t)$ defined in Equation (4.7).

(a) Prove that, for all $t_0 \in [0, 1]$, $H(u, t_0)$ is an injective function for $u \in (0, 2\pi)$ and that $H(0, t_0) = H(2\pi, t_0)$.

(b) Show that the locus of $H(u, 1)$ is a circle.

(c) Show that the locus of $H(u, 0)$ is the curve C.

(d) Use the previous parts of the exercise to conclude that $H(u, t)$ is a homotopy between C and a circle.

4.3.2. Consider the trefoil knot parametrized by $\vec{\alpha}(t) = ((3 + \cos 3t) \cos 2t, (3 + \cos 3t) \sin 2t, \sin 3t)$. Using a CAS, calculate the total curvature of this trefoil knot. Show how to create another simple closed curve that is homotopic to this trefoil knot, with a total curvature of $4\pi + \varepsilon$ for any $\varepsilon > 0$.

4.3.3. Consider the two simple closed curves C_1 and C_2. Prove that $\text{link}(C_1, C_2) = \text{link}(C_2, C_1)$.

4.3.4. Let C_1 and C_2 be the two circles in $\mathbb{R}^3$ parametrized by

$$\vec{\alpha}(t) = (\cos t, \sin t, 0), \qquad \text{for } t \in [0, 2\pi],$$
$$\vec{\beta}(t) = (1 - \cos t, 0, \sin t), \quad \text{for } t \in [0, 2\pi].$$

(a) Using Equation (4.8), write an integral that gives the linking number of these two curves.

(b) Use a computer algebra system to calculate this linking number.

(c) (*) Calculate the integral explicitly.

4.3.5. A rigid motion in $\mathbb{R}^3$ is a transformation $F : \mathbb{R}^3 \to \mathbb{R}^3$ defined by $F(\vec{x}) = A\vec{x} + \vec{b}$, where A is an orthogonal matrix. Prove from Equation (4.8) that the linking number is a geometric invariant, i.e., that if a rigid motion is applied to two simple closed curves C_1 and C_2, then their linking number does not change.

4.3.6. Consider the two simple closed curves

$$\vec{\alpha}(t) = (3 \cos t, 3 \sin t, 0), \qquad \text{for } t \in [0, 2\pi],$$
$$\vec{\beta}(t) = ((3 + \cos(nt)) \cos t, (3 + \cos(nt)) \sin t, \sin(nt)), \quad \text{for } t \in [0, 2\pi].$$

(a) Explain from the definition why the linking number of these two curves is n.

(b) The formula of Gauss in Equation (4.8) is quite difficult to use, but, using a computer algebra system, give support for the above answer.

CHAPTER 5

Regular Surfaces

5.1 Parametrized Surfaces

There are many approaches that one can take to introduce surfaces. Some texts immediately build the formalism of differentiable manifolds, some texts first encounter surfaces in $\mathbb{R}^3$ as the solution set to an algebraic equation $F(x, y, z) = 0$ with three variables, and other texts present surfaces as the images of vector functions of two variables. In this section, we introduce the last of these three options and occasionally refer to the connection with the surfaces as solution sets to algebraic equations. Only later will we show why one requires more technical definitions to arrive at a workable definition that matches what one typically means by a "surface."

We imitate the definition of parametrized curves in $\mathbb{R}^n$ and begin our study of surfaces (yet to be defined) by considering continuous functions $\vec{X} : U \to \mathbb{R}^3$, where U is a subset of $\mathbb{R}^2$. Below are some examples of such functions whose images in $\mathbb{R}^3$ are likely to appear in a multivariable calculus course.

Example 5.1.1 (Planes). If $\vec{a}$ and $\vec{b}$ are linearly independent vectors in $\mathbb{R}^3$, then the plane through the point $\vec{p}$ and parallel to the vector $\vec{a}$ and $\vec{b}$ can be expressed as the image of the following function:

$$\vec{X}(u, v) = \vec{p} + u\vec{a} + v\vec{b}$$

for $(u, v) \in \mathbb{R}^2$. In other words, we can write

$$\vec{X}(u, v) = (p_1 + a_1 u + b_1 v, p_2 + a_2 u + b_2 v, p_3 + a_3 u + b_3 v)$$

for constants p_i, a_i, b_i, with $1 \le i \le 3$ that make (a_1, a_2, a_3) and (b_1, b_2, b_3) noncollinear vectors.

Example 5.1.2 (Graphs of Functions). If $z = f(x, y)$ is a real function of two variables defined over some set $U \subset \mathbb{R}^2$, one can obtain the graph of this function as the image of a function $\vec{X} : U \to \mathbb{R}^3$ by setting

$$\vec{X}(u, v) = (u, v, f(u, v)).$$

Example 5.1.3 (Spheres). One can obtain a sphere of radius R centered at the origin as the image of

$$\vec{X}(u, v) = (R \cos u \sin v, R \sin u \sin v, R \cos v),$$

with $(u, v) \in [0, 2\pi] \times [0, \pi]$. This expression should not appear too mysterious since it merely puts the equation of spherical coordinates $r = R$ into Cartesian coordinates, with $u = \theta$ and $v = \phi$.

Example 5.1.4 (Conics). Besides getting the sphere, one could also obtain an ellipsoid as the image of some vector function $\vec{X} : U \to \mathbb{R}^2$ by modifying the coefficients in front of the x, y, and z coordinate functions of the parametrization for the sphere.

One can obtain all of the other conic surfaces as follows. The circular cone

$$\frac{x^2}{a^2} + \frac{y^2}{a^2} = \frac{z^2}{b^2}$$

is the image of

$$\vec{X}(u, v) = (au \cos v, au \sin v, bu), \quad \text{with } (u, v) \in \mathbb{R} \times [0, 2\pi].$$

The function

$$\vec{X}(u, v) = (\cosh u \cos v, \cosh u \sin v, \sinh u), \quad \text{with } (u, v) \in \mathbb{R} \times [0, 2\pi],$$

traces out the hyperboloid of one sheet. (See Figure 5.1 for two perspectives of a hyperboloid of one sheet.) For the hyperboloid of two sheets, we need two separate functions

$$\vec{X}(u, v) = (\sinh u \cos v, \sinh u \sin v, \cosh u)$$

and

$$\vec{X}(u, v) = (\sinh u \cos v, \sinh u \sin v, -\cosh u).$$

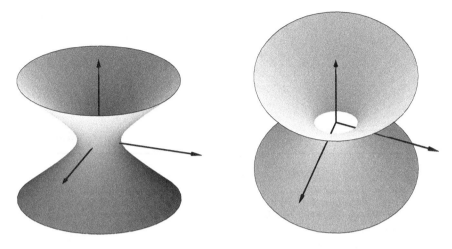

Figure 5.1. Hyperboloid of one sheet.

Example 5.1.5 (Surfaces of Revolution). A *surface of revolution* is a set
S in $\mathbb{R}^3$ obtained by rotating a regular plane curve C in a plane $\mathcal{P}$
about a line L in $\mathcal{P}$ that does not meet C. The curve C is called
the *generating curve*, and the line L is called the *rotation axis*. The
circles described by the rotation locus of the points of C are called
the *parallels* of S, and the various positions of C are called the *meridians* of S. A surface of revolution can be parametrized naturally by
the parameter on the curve C and the angle of rotation u about the
axis.

Let $\vec{x} : I \to \mathbb{R}^2$ be a parametric curve in the xy-plane, with $\vec{x}(t) =$
$(x(t), y(t))$, and suppose we want to find the surface of revolution
of this curve about either the x- or y-axis. About the y-axis, we
take $\vec{X}(t, u) = (x(t)\cos u, y(t), x(t)\sin(u))$, and about the x-axis,
we would take $\vec{X}(t, u) = (x(t), y(t)\cos u, y(t)\sin(u))$. Both of these
options have domains of $I \times [0, 2\pi]$. (See Figure 5.2 for an example
of a surface of revolution based on a semicycloid.)

Looking more closely at Example 5.1.3, one should note that
though the function $\vec{X}(u, v)$ maps onto the sphere of radius R centered at the origin, this function exposes two potential problems with
the intuitive approach. The first problem is that $\vec{X}$ is not injective.

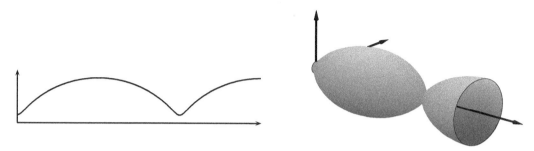

Figure 5.2. Surface of revolution.

In particular, for any point (x, y, z) on this sphere with $y = 0$ and $x > 0$, we have

$$v = \cos^{-1}\left(\frac{z}{R}\right) \qquad \text{and} \qquad u = 0 \text{ or } 2\pi,$$

and so there exist two preimages for such points. Worse still, if we consider the north and south poles $(0, 0, 1)$ and $(0, 0, -1)$, the preimages are the points $(u, 0)$ and (u, π), respectively, for all $u \in [0, 2\pi]$. With the given parametrization, the arc on the sphere defined by $u = 0$ is the set of points with more than one preimage. This latter remark leads to the second problem because the points on this arc have no particular geometric significance, and one would wish to avoid any formulation of a surface that confers special properties on some points that are geometrically ordinary.

Before addressing these and other issues necessary to be able to "do calculus" on a surface, we need to define the class of subsets of $\mathbb{R}^3$ that one can even hope to study with differential geometry. The primary criteria for such a subset is that one must be able to describe its points using continuous functions.

Definition 5.1.6. A subset $S \subseteq \mathbb{R}^3$ is called a *parametrized surface* if for each $p \in S$, there exists an open set $U \subset \mathbb{R}^2$, an open neighborhood V of p in $\mathbb{R}^3$, and a continuous function $\vec{X} : U \to \mathbb{R}^3$ such that $\vec{X}(U) = V \cap S$. Each such $\vec{X}$ is called a *parametrization* of a neighborhood of S. We call a parametrized surface *of class* C^r if it can be covered by parametrizations $\vec{X}$ of class C^r.

(Refer to Section A.2.2 in the appendix of [22] for a clear discussion of open and closed sets on surfaces.) Note that in contrast to

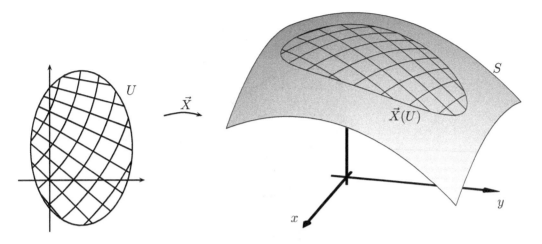

Figure 5.3. Coordinate patch.

space curves, where one uses intervals as the domain for parametrizations, this definition uses open sets. Furthermore, this definition does not insist that S be the image of a single parametrization but, rather that it be covered by images of parametrizations.

For any such function $\vec{X} : U \to V \cap S$, where $V \cap S$ is an open neighborhood of p in S, we call $\vec{X}$ a parametrization of the *coordinate patch* $V \cap S$. (See Figure 5.3 for an illustration of the relation of the domain $U \subset \mathbb{R}^2$ and the coordinate patch on the surface $V \cap S$.)

Definition 5.1.6 states that, for every point $p \in S$, there is an open neighborhood of p on S that is the image of some vector function.

With the tools of curves in the plane or space, one can study properties of a surface S by considering various families of curves on the surface. If one wishes to study S near a point p, one could look for common properties of all the curves on S passing through p.

There are two natural classes of curves on any given parametrized surface S that arise naturally. First, one might study *slices* of S, by which we mean the intersection of S with a family of parallel planes. For example, Figure 5.4 shows the parametric surface

$$\vec{X}(u, v) = (\cos v, \sin 2v \cos u, \sin 2v \sin u),$$

with domain $[0, 2\pi] \times [0, \pi]$. One can understand some of its geometry by considering the intersection of S with planes of the form $x = a$,

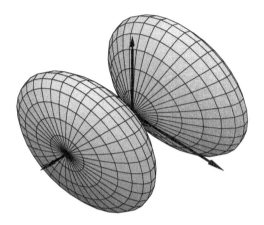

Figure 5.4. Slices of a surface of revolution.

$y = b$, or $z = c$. If we fix $x = a$, with $-1 \le a \le 1$, we get $\cos v = a$, and there exists one value $v_0 \in [0, \pi]$ satisfying this equation. Then the slice of $\vec{X}(u, v)$ intersecting the $x = a$ plane gives a circle parallel to the yz-plane, given by

$$(x, y, z) = (a, \sin(2v_0)\cos u, \sin(2v_0)\sin u).$$

On the other hand, any plane P through the origin containing the x-axis can be written as $by + cz = 0$ for some real constants b, c. The intersection of the surface with the plane P leads to the equation

$$b\sin 2v \cos u + c\sin 2v \sin u = 0 \quad \Longleftrightarrow \quad (\sin 2v)(b\cos u + c\sin u) = 0.$$

The solutions that involve $\sin 2v = 0$ leads to three points (namely, $(0, 0, 0)$, $(1, 0, 0)$, and $(-1, 0, 0)$). Otherwise, if $\sin 2v \ne 0$, then there exists a unique value $u = u_0$ such that

$$b\cos u_0 + c\sin u_0 = 0.$$

Using the basis $\{(1, 0, 0), (0, \cos u_0, \sin u_0)\}$ for the plane P, the slice $S \cap P$ has parametric equations $(\cos v, \sin 2v)$.

Another natural family of curves on parametrized surfaces is what is generically called the *coordinate lines* of the surface S. A coordinate line on S is the image of a space curve defined by fixing one of the variables in a particular parametrization of a coordinate patch

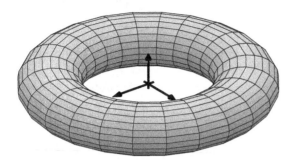

Figure 5.5. Coordinate lines on a torus.

of S. Namely, if $\vec{X} : U \to \mathbb{R}^3$ parametrizes a patch of S, then a curve of the form

$$\vec{\gamma_1}(u) = \vec{X}(u, v_0), \quad \text{respectively} \quad \vec{\gamma_2}(v) = \vec{X}(u_0, v),$$

is called a coordinate line of the variable u, respectively v. (See Figure 5.5 for an example of coordinate lines of a torus parametrized by $\vec{X}(u, v) = ((3 + \cos v) \cos u, (3 + \cos v) \sin u, \sin v)$ for $(u, v) \in [0, 2\pi] \times [0, 2\pi]$.)

Example 5.1.7. With the surfaces of revolution described in Example 5.1.5, the coordinate lines of t and u, respectively, are precisely the parallels and the meridians.

Problems

5.1.1. Find a parametrization $\vec{X}$ for the cylinder $\{(x, y, z) \in \mathbb{R}^3 \,|\, x^2 + y^2 = 1\}$. Can the domain of this parametrization be chosen so that $\vec{X}$ is bijective onto the cylinder? Explain why or why not.

5.1.2. Prove that the following functions provide parametrizations of the hyperbolic paraboloid $x^2 - y^2 = z$. Provide appropriate domains of definitions. Which ones are injective? Which ones are surjective onto the surface?

    (a) $\vec{X}(u, v) = (v \cosh u, v \sinh u, v^2)$.

    (b) $\vec{X}(u, v) = ((u + v), (u - v), 4uv)$.

    (c) $\vec{X}(u, v) = (uv, u(1 - v), u^2(2v - 1))$.

5.1.3. Let $\vec{\alpha} : I \to \mathbb{R}^2$ be a regular, plane curve, and let L be a line that does not intersect the image of $\vec{\alpha}$. Suppose that L goes through the point $\vec{p}$ and has direction vector $\vec{u}$ so that $\vec{p} + t\vec{u}$ is a parametrization of L. Find a formula for the parametrization of the surface of revolution obtained by rotating $\vec{\alpha}(I)$ about L.

5.1.4. Prove that on the unit sphere parametrized by $\vec{X}(\theta, \phi) = (\cos\theta\sin\phi, \sin\theta\sin\phi, \cos\phi)$, the curves given by

$$(\theta, \phi) = (\ln t, 2\tan^{-1} t)$$

intersect every meridian with an angle of $\frac{\pi}{4}$. (Note that a meridian is a curve such that $\theta =$const. A curve on the sphere that intersects the meridians at a constant angle is called a *loxodrome*.)

5.1.5. Consider a curve in the plane $\vec{x}(t) = (x(t), y(t))$ and the surface of revolution obtained by rotating the image of $\vec{x}$ about the x-axis. This surface of revolution is parametrized by $\vec{X}(t, u) = (x(t)\cos(u), y(t), x(t)\sin(u))$.

    (a) Consider u-coordinate lines, with the space curves $\gamma_1(t) = (x(t)\cos(u_0), y(t), x(t)\sin(u_0))$, where u_0 is fixed. Calculate the space curvature and torsion of γ_1.

    (b) Repeat the above question with the t-coordinate lines, i.e., the space curves $\gamma_2(u) = (x(t_0)\cos(u), y(t_0), x(t_0)\sin(u))$, where t_0 is fixed.

5.1.6. Consider the set of points $S = \{(x, y, z) \in \mathbb{R}^3 \,|\, x^4 + y^4 + z^4 = 1\}$. Modify the usual parametrization for a sphere to find parametrizations that cover S.

5.2   Tangent Planes and Regular Surfaces

In the local theory of curves, we called a curve C regular at a point p if there is a parametrization $\vec{x}(t)$ of C near $p = \vec{x}(t_0)$ such that $\vec{x}'(t_0)$ exists and $\vec{x}'(t_0) \neq \vec{0}$. In Section 1.2, we saw that this definition is tantamount to requiring that

$$\lim_{t \to t_0} \frac{\vec{x}'(t)}{\|\vec{x}'(t)\|}$$

exists and, hence, that there exists a tangent line to C at p. Imitating this latter geometric property, we will eventually call a point $p \in S$

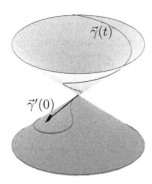

Figure 5.6. Double cone.

regular if one can define a tangent plane to p at S, but we must delay
a precise definition until after we define the tangent plane to S at p.

Definition 5.2.1. Let S be a parametrized surface and p a point in S.
Consider the set of space curves $\vec{\gamma} : (-\varepsilon, \varepsilon) \to \mathbb{R}^3$ such that $\vec{\gamma}(0) = p$,
and the image of $\vec{\gamma}$ lies entirely in S. A *tangent vector* to S at p is
any vector in $\mathbb{R}^3$ that can be expressed as $\vec{\gamma}'(0)$, where $\vec{\gamma}$ is such a
space curve.

From the point of intuition, one often views tangent vectors to
S at p as based at p. The set of tangent vectors, as a subset of $\mathbb{R}^3$,
gives a sort of approximation of the behavior of S at p. By rescaling
the parameter t, one notices that the set of tangent vectors to S at a
point p contains $\vec{0}$ and is closed under scalar multiplication. Hence,
the set of tangent vectors will be a (usually infinite) union of lines,
and, as we shall see, it is often a plane but not always.

Example 5.2.2. As an example of a set of tangent vectors that is not
a plane, consider the cone with equation

$$x^2 + y^2 - z^2 = 0,$$

with parametrization

$$\vec{X}(u, v) = (v \cos u, v \sin u, v), \quad \text{for } (u, v) \in [0, 2\pi] \times \mathbb{R}$$

(see Figure 5.6). We note that $\vec{X}(u, 0) = (0, 0, 0)$ for all $u \in [0, 2\pi]$
and consider the point $p = \vec{X}(u, 0) = (0, 0, 0)$. A curve on the cone

through p has the form $\vec{\gamma}(t) = \vec{X}(u(t), v(t))$, where $v(0) = 0$ and u is any real in $[0, 2\pi]$. For such a curve, using the multivariable chain rule, we have

$$\vec{\gamma}'(t) = \frac{\partial \vec{X}}{\partial u}\frac{du}{dt} + \frac{\partial \vec{X}}{\partial v}\frac{dv}{dt} = u'(t)\frac{\partial \vec{X}}{\partial u} + v'(t)\frac{\partial \vec{X}}{\partial v},$$

and since $\vec{X}_u(u, 0) = (0, 0, 0)$,

$$\vec{\gamma}'(0) = v'(0)\,(\cos u, \sin u, 1).$$

Since $v'(0)$ and u can be any real values, we determine that the set of tangent vectors to the cone at $(0,0,0)$ is precisely the cone itself.

Definition 5.2.3. Let S be a parametrized surface and p a point in S. If the set of tangent vectors to S at p forms a two-dimensional subspace of $\mathbb{R}^3$, we call this subspace the *tangent space* to S at p and denote it by T_pS. If T_pS exists, we call the *tangent plane* the set of points

$$\{p + \vec{v} \mid \vec{v} \in T_pS\}.$$

We wish to find a condition that determines when the set of tangent vectors is a two-dimensional vector subspace.

Let p be a point on a surface S, and let $\vec{X} : U \to \mathbb{R}^3$ be a parametrization of a neighborhood $V \cap S$ of p for some $U \subseteq \mathbb{R}^2$. Suppose that $p = \vec{X}(u_0, v_0)$ and consider the coordinate lines $\vec{\gamma}_1(t) = \vec{X}(u_0 + t, v_0)$ and $\vec{\gamma}_2(t) = \vec{X}(u_0, v_0 + t)$ through p. Both $\vec{\gamma}_1$ and $\vec{\gamma}_2$ lie on the surface and, by the chain rule,

$$\vec{\gamma}_1'(0) = \frac{\partial \vec{X}}{\partial u}(u_0, v_0) \quad \text{and} \quad \vec{\gamma}_2'(0) = \frac{\partial \vec{X}}{\partial v}(u_0, v_0).$$

Furthermore, for any curve $\vec{\gamma}(t)$ on the surface with $\vec{\gamma}(0) = p$, we can write

$$\vec{\gamma}(t) = \vec{X}(u(t), v(t))$$

for some functions $u(t)$ and $v(t)$, with $u(0) = u_0$ and $v(0) = v_0$. By the chain rule,

$$\vec{\gamma}'(t) = u'(t)\frac{\partial \vec{X}}{\partial u}(u(t), v(t)) + v'(t)\frac{\partial \vec{X}}{\partial v}(u(t), v(t)),$$

so at p,

$$\vec{\gamma}'(0) = u'(0)\frac{\partial \vec{X}}{\partial u}(u_0, v_0) + v'(0)\frac{\partial \vec{X}}{\partial v}(u_0, v_0) = u'(0)\vec{\gamma}_1'(0) + v'(0)\vec{\gamma}_2'(0).$$

Consequently, $\vec{\gamma}'(0)$ is a linear combination of $\frac{\partial \vec{X}}{\partial u}(u_0, v_0)$ and $\frac{\partial \vec{X}}{\partial v}(u_0, v_0)$.

Definition 5.1.6 does not assume that the parametrization $\vec{X}$ is injective. Therefore, the set of tangent vectors is a plane if and only if the union of all linear subspaces

$$\text{Span}\left(\frac{\partial \vec{X}}{\partial u}(u_0, v_0), \frac{\partial \vec{X}}{\partial v}(u_0, v_0)\right),$$

where $\vec{X}(u_0, v_0) = p$ is a plane. In particular, if there exists only one (u_0, v_0) such that $\vec{X}(u_0, v_0) = p$, then a tangent plane exists at p if $\frac{\partial \vec{X}}{\partial u}(u_0, v_0)$ and $\frac{\partial \vec{X}}{\partial v}(u_0, v_0)$ are not collinear.

To simplify notations, one often uses the following abbreviated notation for partial derivatives:

$$\vec{X}_u(u, v) = \frac{\partial \vec{X}}{\partial u}(u, v) \qquad \text{and} \qquad \vec{X}_v(u, v) = \frac{\partial \vec{X}}{\partial v}(u, v).$$

One then writes succinctly that if $\{(u_0, v_0)\} = \vec{X}^{-1}(p)$, then S has a tangent plane at p if and only if $\vec{X}_u(u_0, v_0) \times \vec{X}_v(u_0, v_0) \neq \vec{0}$. Furthermore, the tangent plane to S at p is the unique plane through p with normal vector $\vec{X}_u(u_0, v_0) \times \vec{X}_v(u_0, v_0) \neq \vec{0}$. This leads us to a formula for the tangent plane.

Proposition 5.2.4. *Let S be a surface parametrized near a point p by* *$\vec{X} : U \to \mathbb{R}^3$. Suppose that $\vec{X}$ is injective at p (i.e., p has only one preimage) and that $p = \vec{X}(u_0, v_0)$. Then the set of tangent vectors forms a subspace of $\mathbb{R}^3$ if and only if $\vec{X}_u(u_0, v_0) \times \vec{X}_v(u_0, v_0) \neq \vec{0}$. In this case, the tangent plane to S at p satisfies the following equation for position vectors $\vec{x} = (x, y, z)$:*

$$(\vec{x} - \vec{X}(u_0, v_0)) \cdot (\vec{X}_u(u_0, v_0) \times \vec{X}_v(u_0, v_0)) = 0. \qquad (5.1)$$

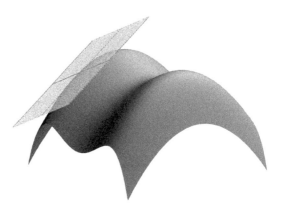

Figure 5.7. A tangent plane.

Through an abuse of language, even if $\vec{X}$ is not injective at p and $\vec{X}_u(u_0, v_0) \times \vec{X}_v(u_0, v_0) \neq 0$ for some $(u_0, v_0) \in U$, we will call Equation (5.1) the equation of the tangent plane to $\vec{X}$ at $q = (u_0, v_0)$. This is an abuse of language since if $p = \vec{X}(u_0, v_0)$ and S is the image of $\vec{X}$, then p might have more than one preimage and, therefore, the set of tangent vectors to S of p might not be a plane but be a union of planes. Problem 5.2.9 gives an example of a parametrized surface that has points where the set of tangent vectors is the union of two planes.

Example 5.2.5. As a first example, consider the vector function $\vec{X}(u, v) = (u, v, 4 + 3u^2 - u^4 - 2v^2)$. We calculate

$$\vec{X}_u \times \vec{X}_v = (1, 0, 6u - 4u^3) \times (0, 1, -4v)$$
$$= (-6u + 4u^3, 4v, 1).$$

Figure 5.7 shows the surface along with the tangent plane at the point $\vec{X}(-1.4, -1)$. At this specific point, we have $p = \vec{X}(-1.4, -1) = (-1.4, -1, 4.0384)$ and $\vec{X}_u \times \vec{X}_v(-1.4, -1) = (-2.576, -4, 1)$. Since $\vec{X}$ is in fact the graph of an injective two-variable function, by Proposition 5.2.4, the tangent plane to $\vec{X}$ at p is

$$-2.576(x+1.4)-4(y+1)+(z-4.0384) = 0 \iff -2.576x-4y+z = 11.6448.$$

 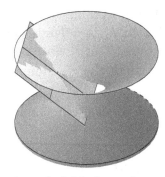

Figure 5.8. A tangent plane on a hyperboloid of one sheet.

Example 5.2.6. As a second example, consider the hyperboloid of one sheet given by the parametrization

$$\vec{X}(u, v) = (\cosh u \cos v, \cosh u \sin v, \sinh u).$$

A calculation produces

$$\vec{X}_u \times \vec{X}_v = (-\cosh^2 u \cos v, -\cosh^2 u \sin v, \cosh u \sinh u).$$

Then at the point $p = \vec{X}(1, 0) = (\cosh 1, 0, \sinh 1)$, the tangent plane exists, and its equation is

$$-\cosh^2 1(x - \cosh 1) + \cosh 1 \sinh 1(z - \sinh 1) = 0 \iff (\cosh 1)x - (\sinh 1)z = 1.$$

Figure 5.8 shows two perspectives of this hyperboloid along with its tangent plane at p. It is interesting to contrast this picture with Figure 5.7, in which the tangent plane does not intersect the surface except at p. The shape of the hyperboloid is such that at any point p the tangent plane at p intersects the hyperboloid in nontrivial curves.

Using the notion of the differential of the parametrization $\vec{X}$, we can summarize Proposition 5.2.4 in another way that is more convenient when we discuss parametrized surfaces in higher dimensions. (See Section 5.2 in [22] for a longer explanation of the differential of a function from $\mathbb{R}^m$ to $\mathbb{R}^n$.)

If $\vec{F}$ is a function from an open set $U \subset \mathbb{R}^n$ to $\mathbb{R}^m$, we write $\vec{F} = (F_1, F_2, \ldots, F_m)^T$ and think of $\vec{F}$ as a column vector of functions

F_i, each in n variables. For all $\vec{a} \in U$, we say that F is *differentiable* at $\vec{a}$ if each F_i is differentiable at $\vec{a}$. Furthermore, we define the *differential* of $\vec{F}$ at $\vec{a}$, written $d\vec{F}_{\vec{a}}$, as the linear transformation $\mathbb{R}^n \to \mathbb{R}^m$ that sends the jth standard vector $\vec{e}_j$ to $\frac{\partial \vec{F}}{\partial x_j}(\vec{a})$. We denote by $[d\vec{F}_{\vec{a}}]$ the matrix of $d\vec{F}_{\vec{a}}$ with respect to the standard bases of $\mathbb{R}^n$ and $\mathbb{R}^m$, which we can write explicitly as

$$[d\vec{F}_{\vec{a}}] = \begin{pmatrix} \dfrac{\partial \vec{F}}{\partial x_1} & \dfrac{\partial \vec{F}}{\partial x_2} & \cdots & \dfrac{\partial \vec{F}}{\partial x_n} \end{pmatrix} = \begin{pmatrix} \frac{\partial F_1}{\partial x_1} & \frac{\partial F_1}{\partial x_2} & \cdots & \frac{\partial F_1}{\partial x_m} \\ \frac{\partial F_2}{\partial x_1} & \frac{\partial F_2}{\partial x_2} & \cdots & \frac{\partial F_2}{\partial x_m} \\ \vdots & \vdots & \ddots & \vdots \\ \frac{\partial F_n}{\partial x_1} & \frac{\partial F_n}{\partial x_2} & \cdots & \frac{\partial F_n}{\partial x_m} \end{pmatrix}. \quad (5.2)$$

By a slight abuse of notation, we will sometimes write $[d\vec{F}]$ or even $d\vec{F}$, without the subscript, to mean the matrix of functions with $\partial \vec{F} / \partial x_j$ as the jth column.

We can now restate Proposition 5.2.4 as follows.

Corollary 5.2.7. *Let S be a surface, where $p \in S$, and suppose that in an open neighborhood of p, the surface S is parametrized by $\vec{X}$: $U \to \mathbb{R}^3$. The tangent space (and tangent plane) to S at p exists if $\vec{X}^{-1}(p) = \{q\}$ (a singleton set), the differential $d\vec{X}_q$ exists, and $d\vec{X}_q$ has maximal rank. Furthermore, the tangent space is given by $T_p S = \text{Im}(d\vec{X}_q)$.*

Proof: Note that with respect to the standard bases in $\mathbb{R}^2$ and $\mathbb{R}^3$, if $q = (u_0, v_0)$, the matrix of $d\vec{X}_q$ is the 3×2 matrix

$$[d\vec{X}_q] = \begin{pmatrix} \frac{\partial X_1}{\partial u}(q) & \frac{\partial X_1}{\partial v}(q) \\ \frac{\partial X_2}{\partial u}(q) & \frac{\partial X_2}{\partial v}(q) \\ \frac{\partial X_3}{\partial u}(q) & \frac{\partial X_3}{\partial v}(q) \end{pmatrix}, \quad (5.3)$$

where we have written $\vec{X}(u,v) = (X_1(u,v), X_2(u,v), X_3(u,v))^T$.

First note that for the tangent plane to exist, the partial derivatives of $\vec{X}$ must exist, so the differential exists. From Proposition 5.2.4, we know that the surface is a plane if $\vec{X}^{-1}(p)$ is a singleton set, which we'll denote by $\{q\} \subset U$, and $\vec{X}_u \times \vec{X}_v(q)$ exists and is nonzero. This is equivalent to saying that $\vec{X}_u(q)$ and $\vec{X}_v(q)$ are linearly independent, which means that $d\vec{X}_q$ has maximal rank. $\quad \square$

The condition in Corollary 5.2.7 that requires $d\vec{X}_q$ to have maximal rank occurs often enough in various contexts that we give it a name, as follows.

Definition 5.2.8. Let F be a function from an open set $U \subset \mathbb{R}^n$ to $\mathbb{R}^m$. A point $q \in U$ is called a *critical point* if dF_q does not have maximal rank, i.e., $\operatorname{rank} dF_q < \min(m, n)$. If q is a critical point of F, we call $F(q) \in \mathbb{R}^m$ a *critical value* of F. If $p \in \mathbb{R}^m$ is not a critical value of F (even if p is not in the image of F), then we call p a *regular value* of F.

Finally, suppose that S is a surface parametrized by $\vec{X} : U \to \mathbb{R}^3$, and let $p = \vec{X}(q)$ be a point of S. We call p a *singular point* of S if q is a critical point of $\vec{X}$ or $\vec{X}^{-1}(p)$ is not a singleton.

With this terminology, according to Corollary 5.2.7, the singular points on a parametrized surface are precisely the points where the set of tangent vectors does not form a subspace.

We have belabored the issue of whether or not one can define a tangent plane to S at a point p because, in an intuitive sense, if a tangent plane does not exist locally near p, then the surface S does not "resemble" a two-dimensional plane and thus does not look like what we would expect in a surface. Just as with curves we introduced the notion of a regular curve (i.e., a curve $\vec{x} : I \to \mathbb{R}^n$ such that $\vec{x}'(t) \neq \vec{0}$ for all $t \in I$), so with surfaces, we would like to define a class of surfaces that is smooth enough so that we can "do calculus" on it. The points in the above discussion lead to the following definition.

Definition 5.2.9. A subset $S \subseteq \mathbb{R}^3$ is a *regular surface* if for each $p \in S$, there exists an open set $U \in \mathbb{R}^2$, an open neighborhood V of p in $\mathbb{R}^3$, and a surjective continuous function $\vec{X} : U \to V \cap S$ such that

1. $\vec{X}$ is differentiable: if we write $\vec{X}(u, v) = (x(u, v), y(u, v), z(u, v))$, then the functions $x(u, v)$, $y(u, v)$, and $z(u, v)$ have continuous partial derivatives of all orders;

2. $\vec{X}$ is a homeomorphism: $\vec{X}$ is continuous and has an inverse $\vec{X}^{-1} : V \cap S \to U$ such that $\vec{X}^{-1}$ is continuous;

3. $\vec{X}$ satisfies the regularity condition: for each $(u, v) \in U$, the differential $d\vec{X}_{(u,v)} : \mathbb{R}^2 \to \mathbb{R}^3$ is a one-to-one linear transformation.

The parametrization $\vec{X}$ is called a *system of coordinates* (in a neighborhood) of p, and the neighborhood $V \cap S$ of p in S is called a *coordinate neighborhood*. One should observe that a regular surface is defined as a subset of $\mathbb{R}^3$ and not as a function from a subset of $\mathbb{R}^2$ to $\mathbb{R}^3$. However, one may view a regular surface as the union of the images of an appropriate set of systems of coordinates.

Example 5.2.10. We can prove directly that the unit sphere $\mathbb{S}^2$ is a regular surface in a few different ways. We'll use rectangular coordinates first. Consider a point $p = (x, y, z) \in \mathbb{S}^2$, and let $U = \{(u, v) \,|\, u^2 + v^2 < 1\}$. If $z > 0$, then the mapping $\vec{X}_{(1)} : U \to \mathbb{R}^3$ defined by $(u, v, \sqrt{1 - u^2 - v^2})$ is clearly a bijection between U and $\mathbb{S}^2 \cap \{(x, y, z) | z > 0\}$. $\vec{X}_{(1)}$ is also a homeomorphism because it is continuous, and its inverse $\vec{X}_{(1)}^{-1}$ is simply the vertical projection of the upper unit sphere onto $\mathbb{R}^2$ and since projection is a linear transformation, it is continuous. Furthermore, $\vec{X}_{(1)}$ satisfies all the conditions in Definition 5.2.9. We leave it to the reader to check Condition 1, while for Condition 3, we calculate the differential

$$
d\vec{X}_{(1)} = \begin{pmatrix} 1 & 0 \\ 0 & 1 \\ -\dfrac{u}{\sqrt{1 - u^2 - v^2}} & -\dfrac{v}{\sqrt{1 - u^2 - v^2}} \end{pmatrix}.
$$

For all $q = (u, v) \in U$, the entries in this matrix are well defined and produce a matrix that defines a one-to-one linear transformation. Consequently, $[d\vec{X}_{(1)q}]$ is of maximal rank.

The mapping $\vec{X}_{(1)} : U \to \mathbb{R}^3$ so far only tells us that the open upper half of the sphere is a regular surface, but we wish to show that the whole sphere is a regular surface.

For other points $p = (x, y, z) \in \mathbb{S}^2$, we use the following parametrizations $\vec{X}_{(i)} : U \to \mathbb{R}^3$ of coordinate patches around p:

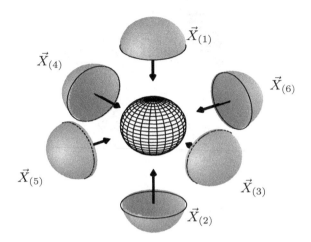

Figure 5.9. Six coordinate patches on the sphere.

if $z > 0$, $\vec{X}_{(1)}(u,v) = (u, v, \sqrt{1 - u^2 - v^2})$,   if $z < 0$, $\vec{X}_{(2)}(u,v) = (u, v, -\sqrt{1 - u^2 - v^2})$,

if $y > 0$, $\vec{X}_{(3)}(u,v) = (u, \sqrt{1 - u^2 - v^2}, v)$,   if $y < 0$, $\vec{X}_{(4)}(u,v) = (u, -\sqrt{1 - u^2 - v^2}, v)$,

if $x > 0$, $\vec{X}_{(5)}(u,v) = (\sqrt{1 - u^2 - v^2}, u, v)$,   if $x < 0$, $\vec{X}_{(6)}(u,v) = (-\sqrt{1 - u^2 - v^2}, u, v)$.

These six parametrizations give six coordinate patches such that for all $p \in \mathbb{S}^2$, p is in at least one of these coordinate patches (see Figure 5.9 for an illustration of the six patches covering the sphere). Since the domain for each parametrization is open, all six patches are necessary.

Example 5.2.11. Carefully using two latitude - longitude parametrizations, we get another way to show that the unit sphere $\mathbb{S}^2$ is a regular surface. Let H_1 be the closed half-plane

$$H_1 = \{(x, y, z) \in \mathbb{R}^3 \mid x \geq 0 \text{ if } y = 0\}.$$

Let $p \in \mathbb{S}^2 - H_1$, and let $U = (0, 2\pi) \times (0, \pi)$. Then $\vec{X}_{(1)} : U \to \mathbb{R}^3$ defined by

$$\vec{X}_{(1)}(u, v) = (\cos u \sin v, \sin u \sin v, \cos v)$$

is a homeomorphism between U and $\mathbb{S}^2 - H_1$ that satisfies the remaining conditions in the definition of a regular surface. (The reader should check this claim.)

Figure 5.10. Not a bijection.

On the other hand, if $p \in \mathbb{S}^2 - H_2$, where H_2 is the closed half-plane

$$H_2 = \{(x, y, z) \in \mathbb{R}^3 | \, x \leq 0 \text{ if } z = 0\},$$

then a homeomorphism between $U \to \mathbb{S}^2 - H_2$ is

$$\vec{X}_{(2)}(u, v) = (-\cos u \sin v, \cos v, -\sin u \sin v).$$

This also satisfies the regularity and smoothness conditions. Furthermore, $(\mathbb{S}^2 - H_1) \cup (\mathbb{S}^2 - H_2) = \mathbb{S}^2$, so the two parametrizations of coordinate patches cover the whole sphere.

The three conditions in Definition 5.2.9 each eliminate various situations we do not wish to let into the class of surfaces of interest in differential geometry. The condition that $\vec{X}$ be differentiable eliminates the possibility of corners or folds. A cube, for example, is not a regular surface because for whatever parametrization is used in the neighborhood of an edge where two faces meet, at least one of the partial derivatives will not exist.

The second condition, that $\vec{X}$ be a homeomorphism, might initially appear the least intuitive. Since we assume that in the neighborhood of each point $p \in S$ there is a parametrization $\vec{X} : U \to V \cap S$ that is a bijection, we already eliminate a surface that intersects itself or degenerates to a curve. Figure 5.10 (see Problem 5.2.9 for the parametrization) shows the image of a parametrized surface $\vec{X}$ that intersects itself along a ray. The tangent surface to this surface at any point along this ray is the union of two planes and not a single plane.

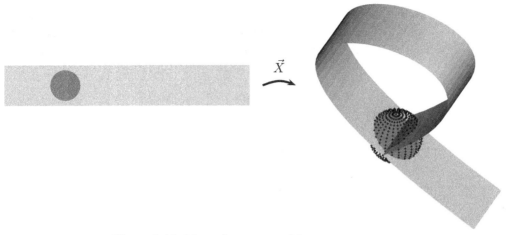

Figure 5.11. Not a homeomorphism.

Requiring not only that each patch $\vec{X} : U \to V \cap S$ of a parametrized surface be a bijection but also bicontinuous (i.e., be a homeomorphism) means intuitively that if $\vec{X}(q_1)$ and $\vec{X}(q_2)$ are arbitrarily close in $\mathbb{R}^3$, then q_1 and q_2 are arbitrarily close. This eliminates situations similar to that depicted in Figure 5.11, where an open strip is twisted back onto itself so that it does not intersect itself but that the distance between one end and another part in the middle of the strip is 0.

We now explain why this is not a homeomorphism. Consider an open disk U inside the open strip, as shown in the figure. Then $\vec{X}(U)$ is not open because a set U' is open in S if and only if $U' = S \cap V$ for some open set V in $\mathbb{R}^3$. But any open set $V \subset \mathbb{R}^3$ that contains $\vec{X}(U)$ must contain other points of S, as shown. Thus, U open in the domain does not imply that $\vec{X}(U)$ is open, which shows that $\vec{X}^{-1}$ is not continuous and, hence, that $\vec{X}$ is not a homeomorphism.

The third condition of Definition 5.2.9 is necessary because, even with the first two conditions satisfied, it is still possible to create a parametrized surface $\vec{X}$ that has a point p such that the tangent space to the surface at p is not a plane but a line. (See Problem 5.2.12 for an example of a surface that satisfies the first two conditions for a regular surface but not the third.)

From the above explanations of the criteria for a regular surface, it would seem hard to determine at the outset whether or not a given surface is regular. However, the next two propositions show under what conditions function graphs and surfaces given as solutions to one equation are regular.

Proposition 5.2.12. *Let $U \subset \mathbb{R}^2$ be open. Then if a function $f : U \to \mathbb{R}$ is differentiable, the subset $S = \{(x, y, x) \in \mathbb{R}^3 : (x, y, z) = (u, v, f(u, v))\}$ is a regular surface.*

Proof: The inverse function $F^{-1} : S \to U$ is the projection onto the xy-plane. F^{-1} is continuous, and since f is differentiable, F is also continuous. Thus, F is a homeomorphism between U and S. The regularity condition is clearly satisfied because

$$dF_{(u,v)} = \begin{pmatrix} 1 & 0 \\ 0 & 1 \\ f_u & f_v \end{pmatrix}$$

is always one-to-one regardless of the values of f_u and f_v. $\square$

Another class of surfaces in $\mathbb{R}^3$ one studies arises as level surfaces of functions $f : \mathbb{R}^3 \to \mathbb{R}$, that is, as points that satisfy the equation $f(x, y, z) = a$. The following proposition actually shows where the terminology "regular surface" comes from in light of the notion of a regular value of multivariable functions presented in Definition 5.2.8.

Proposition 5.2.13. *Let $f : U \subset \mathbb{R}^3 \to \mathbb{R}$ be a differentiable function, and let $a \in \mathbb{R}$ be a regular value of f, i.e., a real number such that for all $p \in f^{-1}(a)$, df_p is not $\vec{0}$. Then the surface defined by $S = \{(x, y, z) \in \mathbb{R}^3 \mid f(x, y, z) = a\}$ is a regular surface. Furthermore, the gradient $df_p = \vec{\nabla} f(p)$ is normal to S at p.*

Proof: Let $p \in S$. Since $df_p \neq \vec{0}$, after perhaps relabeling the axes, we have $f_z(p) \neq 0$. We consider the function $G : \mathbb{R}^3 \to \mathbb{R}^3$ defined by $G(x, y, z) = (x, y, f(x, y, z))$ and notice that

$$dG_p = \begin{pmatrix} 1 & 0 & 0 \\ 0 & 1 & 0 \\ f_x & f_y & f_z \end{pmatrix}.$$

Consequently, $\det(dG_p) = f_z(p) \neq 0$, so dG_p is invertible. The Implicit Function Theorem allows us to conclude that there exists a neighborhood U' of p in U such that G is one-to-one on U',

$V = G(U')$ is open, and the inverse $G^{-1} : V \to U'$ is differentiable. However, since $(u, v, w) = G(x, y, z) = (x, y, f(x, y, z))$, we will be able to write the inverse function as $(x, y, z) = G^{-1}(u, v, w) = (u, v, g(u, v, w))$ for some differentiable function $g : V \to \mathbb{R}$.

Furthermore, if V' is the projection of V onto the uv-plane, then $h : V' \to \mathbb{R}$ defined by $h_a(u, v) = g(u, v, a)$ is differentiable and

$$G(f^{-1}(a) \cap U') = V \cap \{(u, v, w) \in \mathbb{R}^3 \mid w = a\} \Longleftrightarrow$$
$$f^{-1}(a) \cap U' = \{(u, v, h_a(u, v)) \mid (u, v) \in V'\}.$$

Thus, $f^{-1}(a) \cap U'$ is the graph of h_a, and by Proposition 5.2.12, $f^{-1}(a) \cap U'$ is a coordinate neighborhood of p. Thus, every point $p \in S$ has a coordinate neighborhood, and so $f^{-1}(a)$ is a regular surface in $\mathbb{R}^3$.

To prove that df_p is normal to S at p, suppose that $\vec{X}(u, v) = (x(u, v), y(u, v), z(u, v))$ parametrizes regularly a neighborhood of S around p. Then, differentiating the defining equation $f(x, y, z) = a$ with respect to u and v one obtains

$$\begin{cases} \dfrac{\partial f}{\partial x}\dfrac{\partial x}{\partial u} + \dfrac{\partial f}{\partial y}\dfrac{\partial y}{\partial u} + \dfrac{\partial f}{\partial z}\dfrac{\partial z}{\partial u} = 0, \\ \dfrac{\partial f}{\partial x}\dfrac{\partial x}{\partial v} + \dfrac{\partial f}{\partial y}\dfrac{\partial y}{\partial v} + \dfrac{\partial f}{\partial z}\dfrac{\partial z}{\partial u} = 0. \end{cases}$$

Thus, $df_p = (f_x(p), f_y(p), f_z(p))$ is perpendicular to $\vec{X}_u$ and $\vec{X}_v$ at p. $\qquad\square$

(Note: See Section 6.4 in [22] for a statement of the Implicit Function Theorem and some examples.)

We point out that Definition 5.2.9 can be easily modified to provide a definition of regularity for a surface S in $\mathbb{R}^n$. The only required changes are that each parametrization of a neighborhood of S be a continuous function $\vec{X} : \mathbb{R}^2 \to \mathbb{R}^n$ and that at each point $p = \vec{X}(u, v)$, the differential $d\vec{X}_{(u,v)}$ be an injective linear transformation $\mathbb{R}^2$ to $\mathbb{R}^n$.

Problems

5.2.1. Consider the parametrized surface

$$\vec{X}(u, v) = ((v^2+1)\cos u, (v^2+1)\sin u, v), \qquad \text{where } (u, v) \in [0, 2\pi] \times \mathbb{R}.$$

Find the equation of the tangent plane to this surface at $p = (2, 0, 1) = \vec{X}(0, 1)$.

5.2.2. Calculate the equation of the tangent plane to the torus

$$\vec{X}(u, v) = ((2 + \cos v) \cos u, (2 + \cos v) \sin u, \sin v)$$

at the point $(u, v) = \left(\frac{\pi}{6}, \frac{\pi}{3}\right)$.

5.2.3. Show that the equation of the tangent plane at (x_0, y_0, z_0) at a regular surface given by $f(x, y, z) = 0$, where 0 is a regular value of f, is

$$f_x(x_0, y_0, z_0)(x - x_0) + f_y(x_0, y_0, z_0)(y - y_0) + f_z(x_0, y_0, z_0)(z - z_0) = 0.$$

5.2.4. Determine the tangent planes to $x^2 + y^2 - z^2 = 1$ at the points $(x, y, 0)$ and show that they are all parallel to the z-axis.

5.2.5. Show that the cylinder $\{(x, y, z) \in \mathbb{R}^3 \mid x^2 + y^2 = 1\}$ is a regular surface and find parametrizations whose coordinate neighborhoods cover it.

5.2.6. Show that the two-sheeted cone with equation $x^2 + y^2 - z^2 = 0$ is not a regular surface.

5.2.7. *Surfaces of Revolution.*   Consider the surface of revolution S with parametrization $\vec{X}(u, v) = (f(v) \cos u, f(v) \sin u, g(v))$, where $(f(v), g(v))$ is a regular plane curve (in the xz-plane). Prove that S is regular if and only if for all v in the domain of f, $f(v) \neq 0$ and $g'(v) \neq 0$ or $f(v) \neq 0$ and $f'(v) \neq 0$.

5.2.8. Let $f(x, y, z) = (x + 2y + 3z - 4)^2$.

(a) Locate the critical points and the critical values of f.

(b) For what values of c is the set $f(x, y, z) = c$ a regular surface?

(c) Repeat (a) and (b) for the function $f(x, y, z) = xy^2 z^3$.

5.2.9. Consider the surface parametrized by $\vec{X}(u, v) = (uv, u^2 - v^2, u^3 - v^3)$. Prove that $\vec{X}$ intersects itself along the ray $(x, 0, 0)$, with $x \geq 0$. Find the points (u, v) that map to this ray. Use these to prove that the tangent space to $\vec{X}$ at any point on the open ray $(x, 0, 0)$, with $x > 0$, is the union of two planes. (See Figure 5.10 for the plot of a portion of this surface near $(0, 0, 0)$.)

5.2.10. Consider the parametrized $\vec{X}(u, v) = (u, v^3, -v^2)$. Prove that for any point on the surface $p = \vec{X}(q)$ such that $q = (u, 0)$, $d\vec{X}_q$ does not have maximal rank.

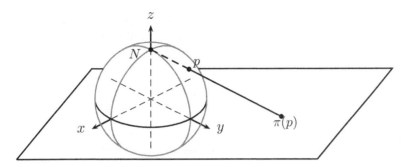

Figure 5.12. Stereographic projection.

5.2.11. Consider $\vec{X}(u,v) = (v \cos u, v \sin u, e^{-v})$ for $(u,v) \in [0, 2\pi] \times [0, \infty)$. Find all points where $\vec{X}$ fails the regularity condition.

5.2.12. Show that the parametrized surface $\vec{X}(u,v) = (u^3 + v + u, u^2 + uv, v^3)$ satisfies the first and second conditions of Definition 5.2.9 but does not satisfy the third at $p = (0,0,0)$.

5.2.13. Consider the hyperbolic paraboloid defined by $S = \{(x,y,z) \in \mathbb{R}^3 \mid z = x^2 - y^2\}$. Check that the following provide systems of coordinates for parts of S.

  (a) $\vec{X}(u,v) = (u + v, u - v, 4uv)$ for $(u,v) \in \mathbb{R}^2$.
  (b) $\vec{X}(u,v) = (u \cosh v, u \sinh v, u^2)$ for $(u,v) \in \mathbb{R} \times \mathbb{R}$.

5.2.14. Find a parametrization for the hyperboloid of two sheets $x^2 + y^2 - z^2 = 1$.

5.2.15. *Stereographic Projection.* One way to define coordinates on the surface of the sphere $\mathbb{S}^2$ given by $x^2 + y^2 + z^2 = 1$ is to use the stereographic projection of $\pi : \mathbb{S}^2 - \{N\} \to \mathbb{R}^2$, where $N = (0,0,1)$, defined as follows. Given any point $p \in \mathbb{S}^2$, the line (pN) intersects the xy-plane at exactly one point, which is the image of the function $\pi(p)$. If (x,y,z) are the coordinates for p in $\mathbb{S}^2$, let us write $\pi(x,y,z) = (u,v)$ (see Figure 5.12).

  (a) Prove that $\pi(x,y,z) = (\frac{x}{1-z}, \frac{y}{1-z})$.
  (b) Prove that
  $$\pi^{-1}(u,v) = \left(\frac{2u}{u^2+v^2+1}, \frac{2v}{u^2+v^2+1}, \frac{u^2+v^2-1}{u^2+v^2+1} \right).$$

  (c) Prove that by using stereographic projections, it is possible to cover the sphere with two coordinate neighborhoods.

5.2.16. *Cones.* Let C be a regular planar curve parametrized by $\vec{\alpha} : (a, b) \to \mathbb{R}^3$ lying in a plane $\mathcal{P}$ that does not contain the origin O. The cone Σ over C is the union of all the lines passing through O and p for all points p on C.

    (a) Find a parametrization $\vec{X}$ of the cone Σ over C.

    (b) Find the points where Σ is not regular.

5.2.17. *Central Projection.* The central projection used in cartography involves a function $f : \mathbb{S}^2 \to (0, 2\pi) \times \mathbb{R}$ defined geometrically as follows. View $(0, 2\pi) \times \mathbb{R}$ as a cylinder wrapping around the unit sphere $\mathbb{S}^2$ in such a way that the slit in the cylinder falls in the half-plane $\mathcal{P}$ with $y = 0$ and $x \geq 0$. The central projection f sends a point $p \in \mathbb{S}^2 - \mathcal{P}$ to $f(p)$ on the cylinder by defining $f(p)$ as the unique point on the cylinder on ray $[O, p)$, where $O = (0, 0, 0)$ is the origin.

    (a) Let (x, y, z) be the coordinates of $p \in \mathbb{S}^2 - \mathcal{P}$ and call $f(x, y, z) = (u, v)$. Find a formula for f.

    (b) Calculate a formula for f^{-1}.

    (c) If one uses central projections for coordinate neighborhoods on the sphere, how many are necessary to cover the sphere?

5.3   Change of Coordinates

A particular patch of a surface can be parametrized in a variety of ways. In particular, we could use a different coordinate system in $\mathbb{R}^2$ to describe the domain U of $\vec{X}$.

Example 5.3.1 (Two Parametrizations of the Sphere). As an example, consider the following two parametrizations of the upper half of the unit sphere in $\mathbb{R}^3$:

1. $\vec{X}(u, v) = (u, v, \sqrt{1 - u^2 - v^2})$, with $(u, v) \in U$ where the domain U is the closed unit disk, i.e., $U = \{(u, v) \in \mathbb{R}^2 \, | \, u^2 + v^2 \leq 1\}$. To be more specific (if we needed to provide intervals for (u, v), say to calculate integrals), we could say

$$U = \{(u, v) \in \mathbb{R}^2 \, | -\sqrt{1 - u^2} \leq v \leq \sqrt{1 - u^2} \text{ and } -1 \leq u \leq 1\},$$

2. $\vec{Y}(\theta, \varphi) = (\cos \theta \sin \varphi, \sin \theta \sin \varphi, \cos \varphi)$, with $(\theta, \varphi) \in U'$ where we have $U' = [0, 2\pi] \times [0, \frac{\pi}{2}]$.

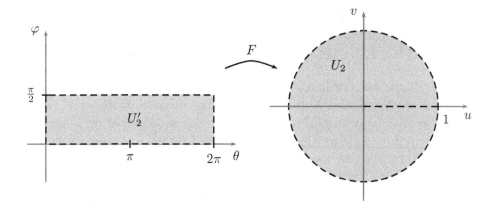

Figure 5.13. Coordinate change by F.

In this case, it's relatively easy to verify that $\vec{X}$ and $\vec{Y}$ parametrize the same surface patch by setting the change of coordinates $(u, v) = F(\theta, \varphi)$, with

$$F : \quad U' \longrightarrow U$$
$$(\theta, \varphi) \longmapsto (\cos\theta \sin\varphi, \sin\theta \sin\varphi).$$

We notice that $F(U) = U'$ (i.e., F is a surjective function) and that, as vector functions, $\vec{Y} = \vec{X} \circ F$.

The function $F : \mathbb{R}^2 \to \mathbb{R}^2$ is not a bijection between U and U', but it is a bijection between

$$U_2 = \{(u, v) \in U \,|\, u^2 + v^2 < 1 \text{ and } v = 0 \text{ implies } u < 0\} \quad \text{and} \quad U_2' = (0, 2\pi) \times \left(0, \frac{\pi}{2}\right),$$

and the inverse function $(\theta, \varphi) = F^{-1}(u, v)$ is given by

$$\theta = \begin{cases} \tan^{-1}\left(\frac{v}{u}\right), & \text{if } u > 0 \text{ and } v \geq 0, \\ \frac{\pi}{2}, & \text{if } u = 0 \text{ and } v > 0, \\ \tan^{-1}\left(\frac{v}{u}\right) + \pi, & \text{if } u < 0, \\ \frac{3\pi}{2}, & \text{if } u = 0 \text{ and } v < 0, \\ \tan^{-1}\left(\frac{v}{u}\right) + 2\pi, & \text{if } u > 0 \text{ and } v \leq 0. \end{cases} \quad \text{and} \quad \varphi = \sin^{-1}(\sqrt{u^2 + v^2}).$$

It is tedious but not hard to verify that both F and F^{-1} are differentiable on the open domains U_2 and U_2'.

We now consider a general change of coordinates F in $\mathbb{R}^2$. Consider two parametrized surfaces defined by $\vec{X} : U \to \mathbb{R}^3$ and $\vec{Y} = \vec{X} \circ F : U' \to \mathbb{R}^3$, where $F : U' \to U$ for subsets U' and U in $\mathbb{R}^2$. The above example illustrates three progressively stringent conditions on the change-of-coordinates function F. First, a sufficient condition for $\vec{Y} = \vec{X} \circ F$ and $\vec{X}$ to have the same image in $\mathbb{R}^3$ is for F to be surjective. Secondly, if one wishes for the coordinates on a surface to be unique or if we are concerned with $\vec{Y}$ being a homeomorphism, we might wish to require that F be a bijection, in which case, if F is already surjective, one merely must restrict the domain of F to make it a bijection. Finally, requiring that both F and F^{-1} be differentiable adds a "smoothness" condition to how the coordinates transform. This latter condition has a name.

Definition 5.3.2. Let U and V be subsets of $\mathbb{R}^n$. We call a function $F : U \to V$ a *diffeomorphism* if it is a bijection, and both F and F^{-1} are differentiable.

Proposition 5.3.3. *Let U and V be subsets of $\mathbb{R}^n$, and let $F : U \to V$ be a diffeomorphism. Suppose that $q = F(q')$. Then the linear transformation $dF_{q'}$ is invertible, and*

$$dF_{q'}^{-1} = d(F^{-1})_q.$$

Proof: Suppose that $F : \mathbb{R}^m \to \mathbb{R}^n$ and $G : \mathbb{R}^n \to \mathbb{R}^s$ are differentiable, multivariable, vector-valued functions. (For what follows, one can also assume that these functions have small domains, just as long as the range of F is a subset of the domain of G.) Using the notion of the differential defined in Equation (5.2), it is an easy exercise to show that the formula for the chain rule on $G \circ F$ in a multivariable context can be written as

$$d(G \circ F)_a = dG_{F(a)} \circ dF_a,$$

where we mean composition of linear transformations.

Applying this to the situation of this proposition, where $F^{-1} \circ F = \mathrm{id}_U$, we get

$$d(F^{-1})_{F(q')} \circ dF_{q'} = I_n,$$

where I_n is the $n \times n$ identity linear transformation. Furthermore, the same reasoning holds with $F \circ F^{-1}$, and we conclude that dF_q is invertible and that $dF_{q'}^{-1} = d(F^{-1})_q$.   □

Suppose again that we have parametrized surfaces $\vec{X} : U \to \mathbb{R}^3$ and $\vec{Y} = \vec{X} \circ F : U' \to \mathbb{R}^3$, where $F : U' \to U$ is a diffeomorphism between open subsets U' and U in $\mathbb{R}^2$. Let $q' \in U'$ and define $q = F(q')$. Then, just as in the above proof, the chain rule in multiple variables gives

$$d\vec{Y}_{q'} = d(\vec{X} \circ F)_{q'} = d\vec{X}_q \circ dF_{q'}. \qquad (5.4)$$

If for the change of coordinates F we write

$$(u, v) = F(s, t) = (f(s, t), g(s, t)),$$

then the matrix of the differential of this coordinate change is

$$[dF] = \begin{pmatrix} \dfrac{\partial u}{\partial s} & \dfrac{\partial u}{\partial t} \\[2mm] \dfrac{\partial v}{\partial s} & \dfrac{\partial v}{\partial t} \end{pmatrix} = \begin{pmatrix} \dfrac{\partial f}{\partial s} & \dfrac{\partial f}{\partial t} \\[2mm] \dfrac{\partial g}{\partial s} & \dfrac{\partial g}{\partial t} \end{pmatrix}. \qquad (5.5)$$

With matrices, the chain rule is written as

$$[d\vec{Y}_{(s,t)}] = [d\vec{X}_{F(s,t)}] \cdot \begin{pmatrix} \dfrac{\partial u}{\partial s} & \dfrac{\partial u}{\partial t} \\[2mm] \dfrac{\partial v}{\partial s} & \dfrac{\partial v}{\partial t} \end{pmatrix}. \qquad (5.6)$$

Writing $\vec{X} = (X_1, X_2, X_3)$ and $\vec{Y} = (Y_1, Y_2, Y_3)$ if necessary, we find that Equation (5.6) is equivalent to the relations

$$\frac{\partial \vec{Y}}{\partial s} = \frac{\partial \vec{X}}{\partial u}\frac{\partial u}{\partial s} + \frac{\partial \vec{X}}{\partial v}\frac{\partial v}{\partial s} = \frac{\partial u}{\partial s}\vec{X}_u + \frac{\partial v}{\partial s}\vec{X}_v,$$
$$\frac{\partial \vec{Y}}{\partial t} = \frac{\partial \vec{X}}{\partial u}\frac{\partial u}{\partial t} + \frac{\partial \vec{X}}{\partial v}\frac{\partial v}{\partial t} = \frac{\partial u}{\partial t}\vec{X}_u + \frac{\partial v}{\partial t}\vec{X}_v. \qquad (5.7)$$

Problems

5.3.1. Prove the claim in Example 5.3.1 that both F and F^{-1} are differentiable on the open domains U_2 and U_2'.

5.3.2. Consider the coordinate change in two variables

$$F(t, s) \rightarrow (t^2 - s^2, 2st).$$

Let $U = \{(s, t) \,|\, s^2 + t^2 \leq 1 \text{ and } t \geq 0\}$. Show that $F(U)$ is the whole unit disk. Prove that if $\vec{X}(x, y) = (x, y, \sqrt{1 - x^2 - y^2})$, then $\vec{X} \circ F : U \rightarrow \mathbb{R}^3$ is a parametrization for the upper half of the unit sphere.

5.3.3. Prove that $F(x, y) = (x^3, y^3)$ is a bijection from $\mathbb{R}^2$ to $\mathbb{R}^2$ but not a diffeomorphism. Prove also that $F(x, y) = (x^3 + x, y^3 + y)$ is a diffeomorphism of $\mathbb{R}^2$ onto itself.

5.3.4. Prove that the function $F : \mathbb{R}^2 \rightarrow \mathbb{R}^2$ defined by

$$\begin{aligned} F(x, y) = (x\cos(x^2 + y^2) - y\sin(x^2 + y^2), \\ x\sin(x^2 + y^2) + y\cos(x^2 + y^2)) \end{aligned}$$

is a diffeomorphism of $\mathbb{R}^2$ onto itself.

5.3.5. Find an example of a diffeomorphism of $\mathbb{R}^2$ onto an open square and show why it is a diffeomorphism.

5.4 The Tangent Space and the Normal Vector

Let S be a regular surface parametrized in the neighborhood of a point p by $\vec{X} : U \rightarrow \mathbb{R}^3$. Suppose that $p = \vec{X}(q)$. Since S is regular, the set of tangent vectors to S at p is a subspace T_pS, and we saw in Corollary 5.2.7 that $T_pS = \text{Im}(d\vec{X}_q)$. The condition that S be regular at p insures that the columns of $d\vec{X}_q$ are linearly independent. Thus, given the parametrization $\vec{X}$ of the neighborhood of p, the set of vectors $\{\vec{X}_u(q), \vec{X}_v(q)\}$ forms a basis of T_pS.

Comparing surfaces to curves, recall that at every regular point on a curve, the unit tangent vector to a curve at a point is invariant (up to a change in sign) under reparametrization of the curve, and the tangent line to the curve at a point is an entirely geometric object, completely unchanged under reparametrizations. Implicit in Definition 5.2.1, the set of tangent vectors to a surface at a point, and hence the tangent space, is a geometric object, invariant under reparametrization.

In contrast, the change of basis formulas Equations (5.7) and (5.6) show that when $\vec{Y} = \vec{X} \circ F$ and if $q = F(q')$ and $p = \vec{X}(q)$, the sets of vectors $\{\vec{X}_u(q), \vec{X}_v(q)\}$ and $\{\vec{Y}_s(q'), \vec{Y}_t(q')\}$ are not necessarily equal. Consequently, though T_pS is invariant under a reparametrization, the basis induced from the parametrization is not invariant.

Proposition 5.4.1. *Let $\vec{X} : U \to \mathbb{R}^3$ be a parametrization of a coordinate patch on a regular surface S. Consider a diffeomorphism $F : U' \subset \mathbb{R}^2 \to U$ such that $\vec{Y} = \vec{X} \circ F$ is a regular parametrization of an open subset of S. Set $q = F(q')$ and $p = \vec{X}(q)$. If $\vec{a} \in T_pS$ has coordinates*

$$\vec{a} = \bar{a}_1 \vec{Y}_s + \bar{a}_2 \vec{Y}_t = a_1 \vec{X}_u + a_2 \vec{X}_v,$$

then

$$\begin{pmatrix} \bar{a}_1 \\ \bar{a}_2 \end{pmatrix} = [dF_{q'}]^{-1} \begin{pmatrix} a_1 \\ a_2 \end{pmatrix} = [d(F^{-1})_q] \begin{pmatrix} a_1 \\ a_2 \end{pmatrix}.$$

Proof: By the chain rule in Equations (5.7) and (5.6), we have

$$\bar{a}_1 \vec{Y}_s + \bar{a}_2 \vec{Y}_t = \bar{a}_1 \left(\frac{\partial u}{\partial s} \vec{X}_u + \frac{\partial v}{\partial s} \vec{X}_v \right) + \bar{a}_2 \left(\frac{\partial u}{\partial t} \vec{X}_u + \frac{\partial v}{\partial t} \vec{X}_v \right)$$

$$= \left(\bar{a}_1 \frac{\partial u}{\partial s} + \bar{a}_2 \frac{\partial u}{\partial t} \right) \vec{X}_u + \left(\bar{a}_1 \frac{\partial v}{\partial s} + \bar{a}_2 \frac{\partial v}{\partial t} \right) \vec{X}_v.$$

Since $\vec{a} = a_1 \vec{X}_u + a_2 \vec{X}_v$ and $\{\vec{X}_u, \vec{X}_v\}$ is a basis of T_pS, we deduce that

$$\begin{pmatrix} a_1 \\ a_2 \end{pmatrix} = \begin{pmatrix} \frac{\partial u}{\partial s} & \frac{\partial u}{\partial t} \\ \frac{\partial v}{\partial s} & \frac{\partial v}{\partial t} \end{pmatrix} \begin{pmatrix} \bar{a}_1 \\ \bar{a}_2 \end{pmatrix}.$$

The last claim of the proposition follows in light of Proposition 5.3.3. $\qquad \square$

Using the language of linear algebra, Proposition 5.4.1 shows that $[d(F^{-1})_q]$ is the change of coordinates matrix between the basis corresponding to the parametrization $\vec{X}$ and the basis corresponding to the reparametrization $\vec{Y} = \vec{X} \circ F$.

At this point, we wish to restate Proposition 5.4.1 using notations that we will adopt later, in particular in Chapter 7. We label the coordinates (u, v) as (x_1, x_2), and we label the coordinates (s, t) as $(\bar{x}_1, \bar{x}_2)$. We write the diffeomorphism that relates the coordinates as

$$(\bar{x}_1, \bar{x}_2) = F(x_1, x_2) \qquad \text{and} \qquad (x_1, x_2) = F^{-1}(\bar{x}_1, \bar{x}_2).$$

Then according to Proposition 5.4.1, the change of coordinates on $T_p S$ satisfies

$$\begin{pmatrix} \bar{a}_1 \\ \bar{a}_2 \end{pmatrix} = \begin{pmatrix} \frac{\partial \bar{x}_1}{\partial x_1} & \frac{\partial \bar{x}_1}{\partial x_2} \\ \frac{\partial \bar{x}_2}{\partial x_1} & \frac{\partial \bar{x}_2}{\partial x_2} \end{pmatrix} \begin{pmatrix} a_1 \\ a_2 \end{pmatrix}. \tag{5.8}$$

This matrix product expression is often written in summation notation as

$$\bar{a}_i = \sum_{k=1}^{2} \frac{\partial \bar{x}_i}{\partial x_k} a_k.$$

According to Proposition 5.2.4, if $\vec{X}$ is a regular parametrization of a neighborhood of $p = \vec{X}(q)$ of a regular surface S, the tangent plane has $\vec{X}_u(q) \times \vec{X}_v(q)$ as a normal vector. Since the tangent plane is invariant under reparametrization, this cross product must only be rescaled under reparametrization.

Proposition 5.4.2. *Let $\vec{X} : U \to \mathbb{R}^3$ be a parametrization of a coordinate patch on a regular surface S. Consider a function $F : U' \subset \mathbb{R}^2 \to U$ such that $\vec{Y} = \vec{X} \circ F$ is a regular parametrization of an open subset of S. Then if $q = F(q')$ and $p = \vec{X}(q)$,*

$$\vec{Y}_s(q') \times \vec{Y}_t(q') = \det(dF_{q'})(\vec{X}_u(q) \times \vec{X}_v(q)).$$

Proof: (Left as an exercise for the reader.) $\qquad\qquad\qquad\qquad\qquad \square$

We remind the reader that the determinant $\det(dF_{q'})$ of the differential matrix of the change of coordinate function is called the *Jacobian* of F at q'. (The reader may recall that the Jacobian appears in the change of variables formula for multiple integrals.)

Proposition 5.4.2 motivates us to define the unit normal to a regular surface S at point p as follows. If $\vec{X}$ is a parametrization of a neighborhood of p and $p = \vec{X}(q)$, then the unit normal vector is

$$\vec{N}(q) = \frac{\vec{X}_u \times \vec{X}_v}{\|\vec{X}_u \times \vec{X}_v\|}(q). \tag{5.9}$$

This vector is invariant under reparametrization except up to a change in sign, namely, the sign of the Jacobian of the corresponding change of coordinates.

This unit normal vector, among other things, allows one to easily answer questions that involve the angle at which two surfaces intersect at a point. In other words, let S_1 and S_2 be two regular surfaces, and let $p \in S_1 \cap S_2$. Then S_1 intersects with S_2 at p, with an angle θ, where $\vec{N}_1 \cdot \vec{N}_2 = \cos \theta$. Of course, the particular parametrization might change the sign of one of the unit normal vectors, but the acute angle between $T_p(S_1)$ and $T_p(S_2)$ is well defined, regardless of signs.

Problems

5.4.1. Calculate the unit normal vector for function graphs $\vec{X}(u, v) = (u, v, f(u, v))$.

5.4.2. Suppose that a coordinate neighborhood of a regular surface can be parametrized by
$$\vec{X}(u, v) = \vec{\alpha}(u) + \vec{\beta}(v),$$
where $\vec{\alpha}$ and $\vec{\beta}$ are regular parametrized curves defined over intervals I and J, respectively. Prove that along coordinate lines (either $u = u_0$ or $v = v_0$) all the tangent planes to the surface are parallel to a fixed line.

5.4.3. Let $\vec{X} : U \to \mathbb{R}^3$ be a parametrization of an open set of a regular surface, and let $\vec{p}$ be a point in $\mathbb{R}^3$ that is not in S. Consider the function $F : U \to \mathbb{R}$ defined by
$$F(u, v) = \|\vec{X}(u, v) - \vec{p}\|.$$
Prove that if $\vec{q} = (u_0, v_0)$ is a critical point of F, then the vector $\vec{X}(\vec{q}) - \vec{p}$ is normal to the surface S at $\vec{X}(\vec{q})$.

5.4.4. *Tangential Surfaces.* Let $\vec{\alpha} : I \to \mathbb{R}^3$ be a regular parametrized curve with curvature $\kappa(t) \neq 0$. Call the tangential surface to $\vec{\alpha}$ the parametrized surface
$$\vec{X}(t, u) = \vec{\alpha}(t) + u\vec{\alpha}'(t), \qquad \text{with } t \in I \text{ and } u \neq 0.$$
Prove that for any fixed $u_0 \in I$, along any curve $\vec{X}(u_0, v)$, the tangent planes are all equal.

5.4.5. *Tubes.* Let $\vec{\alpha} : I \to \mathbb{R}^3$ be a regular parametrized curve. Let r be a positive real constant. Define the *tube* of radius r around $\vec{\alpha}$ as the parametrized surface

$$\vec{X}(t,u) = \vec{\alpha}(t) + r(\cos u)\vec{P}(t) + r(\sin u)\vec{B}(t), \quad \text{with } (t,u) \in I \times [0, 2\pi].$$

(a) Prove that a necessary (though not sufficient) condition for $\vec{X}$ to be a regular surface is that

$$r < 1/(\max_{t \in I} \kappa(t)).$$

(b) Show that when $\vec{X}$ is regular, the unit normal vector is

$$\vec{N}(t,u) = -\cos u \vec{P}(t) - \sin u \vec{B}(t).$$

5.4.6. Prove Proposition 5.4.2.

5.4.7. Let f and g be real functions such that $f(v) > 0$ and $g'(v) \neq 0$. Consider a parametrized surface given by

$$\vec{X}(u,v) = (f(v)\cos u, f(v)\sin u, g(v)).$$

Show that all the normal lines to this surface pass through the z-axis. [Hint: See Problem 5.2.7.]

5.4.8. Two regular surfaces S_1 and S_2 intersect *transversally* if for all $p \in S_1 \cap S_2$, $T_p(S_1) \neq T_p(S_2)$. Prove that if S_1 and S_2 intersect transversally, then the set $S_1 \cap S_2$ is a regular curve.

5.5   Orientable Surfaces

The concept of orientability is a global property of surfaces, and we do not need it again until Section 8.4. Nonetheless, we introduce the notion in this chapter, alongside our careful definitions for a regular surface and the tangent plane to a surface at a point. The reader may choose to skip this section at this point and return to it later when he or she encounters the global Gauss-Bonnet Theorem.

The concept of an orientable surface encapsulates the notion of being able to define an inside and an outside to the surface. At any point p of S, there are two unit normal vectors to the surface. If it makes sense to distinguish between "two sides" of the surface, then specifying the unit normal $\vec{N}_p$ at p eliminates all the options for all the other points on the surface and uniquely defines $\vec{N}_{p'}$ for all other points of S.

Definition 5.5.1. Let S be a regular surface in $\mathbb{R}^3$. We say that S is *orientable* if it can be covered by a collection of coordinate patches R_i, with $i \in I$, given as the images of parametrizations $\vec{X}_{(i)} : U_i \to \mathbb{R}^3$, where $U_i \subset \mathbb{R}^2$, such that if

$$\vec{N}_{(i)}(q) = \frac{\frac{\partial \vec{X}_{(i)}}{\partial u} \times \frac{\partial \vec{X}_{(i)}}{\partial v}}{\left\| \frac{\partial \vec{X}_{(i)}}{\partial u} \times \frac{\partial \vec{X}_{(i)}}{\partial v} \right\|}(q),$$

then $\vec{N}_{(i)}(q) = \vec{N}_{(j)}(q)$ for all $q \in U_i \cap U_j$, for all $i, j \in I$. If for the surface S it is impossible to find such a covering by parametrizations, then we call S *nonorientable*.

By Proposition 5.4.2, we can provide an alternative formulation of this definition.

Proposition 5.5.2. *A regular surface S is orientable if and only it it is possible to cover it with coordinate patches R_i, given as the images of parametrizations $\vec{X}_{(i)} : U_i \to \mathbb{R}^3$, such that if $\vec{X}_{(j)} = \vec{X}_{(i)} \circ F$ for a change of coordinates $F : U_i \cap U_j \to U_i \cap U_j$, then $\det(dF_q) > 0$ for all $q \in U_i \cap U_j$.*

Example 5.5.3. The commonly given example of a nonorientable surface in $\mathbb{R}^3$ is the Möbius strip M. Intuitively, the Möbius strip is a surface obtained by taking a long and narrow strip of paper and gluing the short ends together as though to make a cylinder but putting one twist in the strip before gluing (see Figure 5.14).

By looking at Figure 5.14, one can imagine a normal vector pointing outward on the surface of the Möbius strip, but if one follows

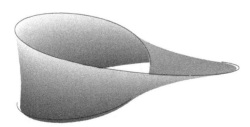

Figure 5.14. Möbius strip.

the direction of this normal vector around on the surface, when it comes back around, it will be pointing inward this time instead of outward.

We wish to be more specific in this case. One can parametrize most (an open dense subset) of the Möbius strip by

$$\vec{X}(u,v) = \left(\left(2 - v \sin \frac{u}{2} \right) \sin u, \left(2 - v \sin \frac{u}{2} \right) \cos u, v \cos \frac{u}{2} \right),$$

with $(u,v) \in (0, 2\pi) \times (-1, 1)$. This coordinate neighborhood omits the boundary of the closed Möbius strip as well as the coordinate line with $u = 0$.

Let p be the point on M given by $\lim_{u \to 0} \vec{X}\left(u, \frac{1}{2}\right) = (0, 2, \frac{1}{2})$. To obtain the same point p as a limit with $u \to 2\pi$, we must have a different value for v, namely, $p = \lim_{u \to 2\pi} \vec{X}\left(u, -\frac{1}{2}\right)$. With some calculations, one can find that

$$\vec{X}_u \times \vec{X}_v = \left(-\frac{v}{2} \cos u - \left(2 - v \sin \frac{u}{2} \right) \sin u \cos \frac{u}{2} \right) \vec{i}$$
$$+ \left(\frac{v}{2} \sin u - \left(2 - v \sin \frac{u}{2} \right) \cos u \cos \frac{u}{2} \right) \vec{j}$$
$$- \left(\left(2 - v \sin \frac{u}{2} \right) \sin \frac{u}{2} \right) \vec{k}$$

and

$$\|\vec{X}_u \times \vec{X}_v\|^2 = \frac{v^2}{4} + \left(2 - v \sin \frac{u}{2} \right)^2.$$

From this, it is not hard to show that

$$\lim_{u \to 0} \vec{N}\left(u, \frac{1}{2}\right) = \left(-\frac{1}{\sqrt{65}}, -\frac{8}{\sqrt{65}}, 0 \right)$$

and

$$\lim_{u \to 2\pi} \vec{N}\left(u, -\frac{1}{2}\right) = \left(\frac{1}{\sqrt{65}}, \frac{8}{\sqrt{65}}, 0 \right).$$

Consequently, no collection of systems of coordinates of M that includes $\vec{X}$ can satisfy the conditions of orientability. However, the specific system of coordinates $\vec{X}$ is not the problem. Using any collection of coordinate patches to cover M, as soon as one tries to "go all the way around" M, there are two overlapping coordinate patches for which the conditions of Definition 5.5.1 fail.

The Möbius strip has a boundary at $v = \pm 1$ in the above parametrization. There are many examples of nonorientable regular surfaces in $\mathbb{R}^3$ that have a boundary. For example, one can attach a "handle" to the Möbius strip, a tube connecting one part of the strip to another. However, for reasons that can only be explained by theorems in topology, there is not "enough room" in $\mathbb{R}^3$ to fit a closed nonorientable surface that does not have a boundary and does not intersect itself. However, because there is "more room" in higher dimensions, there exist many other closed nonorientable surfaces without boundary in $\mathbb{R}^4$ that do not intersect themselves.

Any particular choice of unit normal vector at each point on a regular surface defines a function $n : S \to \mathbb{S}^2$, where $\mathbb{S}^2$ is the two-dimensional sphere. (The notation $\mathbb{S}^k$ refers to the k-dimensional sphere. From a topological perspective, the dimension of the ambient space is irrelevant, but in basic differential geometry, one often pictures $\mathbb{S}^k$ as the unit k-dimensional sphere in $\mathbb{R}^{k+1}$.) We can now state a new criterion for orientability.

Proposition 5.5.4. *Let S be a regular surface in $\mathbb{R}^3$. S is orientable if and only if there exists a continuous function $n : S \to \mathbb{S}^2$ such that $n(p)$ is normal to S at p for all $p \in S$.*

Proof: ($\Rightarrow$) Suppose that S is a regular orientable surface. Suppose that S is covered by a collection of coordinate patches R_i, with $i \in I$, given as the images of parametrizations $\vec{X}_{(i)} : U_i \to \mathbb{R}^3$ that satisfy the condition of Definition 5.5.1. Suppose that p is in the coordinate patch R_k on S and $p = \vec{X}_{(k)}(q)$. Define $n : S \to \mathbb{S}^2$ as $n(p) = \vec{N}_{(k)}(q)$. The criteria for Definition 5.5.1 ensure that n is well defined, regardless of the chosen coordinate patch R_k. Since n is continuous over each R_i and the collection $\{R_i\}_{i \in I}$ covers S, n is continuous.

($\Leftarrow$) Let $n : S \to \mathbb{S}^2$ be a continuous function such that $n(p)$ is normal to S at p for all $p \in S$. Let $\{R_i\}_{i \in I}$ be a collection of coordinate patches of S parametrized by $\vec{X}_{(i)} : U_i \to \mathbb{R}^3$ that covers S and satisfies the definition for a regular surface. Over each open set R_i, the function

$$\vec{N}_{(i)}(u,v) = \frac{\frac{\partial \vec{X}_{(i)}}{\partial u} \times \frac{\partial \vec{X}_{(i)}}{\partial v}}{\left\| \frac{\partial \vec{X}_{(i)}}{\partial u} \times \frac{\partial \vec{X}_{(i)}}{\partial v} \right\|}$$

is defined and continuous. Now for all $i \in I$, if $\vec{X}_{(i)}(q) = p$ for some $p \in R_i$, the normal vector $\vec{N}_{(i)}(q)$ may be equal to $n(p)$ or $-n(p)$. Furthermore, since $\vec{N}_i$ and n are both continuous functions and

$$\|n(p) - (-n(p))\| = 2,$$

we deduce that $\vec{N}_{(i)}(q) = n \circ \vec{X}_{(i)}(q)$ or $\vec{N}_{(i)}(q) = -n \circ \vec{X}_{(i)}(q)$ for all $q \in U_i$.

Define a new collection of coordinate patches as follows: For all $i \in I$, if $\vec{N}_{(i)} = n \circ \vec{X}_{(i)}$ over U_i, then set $\vec{Y}_i = \vec{X}_i$. If $\vec{N}_{(i)} = -n \circ \vec{X}_{(i)}$ over U_i, then define $\vec{Y}_{(i)}(u,v) = \vec{X}_{(i)}(-u,v)$, with the domain of $\vec{Y}_{(i)}$ modified accordingly. Then

$$\vec{N}_{\vec{Y}_{(i)}} = n \circ \vec{Y}_{(i)}$$

for all $i \in I$, and thus the collection of parametrizations $\{\vec{Y}_{(i)}\}_{i \in I}$ satisfies the Definition 5.5.1 for orientability. $\qquad\square$

Definition 5.5.5. Let S be an orientable regular surface in $\mathbb{R}^3$. A choice of continuous function $n : S \to \mathbb{S}^2$ of unit normal vectors on S is called an *orientation* on S. An orientable regular surface equipped with a choice of such a function n is called an *oriented* surface. If S is an oriented regular surface, a pair of vectors $(\vec{v}, \vec{w})$ of the tangent plane $T_p S$ is called a *positively oriented basis* if the vectors form a basis and

$$(\vec{v} \times \vec{w}) \cdot n(p) > 0.$$

A parametrization $\vec{X} : U \to \mathbb{R}^3$ of an open set of S is also called positively oriented if the ordered pair $(\vec{X}_u, \vec{X}_v)$ forms a positively oriented basis of the tangent plane of S at $\vec{X}(u,v)$ for all $(u,v) \in U$.

If we consider surfaces in $\mathbb{R}^k$, the cross product is not available, so Definition 5.5.1 no longer applies. However, Proposition 5.5.2 has

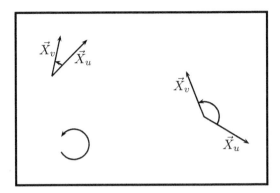

Figure 5.15. Orientation diagram.

no explicit dependency on the dimension of the ambient space and therefore is taken as the definition of orientability for regular surfaces in $\mathbb{R}^k$.

In diagrams depicting orientable regular surfaces, if the regular surface S is in $\mathbb{R}^3$, one can often sketch the surface and indicate a unit normal vector at a point on the surface. Because the surface is orientable, this choice of unit normal vector uniquely determines the unit normal vectors at all points on the surface. However, in a diagram that depicts a neighborhood of a regular surface S in $\mathbb{R}^k$, where k is not necessarily 3, we indicate the orientation of S with an oriented loop. The orientation of the loop indicates that in this diagram, any parametrization of S is such that one must sweep that plane in this direction so that the angle between $\vec{X}_u$ and $\vec{X}_v$ is in the interval $(0, \pi)$. See Figure 5.15.

Problems

5.5.1. Supply the details for Example 5.5.3.

5.5.2. Let S be a regular surface in $\mathbb{R}^3$ given as the set of solutions to the equation $F(x, y, z) = a$, where $F : U \subset \mathbb{R}^3 \to \mathbb{R}$ is differentiable and a is a regular value of F. Prove that S is orientable.

CHAPTER 6

The First and Second Fundamental Forms

Recall that the local geometry of space curves is completely determined by two geometric invariants: the curvature and the torsion. Similarly, as we shall see, the local geometry of a regular surface S in $\mathbb{R}^3$ is determined by the first and second fundamental forms.

The value of restricting one's attention to regular surfaces is that at all points on a regular surface, there is an open neighborhood that is regularly homeomorphic to $\mathbb{R}^2$ via a parametrization $\vec{X}$. Thus, at a point $p \in S$, with $p = \vec{X}(q)$, the differential $d\vec{X}_q$ provides a natural isomorphism between $\mathbb{R}^2$ and T_pS. Whenever one considers vectors on S based at the point p, one must consider these vectors as elements of T_pS, and one can do geometry locally on S by identifying T_pS with $\mathbb{R}^2$.

The first fundamental form establishes a natural direct product on T_pS that ultimately leads to formulas for length of vectors in T_pS, arc length, angle between vectors in T_pS and area formulas on S. The second fundamental form provides a measurement of how the normal vector changes as one moves over the surface S, thereby, in an intuitive sense, describing how S sits in $\mathbb{R}^3$.

6.1 The First Fundamental Form

Definition 6.1.1. Let S be a regular surface and $p \in S$. The *first fundamental form* $I_p(\cdot, \cdot)$ is the restriction of the usual dot product in $\mathbb{R}^3$ to the tangent plane T_pS. Namely, for $\vec{a}, \vec{b}$ in $T_pS = d\vec{X}_q(\mathbb{R}^2)$, $I_p(\vec{a}, \vec{b}) = \vec{a} \cdot \vec{b}$.

Note that for each point $p \in S$, the first fundamental form $I_p(\cdot, \cdot)$ is defined only on the tangent space. There exists a unique matrix that represents $I_p(\cdot, \cdot)$ with respect to the standard basis on T_pS, but it is important to remember that this matrix is a matrix of functions with components depending on the point $p \in S$. For any other point $p_2 \in S$, the first fundamental form is still defined the same way, but the corresponding matrix is most likely different.

This definition draws on the ambient space $\mathbb{R}^3$ for an inner product on each tangent plane. However, if around p the surface S can be parametrized by $\vec{X}$, there is a natural basis on T_pS, namely $\vec{X}_u$ and $\vec{X}_v$. We wish to express the inner product $I_p(\cdot, \cdot)$ in terms of this basis on S. Let

$$\vec{a} = a_1\vec{X}_u(q) + a_2\vec{X}_v(q) \quad \text{and} \quad \vec{b} = b_1\vec{X}_u(q) + b_2\vec{X}_v(q)$$

be tangent vectors. Then we calculate that

$$\vec{a} \cdot \vec{b} = a_1b_1\vec{X}_u(q) \cdot \vec{X}_u(q) + a_1b_2\vec{X}_u(q) \cdot \vec{X}_v(q)$$
$$+ a_2b_1\vec{X}_u(q) \cdot \vec{X}_v(q) + a_2b_2\vec{X}_v(q) \cdot \vec{X}_v(q).$$

This proves the following proposition.

Proposition 6.1.2. *Let $I_p(\cdot, \cdot) : T_pS^2 \to \mathbb{R}$ be the first fundamental form at a point p on a regular surface S. Given a regular parametrization $\vec{X} : U \to \mathbb{R}^3$ of a neighborhood of p, the matrix associated with the first fundamental form $I_p(\cdot, \cdot)$ with respect to the basis $\{\vec{X}_u, \vec{X}_v\}$ is*

$$g = \begin{pmatrix} g_{11} & g_{12} \\ g_{21} & g_{22} \end{pmatrix},$$

where

$$g_{11} = \vec{X}_u \cdot \vec{X}_u \quad and \quad g_{12} = \vec{X}_u \cdot \vec{X}_v,$$
$$g_{21} = \vec{X}_v \cdot \vec{X}_u \quad and \quad g_{22} = \vec{X}_v \cdot \vec{X}_v.$$

The quantity g is a matrix of real functions from the open domain $U \subset \mathbb{R}^2$. With this notation, if $p = \vec{X}(q)$, then the first fundamental form $I_p(\cdot, \cdot)$ at the point p can be expressed as the bilinear form

$$I_p(\vec{a}, \vec{b}) = \vec{a}^T g(q) \vec{b}$$

for all $\vec{a}, \vec{b} \in T_pS$.

One should remark immediately that for any differentiable function $\vec{X} : U \to \mathbb{R}^3$, one has $\vec{X}_u \cdot \vec{X}_v = \vec{X}_v \cdot \vec{X}_u$. Thus, $g_{12} = g_{21}$ and the matrix g is symmetric.

Example 6.1.3 (The xy-Plane). As the simplest possible example, consider the xy-plane. It is a regular surface parametrized by a single system of coordinates $\vec{X}(u,v) = (u,v,0)$. Obviously, we obtain

$$g = \begin{pmatrix} 1 & 0 \\ 0 & 1 \end{pmatrix}.$$

This should have been obvious from the definition of the first fundamental form. The xy-plane is its own tangent space for all p, and the parametrization $\vec{X}$ induces the basis

$$\{\vec{X}_u, \vec{X}_v\} = \{(1,0,0),(0,1,0)\}.$$

Therefore, for any $\vec{v} = \begin{pmatrix} v_1 \\ v_2 \end{pmatrix}$ and $\vec{w} = \begin{pmatrix} w_1 \\ w_2 \end{pmatrix}$, with coordinates given in terms of the standard basis, we have

$$\vec{v} \cdot \vec{w} = v_1 w_1 + v_2 w_2 = \begin{pmatrix} v_1 & v_2 \end{pmatrix} \begin{pmatrix} 1 & 0 \\ 0 & 1 \end{pmatrix} \begin{pmatrix} w_1 \\ w_2 \end{pmatrix}$$

Example 6.1.4 (Cylinder). Consider the right circular cylinder given by $\vec{X}(u,v) = (\cos u, \sin u, v)$, with $(u,v) \in (0, 2\pi) \times \mathbb{R}$. We calculate

$$\vec{X}_u = (-\sin u, \cos u, 0) \quad \text{and} \quad \vec{X}_v = (0,0,1),$$

and thus,

$$g = \begin{pmatrix} 1 & 0 \\ 0 & 1 \end{pmatrix}.$$

Of course, there is no doubt that a cylinder and the xy-plane are not the same surface. As we will see later, a plane and a cylinder share many properties. However, the second fundamental form, which discusses how the normal vector evolves on the cylinder, will differ between the cylinder and the xy-plane.

Example 6.1.5 (Spheres). Consider the following parametrization on the sphere $\vec{X}(u,v) = (\cos u \sin v, \sin u \sin v, \cos v)$, with $(u,v) \in (0, 2\pi) \times (0, \pi)$. The first derivatives of $\vec{X}$ are

$$\vec{X}_u = (-\sin u \sin v, \cos u \sin v, 0),$$
$$\vec{X}_v = (\cos u \cos v, \sin u \cos v, -\sin v).$$

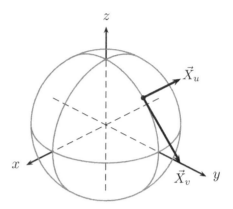

Figure 6.1. Coordinate basis on the sphere.

We deduce that for this parametrization

$$g = \begin{pmatrix} g_{11} & g_{21} \\ g_{12} & g_{22} \end{pmatrix} = \begin{pmatrix} \sin^2 v & 0 \\ 0 & 1 \end{pmatrix}.$$

Figure 6.1 shows the sphere with this parametrization along with the basis $\{\vec{X}_u, \vec{X}_v\}$ at the point $\vec{X}(\pi/2, \pi/4)$.

(We must point out at this time that some classical differential geometry texts use the letters $E = g_{11}$, $F = g_{12}$, and $G = g_{22}$. The classical notation was replaced by the "tensor notation" g_{ij} as the notion of a tensor became more prevalent in differential geometry. Tensor notation is discussed at length in Section 7.1.)

As a first application of how the first fundamental form allows one to do geometry on a regular surface, consider the problem of calculating the arc length of a curve on the surface. Let S be a regular surface with a coordinate neighborhood parametrized by $\vec{X}$: $U \to \mathbb{R}^3$, where $U \subset \mathbb{R}^2$ is open. Consider a curve on S given by $\vec{\gamma}(t) = \vec{X}(u(t), v(t))$, where $(u(t), v(t)) = \vec{\alpha}(t)$ is a differentiable parametrized plane curve in the domain U. The arc length formula over the interval $[t_0, t]$ is

$$s(t) = \int_{t_0}^{t} \|\vec{\gamma}'(\tau)\| \, d\tau.$$

After some reworking, we can rewrite the formula as

$$s(t) = \int_{t_0}^{t} \sqrt{g_{11}(u'(\tau))^2 + 2g_{12}u'(\tau)v'(\tau) + g_{22}(v'(\tau))^2}\, d\tau$$

or, more explicitly,

$$s(t) = \int_{t_0}^{t} \sqrt{g_{11}(u(\tau), v(\tau))(u'(\tau))^2 + 2g_{12}(u(\tau), v(\tau))u'(t)v'(\tau) + g_{22}(u(\tau), v(\tau))(v'(\tau))^2}\, d\tau \ .$$

Using the first fundamental form, we can rewrite the arc length formula as

$$s(t) = \int_{t_0}^{t} \sqrt{I_{\vec{\gamma}(t)}\left(\vec{\alpha}'(\tau), \vec{\alpha}'(\tau)\right)}\, d\tau \ . \tag{6.1}$$

The reader should note that this is an abuse of notation since, for all $p \in S$, $I_p(\cdot, \cdot)$ is a bilinear form on the tangent space T_pS, but $\vec{\alpha}'(t)$ is a vector function in $\mathbb{R}^2$. However, the justification behind this notation is that for all t, the coordinates of $\vec{\alpha}'(t)$ in the standard basis of $\mathbb{R}^2$ are precisely the coordinates of $\vec{\gamma}'(t)$ in the basis $\{\vec{X}_u, \vec{X}_v\}$ based at the point $\vec{\gamma}(t)$. (To be more precise, one could write $I_{\gamma(t)}([\vec{\alpha}'(t)], [\vec{\alpha}'(t)])$ but this notation can become rather burdensome.)

In essence, Equation (6.1) relates the geometry in the particular coordinate neighborhood of S to the plane geometry in the tangent plane T_pS. More precisely, while doing geometry in U, by using $I_p(\cdot, \cdot)$ instead of the usual dot product in $\mathbb{R}^2$, one obtains information about what happens on the regular surface S as opposed to simply what happens on the tangent plane T_pS. This approach does not work only for calculating arc length.

Consider two plane curves $\vec{\alpha}(t)$ and $\vec{\beta}(t)$ in the domain of $\vec{X}$, with $\vec{\alpha}(t_0) = \vec{\beta}(t_0) = q$. Also let $\vec{\gamma} = \vec{X} \circ \vec{\alpha}$ and $\vec{\delta} = \vec{X} \circ \vec{\beta}$ be the corresponding curves on the regular surface S. Then at the point $p = \vec{X}(q)$, the curves $\vec{\gamma}(t)$ and $\vec{\delta}(t)$ form an angle of θ, with

$$\cos\theta = \frac{\vec{\gamma}'(t_0) \cdot \vec{\delta}'(t_0)}{\|\vec{\gamma}'(t_0)\|\, \|\vec{\delta}'(t_0)\|} = \frac{I_p(\vec{\alpha}'(t_0), \vec{\beta}'(t_0))}{\sqrt{I_p(\vec{\alpha}'(t_0), \vec{\alpha}'(t_0))}\sqrt{I_p(\vec{\beta}'(t_0), \vec{\beta}'(t_0))}}.$$

$$\tag{6.2}$$

(Again, the comment following Equation (6.1) applies here as well.)

In the basis $\{\vec{X}_u, \vec{X}_v\}$, the vectors $\vec{X}_u$ and $\vec{X}_v$ have the coordinates $\binom{1}{0}$ and $\binom{0}{1}$, respectively. Therefore, Equation (6.2) implies that the angle φ of the coordinate curves of a parametrization $\vec{X}(u,v)$ is given by

$$\cos\varphi = \frac{\vec{X}_u \cdot \vec{X}_v}{\|\vec{X}_u\| \, \|\vec{X}_v\|} = \frac{I_p((1,0),(0,1))}{\sqrt{I_p((1,0),(1,0))}\sqrt{I_p((0,1),(0,1))}} = \frac{g_{12}}{\sqrt{g_{11}g_{22}}}. \quad (6.3)$$

Thus, all the coordinate curves of a parametrization are orthogonal if and only if $g_{12}(u,v) = 0$ for all (u,v) in the domain of $\vec{X}$. If $\vec{X}$ satisfies this property, we call $\vec{X}$ an *orthogonal parametrization*.

Example 6.1.6 (Loxodromes). Consider the parametrization of the sphere given in Example 6.1.5. Recall that a meridian of the sphere is any curve on the sphere with u fixed and that the coefficients for the first fundamental form are

$$g_{11} = \sin^2 v, \quad g_{12} = g_{21} = 0, \quad g_{22} = 1.$$

A loxodrome on the sphere is a curve that makes a constant angle β with every meridian. Let $\vec{\alpha}(t) = (u(t), v(t))$ be a curve in the domain $(0, 2\pi) \times (0, \pi)$ and $\vec{\gamma}(t) = \vec{X}(\vec{\alpha}(t))$. If $\vec{\gamma}(t)$ makes the same angle β with all the meridians, then by Equation (6.2),

$$\cos\beta = \frac{(u' \ \ v')\begin{pmatrix} \sin^2 v & 0 \\ 0 & 1 \end{pmatrix}\begin{pmatrix} 0 \\ 1 \end{pmatrix}}{\sqrt{(u')^2\sin^2 v + (v')^2}\sqrt{1}} = \frac{v'}{\sqrt{(u')^2\sin^2 v + (v')^2}}.$$

Then

$$(u')^2 \sin^2 v \cos^2\beta + (v')^2\cos^2\beta = (v')^2$$
$$\Longleftrightarrow (v')^2 \sin^2\beta = (u')^2 \sin^2 v \cos^2\beta$$
$$\Longleftrightarrow (u')^2 \sin^2 v = (v')^2 \tan^2\beta$$
$$\Longrightarrow \pm u' \cot\beta = \frac{v'}{\sin v},$$

where the $\pm$ makes sense in that a loxodrome can travel either toward the north pole or toward the south pole. Now integrating both

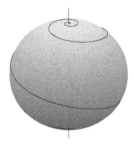

Figure 6.2. Loxodrome on the sphere.

sides of the last equation with respect to t and then performing a substitution, one obtains

$$\int \pm u' \cot \beta \, dt = \int \frac{v'}{\sin v} \, dt \iff \int \pm \cot \beta \, du = \int \frac{1}{\sin v} \, dv$$

$$\iff (\pm \cot \beta) u + C = \ln \tan \left(\frac{v}{2} \right),$$

where we have chosen a form of antiderivative of $1/\sin v$ that suits our calculations best. This establishes an equation between u and v that any loxodrome must satisfy. From this, one can deduce the following parametrization for the loxodrome:

$$(u(t), v(t)) = \left((\pm \tan \beta)(\ln t - C), 2 \tan^{-1} t \right).$$

(Figure 6.2 was plotted using $\tan \beta = 4$ and $C = 0$.)

Not only does the first fundamental form allow us to talk about lengths of curves and angles between curves on a regular surface, but it also provides a way to calculate the area of a region on the regular surface.

Proposition 6.1.7. *Let $\vec{X} : U \subset \mathbb{R}^2 \to S$ be the parametrization for a coordinate neighborhood of a regular surface. Then if Q is a compact (i.e., closed and bounded) subset of U and $R = \vec{X}(Q)$ is a region of S, then the area of R is given by*

$$A(R) = \iint_Q \sqrt{\det(g)} \, du \, dv. \tag{6.4}$$

Proof: Recall the formula for surface area introduced in multivariable calculus that gives the surface area of the region R on S as

$$A(R) = \iint_R dA = \iint_Q \|\vec{X}_u \times \vec{X}_v\| \, du \, dv. \tag{6.5}$$

(See [23, Section 7.4] or [28, Section 17.7] for an explanation of Equation (6.5).)

By Problem 3.1.3, for any two vectors $\vec{v}$ and $\vec{w}$ in $\mathbb{R}^3$, the following identity holds:

$$(\vec{v} \times \vec{w}) \cdot (\vec{v} \times \vec{w}) = (\vec{v} \cdot \vec{v})(\vec{w} \cdot \vec{w}) - (\vec{v} \cdot \vec{w})^2.$$

Therefore,

$$\|\vec{X}_u \times \vec{X}_v\|^2 = (\vec{X}_u \cdot \vec{X}_u)(\vec{X}_v \cdot \vec{X}_v) - (\vec{X}_u \cdot \vec{X}_v)^2 = g_{11}g_{22} - g_{12}^2, \tag{6.6}$$

and the proposition follows because $g_{12} = g_{21}$. $\qquad\square$

Proposition 6.1.7 is interesting in itself as it leads to another property of the first fundamental form. It is obvious from the definition of the first fundamental form that $g_{11}(u,v) \geq 0$ and $g_{22}(u,v) \geq 0$ for all $(u,v) \in U$. However, since $\|\vec{X}_u \times \vec{X}_v\|^2 = \det(g)$ and since $\vec{X}_u \times \vec{X}_v$ is never $\vec{0}$ on a regular surface, we deduce that $\det(g) = g_{11}g_{22} - g_{12} > 0$ for all $(u,v) \in U$. In other words, the matrix of functions g is always a *positive definite* matrix.

Example 6.1.8. Consider the regular surface S parametrized by $\vec{X}(u,v)$ $= (v \cos u, v \sin u, \ln v)$, with $(u,v) \in [0, 2\pi] \times (0, \infty) = U$. (From a practical standpoint, it is not so important that U is not an open subset of $\mathbb{R}^2$, but it should be understood that S requires two coordinate neighborhoods to satisfy the criteria for being a regular surface. See Figure 6.3.) Let $Q = [0, 2\pi] \times [1, 2]$, and let's calculate the area of $R = \vec{X}(Q)$.

It's not hard to show that

$$(g_{ij}) = \begin{pmatrix} v^2 & 0 \\ 0 & 1 + \frac{1}{v^2} \end{pmatrix}.$$

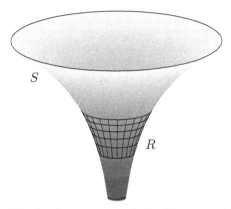

Figure 6.3. Surface area patch for Example 6.1.8.

Then Proposition 6.1.7 gives

$$A(R) = \int_0^{2\pi} \int_1^2 \sqrt{v^2 + 1}\, dv\, du = 2\pi \left[\frac{1}{2}v\sqrt{v^2 + 1} + \frac{1}{2}\sinh^{-1} v \right]_1^2$$

$$= \pi \left(2\sqrt{5} - \sqrt{2} + \ln\left(\frac{2 + \sqrt{5}}{1 + \sqrt{2}} \right) \right).$$

As we have done for many quantities on curves and surfaces, we wish to see how the functions in the first fundamental form change under a coordinate transformation. In order to avoid confusion with various coordinate systems, we introduce some notations that will become more common in future topics. Suppose that an open subset V of a regular surface S can be parametrized by two different sets of coordinates (x_1, x_2) and $(\bar{x}_1, \bar{x}_2)$. If we write $\vec{X}(x_1, x_2)$ and $\vec{Y}(\bar{x}_1, \bar{x}_2)$ for the specific parametrizations, then the function $F = \vec{Y}^{-1} \circ \vec{X}$ is a diffeomorphism between two open subsets of $\mathbb{R}^2$ and

$$(\bar{x}_1, \bar{x}_2) = F(x_1, x_2).$$

For convenience, we will often simply write

$$\begin{cases} \bar{x}_1 = \bar{x}_1(x_1, x_2), \\ \bar{x}_2 = \bar{x}_2(x_1, x_2) \end{cases}$$

to indicate the dependency of one set of variables on the other. Vice versa, we can write

$$(x_1, x_2) = F^{-1}(\bar{x}_1, \bar{x}_2) \quad \text{or simply} \quad \begin{cases} x_1 = x_1(\bar{x}_1, \bar{x}_2), \\ x_2 = x_2(\bar{x}_1, \bar{x}_2). \end{cases}$$

Now $\vec{Y}(\bar{x}_1, \bar{x}_2) = \vec{X}(x_1(\bar{x}_1, \bar{x}_2), x_2(\bar{x}_1, \bar{x}_2))$ for the respective parametrizations. Let g_{ij} be the coefficient functions in the first fundamental form for S parametrized by x_1, x_2, and let $\bar{g}_{ij}$ be the coefficient functions for S parametrized by $\bar{x}_1, \bar{x}_2$. Then by the chain rule,

$$\bar{g}_{ij} = \vec{Y}_{\bar{x}_i} \cdot \vec{Y}_{\bar{x}_j} = \left(\vec{X}_{x_1} \frac{\partial x_1}{\partial \bar{x}_i} + \vec{X}_{x_2} \frac{\partial x_2}{\partial \bar{x}_i} \right) \cdot \left(\vec{X}_{x_1} \frac{\partial x_1}{\partial \bar{x}_j} + \vec{X}_{x_2} \frac{\partial x_2}{\partial \bar{x}_j} \right).$$

After some reorganization, we find

$$\bar{g}_{ij} = \sum_{k=1}^{2} \sum_{l=1}^{2} \frac{\partial x_k}{\partial \bar{x}_i} \frac{\partial x_l}{\partial \bar{x}_j} g_{kl}. \tag{6.7}$$

This type of transformation of coordinates is our first (nontrivial) encounter with tensors. In modern terminology, the collection of the quantities g_{ij} defined at each point of $p \in S$ is referred to as the components of the *metric tensor* of S. Though the first fundamental form, as a bilinear form on $T_p S$, is independent of any particular parametrization, the specific functions g_{ij} depend on the parametrization, but one knows that they behave according to Equation (6.7) under coordinate changes. Because the g_{ij} functions satisfy this particular identity, we call the matrix of functions $g = (g_{ij})$ a tensor of type $(0, 2)$. Later on, as further study will lead to more complicated geometric objects, we will develop a comprehensive theory of tensors in differential geometry.

The value of the first fundamental form is that, given the metric tensor, one can use the appropriate formulas for arc length, for angles between curves and for areas of regions without knowing the specific parametrization of the surface. An additional benefit of the first fundamental form and the metric tensor is that one can still use them to study the geometry of surfaces in higher dimensions.

Suppose that S is a regular surface in $\mathbb{R}^n$ and consider a parametrization of a coordinate patch $\vec{X} : U \to \mathbb{R}^n$, where U is an open subset of $\mathbb{R}^2$. Define the functions $g_{ij} = \vec{X}_i \cdot \vec{X}_j$ as before. Then Equations (6.1) and (6.2) still provide the appropriate formulas for arc length and the angle between vectors in T_pS. In addition, Proposition 6.1.7 generalizes to the following.

Proposition 6.1.9. *Let S be a regular surface in $\mathbb{R}^n$, with $n \geq 3$, and let $\vec{X} : U \to \mathbb{R}^n$ be a parametrization of a coordinate patch of S, where U is an open subset of $\mathbb{R}^2$. Let Q be a compact subset of U and call $R = \vec{X}(Q)$ the compact subset of S. The area of R is given by*

$$A(R) = \iint\limits_Q \sqrt{\det(g)}\, du\, dv. \tag{6.8}$$

Proof: The proof of Equation (6.4) relied on the fact that S is a regular surface in $\mathbb{R}^3$ and made reference to the cross product of $\vec{X}_u$ and $\vec{X}_v$, namely, using the formula

$$\iint\limits_Q \|\vec{X}_u \times \vec{X}_v\|\, du\, dv$$

from multivariable calculus. The Riemann sum behind this integral involves approximating the surface area traced out by $\vec{X}$ over $[u_i, u_{i+1}] \times [v_i, v_{i+1}]$ by a parallelogram spanned on two sides by $\vec{X}_u(u^*, v^*)$ and $\vec{X}_v(u^*, v^*)$, where (u^*, v^*) is a selection point in $[u_i, u_{i+1}] \times [v_i, v_{i+1}]$.

In order to find a formula for the surface area of a regular surface in $\mathbb{R}^n$, one cannot refer to any formula using the cross product since the cross product is only defined in $\mathbb{R}^3$. However, the approach used by the Riemann sum is still the right one. Therefore, one must find a convenient formula for the area of the parallelogram spanned by $\vec{X}_u$ and $\vec{X}_v$ in $\mathbb{R}^n$.

Regardless of the dimension of the ambient Euclidean space, we can calculate the area of these parallelograms as $\|\vec{X}_u\| \|\vec{X}_v\| \sin\theta$, where θ is the angle between the two vectors. One cannot get $\sin\theta$ directly but one can obtain $\cos\theta$ from $\vec{X}_u \cdot \vec{X}_v = \|\vec{X}_u\| \|\vec{X}_v\| \cos\theta$.

Thus,

$$(\vec{X}_u \cdot \vec{X}_v)^2 = (\vec{X}_u \cdot \vec{X}_u)(\vec{X}_v \cdot \vec{X}_v)\cos^2\theta$$

$$\Longrightarrow \qquad (\vec{X}_u \cdot \vec{X}_v)^2 = (\vec{X}_u \cdot \vec{X}_u)(\vec{X}_v \cdot \vec{X}_v)(1 - \sin^2\theta)$$

$$\Longrightarrow \qquad (\vec{X}_u \cdot \vec{X}_u)(\vec{X}_v \cdot \vec{X}_v) - (\vec{X}_u \cdot \vec{X}_v)^2 = (\vec{X}_u \cdot \vec{X}_u)(\vec{X}_v \cdot \vec{X}_v)\sin^2\theta$$

$$\Longleftrightarrow \qquad \sqrt{(\vec{X}_u \cdot \vec{X}_u)(\vec{X}_v \cdot \vec{X}_v) - (\vec{X}_u \cdot \vec{X}_v)^2} = \|\vec{X}_u\| \, \|\vec{X}_v\| \sin\theta.$$

Thus, at each point $(u, v) \in U$, we have $\sqrt{\det(g)} = \|\vec{X}_u\| \, \|\vec{X}_v\| \sin\theta$. Taking the limit of the Riemann sum as the norm of a mesh that goes to 0 gives Equation (6.8). $\qquad\square$

Because of Equation (6.1), a few mathematicians and most physicists use an alternative notation for the metric tensor. These authors describe the metric tensor by saying that the "line element" in a coordinate patch is given by

$$ds^2 = g_{11}(u, v)\, du^2 + 2g_{12}(u, v)\, du\, dv + g_{22}(u, v)\, dv^2. \qquad (6.9)$$

This notation has the advantage that one can write it concisely on a single line instead of in matrix format. However, one should not forget that it not only leads to Equation (6.1) for arc length of a curve on a surface but that it also leads to the Equation (6.2) for angles between curves on the surface and the formula in Proposition 6.1.9 that gives the area of regions on the surface.

Problems

6.1.1. Calculate the metric tensor for the following regular surfaces:

(a) The plane parametrized by $\vec{X}(s, t) = \vec{u} + s\vec{v} + t\vec{w}$ with $(s, t) \in \mathbb{R}^3$.

(b) The surface with parametrization $\vec{X}(u, v) = (u^2 - v, uv, u)$.

(c) The helicoid parametrized by $\vec{X}(u, v) = (v \cos u, v \sin u, au)$ for some $a \in \mathbb{R}$.

6.1.2. Calculate the metric tensor for the following families of conic surfaces:

(a) The ellipsoid $\vec{X}(u, v) = (a \cos u \sin v, b \sin u \sin v, c \cos v)$.

(b) The elliptic paraboloid $\vec{X}(u, v) = (av \cos u, bv \sin u, v^2)$.

(c) The hyperbolic paraboloid $\vec{X}(u,v) = (au\cosh v, bu\sinh v, u^2)$.

(d) The hyperboloid of one sheet

$$\vec{X}(u,v) = (a\cosh v \cos u, b\cosh v \sin u, c\sinh v).$$

(e) The hyperboloid of two sheets

$$\vec{X}(u,v) = (a\cos u \sinh v, b\sin u \sinh v, c\cosh v).$$

6.1.3. Let $0 < r < R$ be real numbers and consider the torus parametrized by

$$\vec{X}(u,v) = ((R + r\cos u)\cos v, (R + r\cos u)\sin v, r\sin u).$$

(a) Show that if the domain of $\vec{X}$ is $U = (0, 2\pi) \times (0, 2\pi)$, then the parametrization is regular and that if $U = [0, 2\pi] \times [0, 2\pi]$ the parametrization is surjective.

(b) Calculate the metric tensor of this torus.

(c) Use the metric tensor to calculate the area of the torus.

6.1.4. Let U be an open subset of $\mathbb{R}^2$, and let $f : U \to \mathbb{R}$ be a two-variable function. Explicitly calculate the metric tensor for the graph of f, which can be parametrized by $\vec{X} : U \to \mathbb{R}^3$, with $\vec{X}(u,v) = (u, v, f(u,v))$. Prove that the coordinate lines are orthogonal if and only if $f_u f_v = 0$.

6.1.5. Let γ be a loxodrome on the sphere as described in Example 6.1.6. Prove that the arc length of the loxodrome between the north and south pole is $\pi \sec \beta$ (regardless of the constant of integration C).

6.1.6. Consider the surface of revolution parametrized by $\vec{X}(u,v) = (f(u)\cos v, f(u)\sin v, g(u))$, with f and g chosen so that the surface is regular (see Problem 5.2.7). Calculate the metric tensor.

6.1.7. Let $\vec{\alpha}(t)$ be a regular space curve and consider the tangential surface S parametrized by $\vec{X}(t,u) = \vec{\alpha}(t) + u\vec{T}(t)$. Calculate the metric tensor for $\vec{X}$.

6.1.8. *Tubes.* Let $\vec{\alpha}(t)$ be a regular space curve. We call the *tube* of radius r around $\vec{\alpha}$ the surface that is parametrized by $\vec{X}(t,u) = \vec{\alpha}(t) + (r\cos u)\vec{P}(t) + (r\sin u)\vec{B}(t)$. Calculate the metric tensor for $\vec{X}$. Supposing that the tube is regular, prove that the area of the tube is $2\pi r$ times the length of $\vec{\alpha}$.

6.1.9. Consider the right circular cone parametrized by $\vec{X}(u,v) = (v\cos u, v\sin u, v)$.

(a) Let $\vec{\alpha}(t)$ be a curve in the plane such that for all t, the vectors $\vec{\alpha}(t)$ and $\vec{\alpha}'(t)$ make a constant angle of β. Prove that $\vec{\alpha}(t) = (f(t)\cos t, f(t)\sin t)$, where $f(t) = Re^{(\cot \beta)t}$ for any real $R > 0$.

(b) Consider the spiral curve $\vec{\gamma}(t) = \vec{X}(t, f(t))$ on the cone. Calculate the arc length of $\vec{\gamma}$ for $0 \leq t \leq 2\pi$. [Hint:

$$R\sqrt{2 + \tan^2 \beta}(e^{2\pi \cot \beta} - 1).]$$

6.1.10. Consider the parametrization $\vec{Y} : \mathbb{R}^2 \to \mathbb{R}^3$ of the unit sphere by stereographic projection (see Problem 5.2.15 and use the parametrization $\vec{Y}(u, v) = \pi^{-1}(u, v)$).

(a) Calculate the corresponding metric tensor component functions g_{ij}.

(b) Consider the usual parametrization $\vec{X}$ of the sphere (as presented in Example 6.1.5.) Give appropriate domains for and explicitly determine the formula for the change of coordinate system given by $F^{-1} = \vec{Y}^{-1} \circ \vec{X} = \pi \circ \vec{X}$.

(c) Explicitly verify Equation (6.7) in this context.

6.1.11. Suppose that the first fundamental form has a matrix of the form

$$g = \begin{pmatrix} 1 & 0 \\ 0 & f(u, v) \end{pmatrix}.$$

Prove that all the v-coordinate lines have equal arc length over any interval $u \in [u_1, u_2]$. In this case, the v-coordinate lines are called parallel.

6.1.12. Let (g_{ij}) be the metric tensor of some surface S parametrized by the coordinates $(x_1, x_2) \in U$, where U is some open subset in $\mathbb{R}^2$. Suppose that we reparametrize the surface with coordinates $(\bar{x}_1, \bar{x}_2)$ in such a way that

$$\begin{cases} x_1 = f(\bar{x}_1, \bar{x}_2), \\ x_2 = \bar{x}_2, \end{cases}$$

where $f : V \to U$ is a function with $V \subset \mathbb{R}$. Let $(\bar{g}_{kl})$ be the metric tensor to S under this parametrization.

(a) Prove that $(\bar{g}_{kl})$ is a diagonal matrix if and only if the function f satisfies the differential equation

$$\frac{\partial f}{\partial \bar{x}_2} = -\frac{g_{12}\left(f(\bar{x}_1, \bar{x}_2), \bar{x}_2\right)}{g_{11}\left(f(\bar{x}_1, \bar{x}_2), \bar{x}_2\right)}.$$

(b) (ODE) Use the above result and the existence theorem for differential equations to prove that every regular surface admits an orthogonal parametrization.

6.1.13. Let C be a regular value of the function $F(x, y, z)$ and consider the regular surface S defined by $F(x, y, z) = C$. Without loss of generality, suppose that the variables x and y can be used to parametrize a neighborhood U of some point $p \in S$. Use implicit differentiation to calculate the metric tensor functions g_{ij} in terms of derivatives of F.

6.1.14. Consider the parametrization $\vec{X}(u, v) = (\cos u \sin v, \sin u \sin v, \cos v)$ of the unit sphere. It is easy to show that the unit normal vector $\vec{N}$ to the sphere at (u, v) is again the vector $\vec{N}(u, v) = \vec{X}(u, v)$. Let $f(u, v)$ be a nonnegative real function in two variables such that $f(u, 0)$ is constant, $f(u, \pi)$ is constant, and for all fixed v_0, $f(u, v_0)$ is periodic 2π. A *normal variation* to a given surface S is a surface created by going out a distance of $f(u, v)$ along the normal vector $\vec{N}$ of S, given by

$$\vec{Y}(u, v) = \vec{X}(u, v) + f(u, v)\vec{N}(u, v).$$

A normal variation of the unit sphere is therefore given by

$$\vec{Y}(u, v) = (1 + f(u, v))\vec{X}(u, v).$$

(a) Calculate the metric tensor for the parametrization $\vec{Y}$.

(b) Use the explicit function $f(u, v) = \cos^2(2u) \cos^2(2v - \pi/2)$. Calculate the metric tensor for $\vec{Y}$.

6.1.15. Consider the following parametrization of the torus in $\mathbb{R}^4$:

$$\vec{X}(u, v) = (\cos u, \sin u, \cos v, \sin v).$$

(The topological definition of a torus is any set that is homeomorphic to $\mathbb{S}^1 \times \mathbb{S}^1$, where $\mathbb{S}^1$ is a circle. The set traced out in $\mathbb{R}^4$ by the function $\vec{X}$ over $[0, 2\pi] \times [0, 2\pi]$ obviously satisfies this topological definition since the image of $\vec{X}$ is exactly $\mathbb{S}^1 \times \mathbb{S}^1$.) Prove that the first fundamental form of $\vec{X}$ is everywhere the usual dot product, i.e., prove that g is the identity matrix for all (u, v). (This is called the *flat torus*.)

6.2   The Gauss Map

In Section 2.2, given any closed, simple, regular curve $\gamma : I \to \mathbb{R}^2$ in the plane, we defined the tangential indicatrix to be the curve given by $\vec{T} : I \to \mathbb{R}^2$, the unit tangent vector of $\gamma(t)$. The image of $\vec{T}$ lies on the unit circle, and though $\vec{T}$ is a closed curve, it need not be regular as its locus may stop and double back. Regardless of the parametrization, the tangential indicatrix is well defined up to a change in sign.

On any regular surface S in $\mathbb{R}^3$, the tangent plane T_pS at a point $p \in S$ is a two-dimensional subspace of $\mathbb{R}^2$, and hence the vectors that are normal to S at p form a one-dimensional subspace. Thus, there exist exactly two possible choices for a unit normal vector. Proposition 5.5.4 implies that if the surface is orientable, we can specify its orientation by a continuous function $n : S \to \mathbb{S}^2$, where $\mathbb{S}^2$ means the unit sphere in $\mathbb{R}^3$.

Definition 6.2.1. Let S be an oriented regular surface in $\mathbb{R}^3$ with an orientation $n : S \to \mathbb{S}^2$. In the classical theory of surfaces, the function n is also called the *Gauss map*.

It is important to remain aware of the distinction between the functions n and $\vec{N}$. Let S be an oriented surface with orientation $n : S \to \mathbb{S}^2$. Suppose that $\vec{X} : U \to \mathbb{R}^3$, where U is an open subset of $\mathbb{R}^2$, parametrizes a neighborhood of S, then the function $\vec{N} : U \to \mathbb{S}^2$ is defined in reference to $\vec{X}$ by Equation (5.9). The functions n and $\vec{N}$ are related by the fact that if $\vec{X}$ is a positively oriented parametrization, then

$$\vec{N} = n \circ \vec{X}.$$

Example 6.2.2. Consider as an example a sphere S in $\mathbb{R}^3$ equipped with the orientation n, with the unit vectors normal to S pointing away from the center $\vec{c}$ of the sphere. The sphere can be given as the solution to the equation

$$\|\vec{x} - \vec{c}\|^2 = R^2. \tag{6.10}$$

Recall that $\|\vec{v}\|^2 = \vec{v} \cdot \vec{v}$. If $\vec{x}(t)$ is any curve on the sphere, then by differentiating the relationship in Equation (6.10), one obtains

$$\vec{x}'(t) \cdot (\vec{x}(t) - \vec{c}) = 0.$$

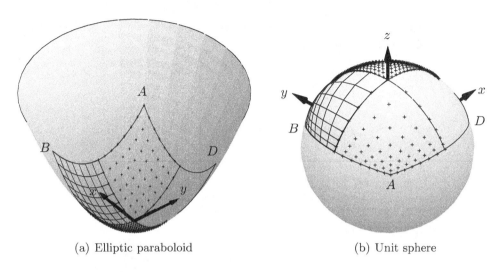

(a) Elliptic paraboloid   (b) Unit sphere

Figure 6.4. Gauss map on the elliptic paraboloid.

The tangent plane to the sphere S at a point $\vec{p}$ consists of all possible vectors $\vec{x}'(t_0)$, where $\vec{x}(t)$ is a curve on S, with $\vec{x}(t_0) = \vec{p}$. Therefore, at any point $\vec{p}$ on the sphere, there are two options for unit normal vectors:

$$\frac{\vec{p} - \vec{c}}{\|\vec{p} - \vec{c}\|} \qquad \text{or} \qquad -\frac{\vec{p} - \vec{c}}{\|\vec{p} - \vec{c}\|}$$

However, it is the former that provides the outward-pointing orientation n. Thus the Gauss map for the sphere of radius R and center $\vec{c}$ is explicitly

$$n(\vec{p}) = \frac{\vec{p} - \vec{c}}{R}.$$

Furthermore, if S is itself the unit sphere centered at the origin, then the Gauss map is the identity function.

Example 6.2.3. Consider the elliptic paraboloid S defined by the equation $z = x^2 + y^2$. This is an orientable surface, and suppose it is oriented with the unit normal always pointing in a positive z-direction. (Figure 6.4 shows this is possible.) Using the function $F(x, y, z) = z - x^2 - y^2$, the elliptic paraboloid is given by the equation $F(x, y, z) = 0$. By Proposition 5.2.13, for all $(x, y, z) \in S$, the gradient $\vec{\nabla}F(x, y, z)$ is a normal vector, so (since the z-direction is positive) the Gauss map is

$$n(x, y, z) = \frac{(-2x, -2y, 1)}{\sqrt{4x^2 + 4y^2 + 1}}.$$

Figure 6.4 shows how the Gauss map acts on the patch

$$\{(x, y, z) \in S \mid -1 \leq x \leq 1 \text{ and } -1 \leq y \leq 1\}.$$

It is not hard to show that the image of the Gauss map on the unit sphere is the upper hemisphere. Indeed, using cylindrical coordinates, the Gauss map is

$$n(r, \theta, z) = \frac{(-2r \cos \theta, -2r \sin \theta, 1)}{\sqrt{4r^2 + 1}}.$$

The function $f(r) = 1/\sqrt{4r^2 + 1}$ is a decreasing bijection from $[0, +\infty)$ to $(0, 1]$. Thus, for any $z_0 \in (0, 1]$, there exists a unique $r_0 \in [0, +\infty)$ with $f(r_0) = z_0$. Then, with $\theta \in [0, 2\pi]$, the image of $n(r_0, \theta, z_0)$ is a circle on the unit sphere at height z_0. Thus, the image of n is

$$n(S) = \{(x, y, z) \in \mathbb{S}^2 \mid z > 0\}.$$

 Example 6.2.4. In contrast to Example 6.2.3, consider the hyperbolic paraboloid given by the parametrization $\vec{X} : \mathbb{R}^2 \to \mathbb{R}^3$, with

$$\vec{X}(u, v) = (u, v, u^2 - v^2).$$

We calculate

$$\vec{X}_u = (1, 0, 2u) \qquad \text{and} \qquad \vec{X}_v = (0, 1, -2v),$$

and thus,

$$\vec{N} = \frac{\vec{X}_u \times \vec{X}_v}{\|\vec{X}_u \times \vec{X}_v\|} = \frac{(-2u, 2v, 1)}{\sqrt{4u^2 + 4v^2 + 1}}.$$

Like Figure 6.4, Figure 6.5 shows the Gauss map for a "square" on the hyperbolic paraboloid. These two examples manifest a central behavior of the Gauss map. On the elliptic paraboloid, the Gauss

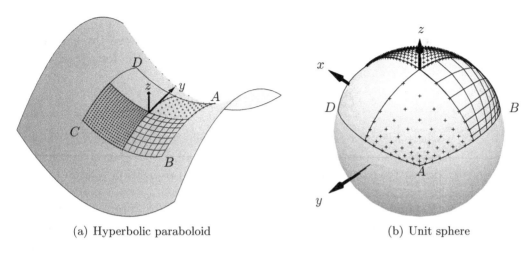

(a) Hyperbolic paraboloid (b) Unit sphere

Figure 6.5. Gauss map on the hyperbolic paraboloid.

map preserves the orientation of the square in the sense that if one travels along the boundary in a clockwise sense on the surface, one also travels along the boundary of the image of the square mapped by n in a clockwise sense. On the other hand, with the hyperbolic paraboloid, the Gauss map reverses the orientation of the square.

Proposition 6.2.5. *Let S be a regular surface in $\mathbb{R}^3$, and let $\vec{X} : U \to \mathbb{R}^3$ be a parametrization of a coordinate patch $V = \vec{X}(U)$, where U is an open set in $\mathbb{R}^2$. Define the vector function $\vec{N} : U \to \mathbb{R}^3$ by*

$$\vec{N} = \frac{\vec{X}_u \times \vec{X}_v}{\|\vec{X}_u \times \vec{X}_v\|}.$$

If $\vec{X}$ is of class C^r, then $\vec{N}$ is of class C^{r-1}.

Proof: By Problem 3.1.10, we deduce that if $\vec{F} : U \to \mathbb{R}^3$, $\vec{G} : U \to \mathbb{R}^3$, and $h : U \to \mathbb{R}$ are of class C^1, then the three functions $\vec{F} \cdot \vec{G} : U \to \mathbb{R}$, $\vec{F} \times \vec{G} : U \to \mathbb{R}^3$, and $h\vec{F} : U \to \mathbb{R}^3$ are also of class C^1, and their partial derivatives follow appropriate product rules. By definition,

$$\vec{N} = \frac{\vec{X}_u \times \vec{X}_v}{\sqrt{(\vec{X}_u \times \vec{X}_v) \cdot (\vec{X}_u \times \vec{X}_v)}}.$$

One can see that any partial derivative of any order of $\vec{N}$ will involve the partial derivative of a combination of the above three types of products and the derivative of an expression of the form

$$f(u,v) = \left((\vec{X}_u \times \vec{X}_v) \cdot (\vec{X}_u \times \vec{X}_v) \right)^{-k/2},$$

where k is an odd positive integer. Since S is regular, $\|\vec{X}_u \times \vec{X}_v\|$ is never 0, so a particular higher derivative of $\vec{N}$ will exist if that higher derivative exists for $\vec{X}_u$ and for $\vec{X}_v$. The proposition follows. $\square$

One would be correct to think that the different behaviors in Figures 6.4 and 6.5 can be illustrated and quantified by a function between tangent spaces to S and $\mathbb{S}^2$, respectively. This perspective is encapsulated in the notion of the differential of functions between regular surfaces. We develop the more general theory of differentials of functions between manifolds in Chapter 3 of [22]. However, the differential of the Gauss map, though only a particular case of the differentials of functions between surfaces, serves a central role in the rest of this chapter.

Let S be a regular oriented surface with orientation n, and let $\vec{X} : U \to \mathbb{R}^3$ be a regular positively oriented parametrization of a coordinate patch $\vec{X}(U)$ of S. Suppose also that S is of class C^2, which, according to Proposition 6.2.5, implies that $\vec{N}$ is of class C^1. Since $\vec{N} \cdot \vec{N} = 1$ for all $(u,v) \in U$, we have

$$\vec{N} \cdot \vec{N}_u = 0 \qquad \text{and} \qquad \vec{N} \cdot \vec{N}_v = 0 \tag{6.11}$$

for all $(u,v) \in U$. Therefore, both derivatives $\vec{N}_u$ and $\vec{N}_v$ are vector functions such that $\vec{N}_u(q)$ and $\vec{N}_v(q)$ are in T_pS when $\vec{X}(q) = p$.

The vector function $\vec{N} : U \to \mathbb{R}^3$ is itself a parametrized surface, with its image lying in the unit sphere, though not necessarily giving a regular parametrization of $\mathbb{S}^2$. The simple fact in Equation (6.11) indicates that if $\vec{N}$ admits a tangent plane at $q \in U$, then

$$T_pS = T_{\vec{N}(q)}(\mathbb{S}^2) \tag{6.12}$$

as subspaces of $\mathbb{R}^3$. In other words, the sets of vectors $\{\vec{X}_u, \vec{X}_v\}$ and $\{\vec{N}_u, \vec{N}_v\}$ span the same subspace T_pS.

The differential of the Gauss map at a point $p \in S$ is a linear transformation $dn_p : T_pS \to T_{n(p)}(\mathbb{S}^2)$, but by virtue of Equation (6.12), one identifies it as a linear transformation $dn_p : T_pS \to T_pS$. For all $\vec{v} \in T_pS$, one defines $dn_p(\vec{v}) = \vec{w}$ if there exists a curve $\vec{\alpha} : I \to U$ such that

$$\begin{cases} \vec{\alpha}(0) = q, \text{ with } \vec{X}(q) = p, \\ \dfrac{d}{dt}\left(\vec{X} \circ \vec{\alpha}\right)(t)\big|_{t=0} = \vec{v}, \text{ and} \\ \dfrac{d}{dt}\left(\vec{N} \circ \vec{\alpha}\right)(t)\big|_{t=0} = \vec{w}. \end{cases} \qquad (6.13)$$

In other words, by writing $\vec{X}(t) = \vec{X} \circ \vec{\alpha}(t)$ and $\vec{N}(t) = \vec{N} \circ \vec{\alpha}(t)$, the differential of the Gauss map satisfies $dn_p(\vec{X}'(0)) = \vec{N}'(0)$.

Using coordinate lines $\vec{X}(t, v_0)$ or $\vec{X}(u_0, t)$ for the curve $\vec{\alpha}(t)$ in the above definition, it is easy to see that

$$\begin{aligned} dn_p(\vec{X}_u) &= \vec{N}_u, \\ dn_p(\vec{X}_v) &= \vec{N}_v. \end{aligned} \qquad (6.14)$$

Though the Gauss map $n : S \to \mathbb{S}^2$ is independent of any coordinate systems on S, the differential dn_p requires reference to a regular positively oriented parametrization of a neighborhood of p. However, though different parametrizations of a neighborhood of p induce different coordinate bases on T_pS, the definition in Equation 6.13 remains unchanged and consequently, as a linear transformation of T_pS to itself, is independent of the parametrization.

Problems

6.2.1. Describe the region of the unit sphere covered by the image of the Gauss map for the following surfaces:

(a) The hyperbolic paraboloid given by $z = x^2 - y^2$.

(b) The hyperboloid of one sheet given by $x^2 + y^2 - z^2 = 1$.

(c) The cone with opening angle α given by $z^2 \tan^2 \alpha = x^2 + y^2$.

(d) The right circular cylinder $x^2 + y^2 = R^2$, with R a constant.

6.2.2. What regular surface S has a single point on $\mathbb{S}^2$ as the image of the Gauss map?

6.2.3. Prove that the image of the Gauss map of a regular surface S is an arc of a great circle if and only if S is a cylinder.

6.2.4. Consider the surface S obtained as a one-sheeted cone over a regular plane curve C (see Problem 5.2.16). Prove that the image of the Gauss map for S is a curve on the unit sphere. Find a parametrization for the image of the Gauss map in terms of a parametrization of C. [Hint: Without loss of generality, suppose that C lies in the plane $z = a$.]

6.3 The Second Fundamental Form

The examples in the previous section illustrate how the differential of the Gauss map dn_p qualifies how much the surface S is curving at the point p. Despite the abstract definition for dn_p, it is not difficult to calculate the matrix for $dn_p : T_pS \rightarrow T_pS$. However, in order for the differential dn_p to exist at all points $p \in S$, we will need to assume that S is a surface of class C^2.

Let S be a regular oriented surface of class C^2 with orientation n, and let $\vec{X} : U \rightarrow \mathbb{R}^3$ be a positively oriented parametrization of a neighborhood V of a point p on S. Since $\vec{N}$ satisfies

$$\vec{N} \cdot \vec{X}_i = 0 \qquad \text{for } i = 1, 2,$$

then by differentiating with respect to another variable, the product rule gives

$$\vec{N}_j \cdot \vec{X}_i + \vec{N} \cdot \vec{X}_{ij} = 0 \qquad \text{for } i, j = 1, 2.$$

Note that since $\vec{X}_{ij} = \vec{X}_{ji}$, by interchanging i and j, one deduces that

$$\vec{N}_i \cdot \vec{X}_j = -\vec{N} \cdot \vec{X}_{ij} = \vec{N}_j \cdot \vec{X}_i. \tag{6.15}$$

(Recall that the notation $\vec{f}_i$ in Equation (6.15) and in what follows refers to taking the derivative of the multivariable vector function $\vec{f}$ with respect to the ith variable.) This leads to the following proposition.

Proposition 6.3.1. *Using the above setup, the linear map dn_p is a self-adjoint operator with respect to the first fundamental form $I_p(\cdot, \cdot)$.*

Proof: Let $\vec{v}, \vec{w} \in T_pS$ and write $\vec{v} = v_1\vec{X}_u + v_2\vec{X}_v$ and $\vec{w} = w_1\vec{X}_u + w_2\vec{X}_v$. Recall that the first fundamental form $I_p(\vec{v}, \vec{w})$ is the dot product $\vec{v} \cdot \vec{w}$ when viewing T_pS as a subset of $\mathbb{R}^3$. Also recall from Equation (6.14) that $dn_p(\vec{X}_u) = \vec{N}_u$ and similarly for v. Hence,

$$I_p(dn_p(\vec{v}), \vec{w}) = I_p(v_1\vec{N}_u + v_2\vec{N}_v, \vec{w}) = (v_1\vec{N}_u + v_2\vec{N}_v) \cdot (w_1\vec{X}_u + w_2\vec{X}_v)$$

$$= \sum_{i,j=1}^{2} v_iw_j\vec{N}_i \cdot \vec{X}_j = \sum_{i,j=1}^{2} v_iw_j\vec{N}_j \cdot \vec{X}_i \qquad \text{(by Equation (6.15))}$$

$$= (v_1\vec{X}_u + v_2\vec{X}_v) \cdot (w_1\vec{N}_u + w_2\vec{N}_v)$$

$$= I_p(\vec{v}, dn_p(\vec{w})). \qquad \square$$

Definition 6.3.2. Let S be an oriented regular surface of class C^2 with orientation n, and let p be a point of S. We define the *second fundamental form* as the quadratic form on T_pS defined by

$$II_p(\vec{v}) = -dn_p(\vec{v}) \cdot \vec{v} = -I_p(dn_p(\vec{v}), \vec{v}).$$

The first fundamental form allows one to measure lengths, angles, and area of regions on a parametrized surface. The second fundamental form provides a measure for how much the normal vector changes if one travels away from p in a particular direction $\vec{v}$, with $\vec{v} \in T_pS$.

Recall from linear algebra that every quadratic form Q on a vector space V of dimension n is of the form

$$Q(\vec{v}) = \vec{v}^t M v$$

for some $n \times n$ matrix. Define the functions $L_{ij} : U \to \mathbb{R}$ by

$$II_p(\vec{v}) = \vec{v}^T \begin{pmatrix} L_{11}(q) & L_{12}(q) \\ L_{21}(q) & L_{22}(q) \end{pmatrix} \vec{v},$$

where $p = \vec{X}(q)$, for all $\vec{v} \in T_pS$. Since

$$II_p(\vec{v}) = -dn_p(\vec{v}) \cdot \vec{v} = -\vec{v}^T dn_p(\vec{v}),$$

by writing $\vec{v} = \binom{a}{b}$ in the coordinate basis $\{\vec{X}_u, \vec{X}_v\}$, one obtains

$$II_p\binom{a}{b} = -(a\vec{N}_1 + b\vec{N}_2) \cdot (a\vec{X}_1 + b\vec{X}_2)$$

$$= -(a^2\vec{N}_1 \cdot \vec{X}_1 + ab\vec{N}_1 \cdot \vec{X}_2 + ab\vec{N}_2 \cdot \vec{X}_1 + b^2\vec{N}_2 \cdot \vec{X}_2).$$

Therefore,

$$L_{ij} = -\vec{N}_j \cdot \vec{X}_i \qquad \text{for all } 1 \le i, j \le 2, \tag{6.16}$$

and by Equation (6.15),

$$L_{ij} = \vec{N} \cdot \vec{X}_{ij}. \tag{6.17}$$

This provides a convenient way to calculate the L_{ij} functions and, hence, the second fundamental form. By Proposition 6.3.1, the matrix (L_{ij}) is a symmetric matrix, so $L_{12} = L_{21}$. In classical differential geometry texts, authors often refer to the coefficients of the second fundamental form using the letters e, f, and g, as follows:

$$e = L_{11}, \qquad f = L_{12} = L_{21}, \qquad g = L_{22}.$$

 Example 6.3.3 (Spheres). Let S be the sphere of radius R with outward orientation and consider the coordinate patch parametrized by $\vec{X}(u, v) = (R \cos u \sin v, R \sin u \sin v, R \cos v)$. The unit normal vector is

$$\vec{N} = (\cos u \sin v, \sin u \sin v, \cos v),$$

and the second derivatives are

$$\vec{X}_{11} = (-R \cos u \sin v, -R \sin u \sin v, 0),$$
$$\vec{X}_{12} = \vec{X}_{21} = (-R \sin u \cos v, R \cos u \cos v, 0),$$
$$\vec{X}_{22} = (-R \cos u \sin v, -R \sin u \sin v, -R \cos v).$$

Thus, the matrix for the second fundamental form is

$$(L_{ij}) = \begin{pmatrix} -R \sin^2 v & 0 \\ 0 & -R \end{pmatrix}.$$

One should have expected the negative signs since along any curve through a point p with direction $\vec{v}$, the unit normal vector $\vec{N}$ changes with a differential in the direction of $\vec{v}$ and not opposite to it, so by Definition 6.3.2, the second fundamental form is always negative.

At a point $p \in S$ in a coordinate neighborhood parametrized positively by $\vec{X}(u, v)$, if $\vec{X}(q) = p$, the function

$$f(s, t) = \frac{1}{2}II_p \begin{pmatrix} s \\ t \end{pmatrix} = \frac{1}{2}(L_{11}(q)s^2 + 2L_{12}st + L_{22}(q)t^2)$$

is called the *osculating paraboloid*. It provides the second-order approximation of the surface near p in reference to the normal vector $\vec{N}(q)$ in the following sense. Setting $q = (u_0, v_0)$, the second-order Taylor approximation of $\vec{X}$ is

$$\vec{X}(u, v) \approx \vec{X}(u_0, v_0) + \vec{X}_u(u_0, v_0)(u - u_0) + \vec{X}_v(u_0, v_0)(v - v_0)$$
$$+ \frac{1}{2}\vec{X}_{uu}(u_0, v_0)(u - u_0)^2 + \vec{X}_{uv}(u_0, v_0)(u - u_0)(v - v_0)$$
$$+ \vec{X}_{vv}(u_0, v_0)(v - v_0)^2. \tag{6.18}$$

Since $\vec{N}$ is perpendicular to $\vec{X}_u$ and $\vec{X}_v$, setting $s = u - u_0$ and $t = v - v_0$,

$$f(s, t) = \left(\vec{X}(s + u_0, t + v_0) - \vec{X}(u_0, v_0) \right) \cdot \vec{N}(u_0, v_0).$$

Furthermore, from the second derivative test in multivariable calculus, one knows that the point $(s, t) = (0, 0)$ is a local extremum if $L_{11}L_{22} - L_{12}^2 > 0$ and is a saddle point if $L_{11}L_{22} - L_{12}^2 < 0$. This leads to the following definition.

Definition 6.3.4. Let S be a regular orientable surface of class C^2. A point p on S is called

1. *elliptic* if $\det(L_{ij}) > 0$;

2. *hyperbolic* if $\det(L_{ij}) < 0$;

3. *parabolic* if $\det(L_{ij}) = 0$ but not all $L_{ij} = 0$; and

4. *planar* if $L_{ij} = 0$ for all i, j.

It is not hard to check (Problem 6.3.5) that if (x_1, x_2) and $(\bar{x}_1, \bar{x}_2)$ are two coordinate systems for the same open set of an oriented surface S, then

$$\det(\bar{L}_{kl}) = \left(\frac{\partial(x_1, x_2)}{\partial(\bar{x}_1, \bar{x}_2)} \right)^2 \det(L_{ij}). \tag{6.19}$$

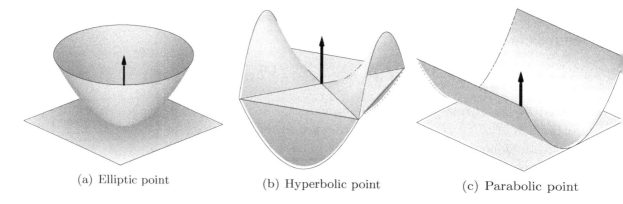

(a) Elliptic point   (b) Hyperbolic point   (c) Parabolic point

Figure 6.6. Elliptic, hyperbolic and parabolic points.

Furthermore, the change of coordinates between (x_1, x_2) and $(\bar{x}_1, \bar{x}_2)$ is a diffeomorphism between two open sets in $\mathbb{R}^2$. Therefore, the Jacobian in Equation (6.19) is never 0 and Definition 6.3.4 is independent of any particular parametrization of S.

Example 6.3.5. Consider the torus parametrized by

$$\vec{X}(u, v) = ((2 + \cos v) \cos u, (2 + \cos v) \sin u, \sin v).$$

The unit normal vector and the second derivatives of the vector function $\vec{X}$ are

$$\vec{N} = (-\cos u \cos v, -\sin u \cos v, -\sin v),$$
$$\vec{X}_{11} = (-(2 + \cos v) \cos u, -(2 + \cos v) \sin u, 0),$$
$$\vec{X}_{12} = (\sin u \sin v, -\cos u \sin v, 0),$$
$$\vec{X}_{22} = (-\cos v \cos u, -\cos v \sin u, -\sin v).$$

Thus,

$$L_{11} = (2 + \cos v) \cos v, \qquad L_{12} = 0, \qquad L_{22} = 1.$$

Thus, in this example it is easy to see that $\det(L_{ij}) = (2 + \cos v) \cos v$. Since $-1 \le \cos v \le 1$, we know $2 + \cos v \ge 1$, so the sign of $\det(L_{ij})$ is the sign of $\cos v$. Hence, the parabolic points on the torus are where $v = \pm\frac{\pi}{2}$, the elliptic points are where $-\frac{\pi}{2} < v < \frac{\pi}{2}$, and the hyperbolic points are where $\frac{\pi}{2} < v < \frac{3\pi}{2}$ (see Figure 6.7).

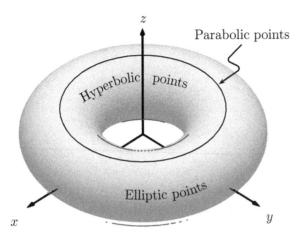

Figure 6.7. Torus.

As Figure 6.6 implies, the only quadratic surface possessing at least one planar point is a plane. However, similar to the "undecided" case in the second derivative test from multivariable calculus, on a general surface S, the existence of a planar point does not imply that S is a plane; it merely implies that third-order behavior, as opposed to second order, governs the local geometry of the surface with respect to the normal vector. In this case, a variety of possibilities can occur.

Example 6.3.6 (Monkey Saddle). The simplest surface that illustrates third-order behavior is the *monkey saddle* parametrized by $\vec{X}(u,v) = (u, v, u^3 - 3uv^2)$. We calculate that

$$\vec{X}_{uu} = (0, 0, 6u),$$
$$\vec{X}_{uv} = (0, 0, -6v),$$
$$\vec{X}_{vv} = (0, 0, -6u).$$

and hence, even without calculating the unit normal vector function, we deduce that $L_{ij}(0,0) = 0$ for all i, j. Consequently, $(0,0,0)$ is a planar point (see Figure 6.8 for a picture). Near $(0,0,0)$, the best approximating quadratic to the surface is in fact the plane $z = 0$, but obviously such an approximation describes the surface poorly. (The terminology "monkey saddle" comes from the fact that a monkey

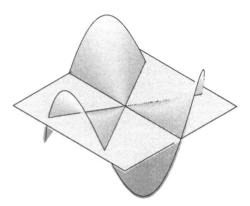

Figure 6.8. Monkey saddle with its tangent plane at $(0,0,0)$.

riding a bicycle would need a saddle with three depressions, one for each leg and one for the tail.)

Example 6.3.5 illustrates a general fact about the local shape of a surface encapsulated in the following proposition.

Proposition 6.3.7. *Let S be a regular surface of class C^2, and let V be a coordinate neighborhood on S.*

  1. *If $p \in V$ is an elliptic point, then there exists a neighborhood V' of p such that T_pS does not intersect $V' - \{p\}$.*

  2. *If $p \in V$ is a hyperbolic point, then T_pS intersects every deleted neighborhood V' of p.*

Proof: Let $\vec{X} : U \to \mathbb{R}^3$ be a parametrization of the coordinate neighborhood V and suppose that $\vec{X}(0,0) = p$. Consider the real-valued function on U

$$h(u,v) = (\vec{X}(u,v) - \vec{X}(0,0)) \cdot \vec{N}(0,0).$$

Since $\vec{N}(0,0)$ is a unit vector perpendicular to T_pS, then $h(u,v)$ is the height function of signed distance between the surface S at $p' = \vec{X}(u,v)$ and the tangent plane T_pS.

The second-order Taylor series of $\vec{X}$ near $(0,0)$ is given by Equation (6.18), with a remainder function $\vec{R}(u,v)$ such that

$$\lim_{(u,v)\to(0,0)} \frac{\vec{R}(u,v)}{u^2 + v^2} = \vec{0}. \qquad (6.20)$$

Thus,

$$h(u,v) = \frac{1}{2}\vec{X}_{uu}(0,0) \cdot \vec{N}(0,0)u^2 + \vec{X}_{uv}(0,0) \cdot \vec{N}(0,0)uv$$

$$+ \frac{1}{2}\vec{X}_{vv}(0,0) \cdot \vec{N}(0,0)v^2 + \vec{R}(u,v) \cdot \vec{N}(0,0)$$

$$= \frac{1}{2}(L_{11}(0,0)u^2 + 2L_{12}(0,0)uv + L_{22}(0,0)v^2) + \vec{R}(u,v) \cdot \vec{N}(0,0)$$

$$= \frac{1}{2}II_p(\begin{pmatrix} u \\ v \end{pmatrix}) + \vec{R}(u,v) \cdot n(p).$$

Solving the quadratic equation

$$L_{11}u^2 + 2L_{12}uv + L_{22}v^2 = 0 \tag{6.21}$$

for u in terms of v leads to

$$L_{11}u = \left(-L_{12} \pm \sqrt{(L_{12})^2 - L_{11}L_{22}}\right)v.$$

If we assume that $(u,v) \neq (0,0)$, Equation(6.21) has no solutions if $\det(L_{ij}) > 0$ and has solutions if $\det(L_{ij}) < 0$. Hence, if p is hyperbolic, $II_p(\begin{pmatrix} u \\ v \end{pmatrix})$ changes sign in a neighborhood of $(0,0)$, and if p is elliptic, it does not.

Define functions $\bar{R}_1$ and $\bar{R}_2$ by

$$\bar{R}_i(u,v) = \frac{R_i(u,v)}{u^2 + v^2} \qquad \text{for } i = 1,2.$$

Then solving $h(u,v) = 0$ amounts to solving

$$(L_{11} + \bar{R}_1(u,v))u^2 + 2L_{12}uv + (L_{22} + \bar{R}_2(u,v))v^2 = 0.$$

Since both $\bar{R}_1$ and $\bar{R}_2$ have a limit of 0 as (u,v) approaches $(0,0)$,

$$\lim_{(u,v) \to (0,0)} (L_{11}(0,0) + \bar{R}_1(u,v))(L_{22}(0,0) + \bar{R}_2(u,v)) - L_{12}(0,0)^2 = L_{11}(0,0)L_{22}(0,0) - L_{12}(0,0)^2.$$

Thus, if p is an elliptic point, there is a neighborhood of $(0,0)$ in which $h(u,v) = 0$ does not have solutions except for $(u,v) = (0,0)$, and if p is hyperbolic, every neighborhood of $(0,0)$ has points through which $h(u,v)$ changes sign. $\qquad\qquad\square$

Let us return to considering the differential of the Gauss map. With Equation (6.17), it is now possible to explicitly calculate the matrix for $dn_p : T_pS \to T_pS$ in terms of the oriented basis $(\vec{X}_u, \vec{X}_v)$, where $\vec{X} : U \to \mathbb{R}^3$ is a positively oriented parametrization of a neighborhood around p. Since $\vec{N}_u$ and $\vec{N}_v$ lie in T_pS, there exist functions $a_j^i(u, v)$ defined on U such that

$$\begin{aligned} \vec{N}_u &= a_1^1 \vec{X}_u + a_1^2 \vec{X}_v, \\ \vec{N}_v &= a_2^1 \vec{X}_u + a_2^2 \vec{X}_v. \end{aligned} \tag{6.22}$$

Using numerical indices to represent corresponding derivatives, Equation (6.22) can be written as

$$\vec{N}_j = \sum_{i=1}^{2} a_j^i \vec{X}_i, \tag{6.23}$$

where we denote $\vec{X}_1 = \vec{X}_u$ and $\vec{X}_2 = \vec{X}_v$. (It is important to remember that the superscripts in a_j^i also correspond to indices and not to powers. This notation, though perhaps awkward at first, is the standard notation for components of a tensor, a language that this book develops more and more through subsequent chapters. This notation is also entirely standard in the theory of manifolds as developed in [22].)

For any vector $\vec{w} \in T_pS$ with coordinates $\vec{w} = \binom{w_1}{w_2} = w_1 \vec{X}_u + w_2 \vec{X}_v$, the differential of the Gauss map satisfies

$$\begin{aligned} dn_p(\vec{w}) &= w_1 \vec{N}_1 + w_2 \vec{N}_2 \\ &= (a_1^1 w_1 + a_2^1 w_2) \vec{X}_1 + (a_1^2 w_1 + a_2^2 w_2) \vec{X}_2 \\ &= \begin{pmatrix} a_1^1 & a_2^1 \\ a_1^2 & a_2^2 \end{pmatrix} \begin{pmatrix} w_1 \\ w_2 \end{pmatrix}. \end{aligned}$$

However, from Equations (6.15), (6.17) and (6.23),

$$-L_{ij} = \vec{N}_i \cdot \vec{X}_j = \left(\sum_{k=1}^{2} \vec{X}_k \right) \cdot \vec{X}_j,$$

and so

$$-L_{ij} = \sum_{k=1}^{2} a_i^k g_{kj}. \tag{6.24}$$

In matrix notation, Equation (6.24) means that

$$-\begin{pmatrix} L_{11} & L_{12} \\ L_{21} & L_{22} \end{pmatrix} = \begin{pmatrix} a_1^1 & a_1^2 \\ a_2^1 & a_2^2 \end{pmatrix} \begin{pmatrix} g_{11} & g_{12} \\ g_{21} & g_{22} \end{pmatrix}. \tag{6.25}$$

Since the metric tensor is a positive definite matrix at any point on a regular surface, one can multiply both sides of Equation (6.25) by the inverse of (g_{ij}), and conclude that the matrix for dn_p given in terms of the basis $\{\vec{X}_u, \vec{X}_v\}$ is

$$\begin{pmatrix} a_1^1 & a_1^2 \\ a_2^1 & a_2^2 \end{pmatrix} = -\begin{pmatrix} L_{11} & L_{12} \\ L_{21} & L_{22} \end{pmatrix} \begin{pmatrix} g_{11} & g_{12} \\ g_{21} & g_{22} \end{pmatrix}^{-1}. \tag{6.26}$$

This matrix formula is more common in modern texts but classical differential geometry texts refer to the individual component equations implicit in the above formula as the *Weingarten equations*.

Since all of the above matrices are 2×2, $\det(L_{ij}) = \det(-L_{ij})$ and since $\det(g_{ij}) > 0$, the determinant $\det(L_{ij})$ has the same sign as $\det(a_j^i)$. Therefore, one deduces the following reformulation of Definition 6.3.4.

Proposition 6.3.8. *Let S be an oriented regular surface of class C^2 with orientation n. Then, a point $p \in S$ is called*

1. *elliptic if* $\det(dn_p) > 0$;

2. *hyperbolic if* $\det(dn_p) < 0$;

3. *parabolic if* $\det(dn_p) = 0$ *but* $dn_p \neq 0$;

4. *planar if* $dn_p = 0$.

Problems

6.3.1. Calculate the second fundamental form (i.e., the matrix of functions (L_{ij})) for the following surfaces:

    (a) The ellipsoid $\vec{X}(u, v) = (a \cos u \sin v, b \sin u \sin v, c \cos v)$.

    (b) The parabolic hyperboloid $\vec{X}(u, v) = (au, bv, uv)$.

    (c) The catenoid $\vec{X}(u, v) = (\cosh v \cos u, \cosh v \sin u, v)$.

6.3.2. Calculate the second fundamental form (i.e., the matrix of functions (L_{ij})) using the following parametrizations of the right cylinder:

(a) $\vec{X}(u,v) = (\cos u, \sin u, v)$.

(b) $\vec{Y}(u,v) = (\cos(u+v), \sin(u+v), v)$.

Using the parametrization $\vec{Y}$, find all vectors $\vec{v} \in T_p S$ such that $II_p(\vec{v}) = 0$. Show that these correspond to the straight lines on the cylinder.

6.3.3. Consider Enneper's surface parametrized by

$$\vec{X}(u,v) = \left(u - \frac{u^3}{3} + uv^2, v - \frac{v^3}{3} + vu^2, u^2 - v^2 \right).$$

Show that

(a) the coefficients of the first fundamental form are

$$g_{11} = g_{22} = (1 + u^2 + v^2)^2, \qquad g_{12} = 0;$$

(b) the coefficients of the second fundamental form are

$$L_{11} = 2, \qquad L_{12} = 0, \qquad L_{22} = -2.$$

6.3.4. Suppose that an open set V of a regular oriented surface has two systems of coordinates (x_1, x_2) and $(\bar{x}_1, \bar{x}_2)$ and suppose that V is parametrized by $\vec{X}(x_1, x_2)$ in terms of $(x_1, x_2) \in U$ and by $\vec{Y}(\bar{x}_1, \bar{x}_2)$ in terms of $(\bar{x}_1, \bar{x}_2)$. Call L_{ij} the terms of the second fundamental form in terms of the (x_1, x_2) coordinates and call $\bar{L}_{ij}$ the terms of the second fundamental form in terms of the $(\bar{x}_1, \bar{x}_2)$ coordinates. Prove that

$$\bar{L}_{kl} = \sum_{i,j=1}^{2} \frac{\partial x_i}{\partial \bar{x}_k} \frac{\partial x_j}{\partial \bar{x}_l} L_{ij}.$$

6.3.5. Use the previous exercise to show that under the same conditions

$$\det(\bar{L}_{kl}) = \left(\frac{\partial(x_1, x_2)}{\partial(\bar{x}_1, \bar{x}_2)} \right)^2 \det(L_{ij}).$$

6.3.6. Suppose we are under the same conditions as in Problem 6.3.4. Call $[dF]$ the 2×2 matrix of the differential of the coordinate change, that is

$$[dF] = \left(\frac{\partial x_i}{\partial \bar{x}_j} \right)_{i,j=1,2}.$$

Call $\bar{a}_i^j$ the coefficients of dn_p in terms of the coordinate system $(\bar{x}_1, \bar{x}_2)$. Prove that, as matrices,

$$(\bar{a}_i^j) = (dF)\,(a_k^l)\,(dF)^{-1}.$$

6.3.7. Calculate the second fundamental form (L_{ij}) and the matrix for dn_p for function graphs $\vec{X}(u, v) = (u, v, f(u, v))$. Determine which points are elliptic, hyperbolic, or parabolic. (This exercise coupled with Proposition 6.3.8 is a proof of the second derivative test from multivariable calculus.)

6.3.8. Prove that all points on the tangential surface of a regular space curve are parabolic.

6.3.9. Give an example of a surface with an isolated parabolic point.

6.3.10. Consider a regular surface S given by the equation $F(x, y, z) = 0$. Use implicit differentiation to provide a criterion for determining whether points are elliptic, hyperbolic, parabolic, or planar.

6.4   Normal and Principal Curvatures

One way to analyze the shape of a regular surface S near a point p is to consider curves on S through p and analyze the normal component of their principal curvature vector. In order to use the techniques of differential geometry, in this section we will always assume that the surface S is of class C^2 and that the curves on S are regular curves also of class C^2.

Definition 6.4.1. Let S be a regular surface of class C^2, and let $\vec{X} :$ $U \to \mathbb{R}^3$ be a parametrization of a coordinate neighborhood V of S. Let $\vec{\gamma} : I \to \mathbb{R}^3$ be a parametrization of class C^2 for a curve C that lies on S in V. The *normal curvature* of S along C is the function

$$\kappa_n(t) = \frac{1}{s'}\vec{T}' \cdot \vec{N} = \kappa(\vec{P} \cdot \vec{N}) = \kappa \cos\theta,$$

where θ is the angle between the principal normal vector $\vec{P}$ of the curve and the normal vector $\vec{N}$ of the surface.

Interestingly enough, though the curvature of a space curve in general depends on the second derivative of the curve, once one has the second fundamental form for a surface, the normal curvature at a point depends only on the direction.

Proposition 6.4.2. *Let p be a point on S in the neighborhood V and suppose that $\vec{\gamma}$ is the parametrization of a curve C on S such that $\vec{\gamma}(0) = p$ and $\vec{T}(0) = \vec{w} \in T_pS$. Then at p, we have*

$$\kappa_n(0) = II_p(\vec{w}).$$

Proof: Suppose that $\vec{\gamma} = \vec{X} \circ \vec{\alpha}$, where $\vec{\alpha}(t) = (u(t), v(t))$ is a curve in the domain U. The normal vector of S along the curve is given by the function $\vec{N}(t) = \vec{N}(\vec{\alpha}(t))$. Since $\vec{T}(t) \cdot \vec{N}(t) = 0$ for all $t \in I$, then

$$\vec{T}' \cdot \vec{N} = -\vec{T} \cdot \vec{N}',$$

and, hence,

$$\kappa_n(t) = -\frac{1}{s'(t)}\vec{T} \cdot \vec{N}'.$$

Since $\vec{N}(t) = \vec{N}(u(t), v(t))$, then $\vec{N}'(t) = \vec{N}_u u'(t) + \vec{N}_v v'(t)$, and therefore, $\vec{N}'(0) = dn_p\binom{u'(0)}{v'(0)} = dn_p(s'(0)\vec{w})$. Thus, at the point p, the normal curvature of S along C is

$$\kappa_n = -\frac{1}{s'(0)}\vec{T}(0) \cdot dn_p(s'(0)\vec{w}) = -\vec{w} \cdot dn_p(\vec{w}) = II_p(\vec{w}). \qquad \square$$

Corollary 6.4.3. *All curves C on S passing through the point p with direction $\vec{w} \in T_pS$ have the same normal curvature at p.*

Because of Corollary 6.4.3, given a point p on a regular surface S, one knows everything about the local change in the Gauss map by knowing the values of the second fundamental form on the unit circle in T_pS around p. In particular, most interesting are the optimal values of $II_p(\vec{w})$ for $\|\vec{w}\| = 1$ in T_pS. To find these optimal values, set $\vec{w} = \binom{a}{b}$ in the coordinate basis $\{\vec{X}_u, \vec{X}_v\}$ and optimize

$$II_p(\vec{w}) = L_{11}a^2 + 2L_{12}ab + L_{22}b^2,$$

with variables a, b subject to the constraint

$$I_p(\vec{w}, \vec{w}) = g_{11}a^2 + 2g_{12}ab + g_{22}b^2 = 1.$$

Using Lagrange multipliers, one finds that there exists λ such that the optimization problem is solved when

$$\begin{cases} L_{11}a + L_{12}b = \lambda(g_{11}a + g_{12}b), \\ L_{21}a + L_{22}b = \lambda(g_{21}a + g_{22}b), \end{cases}$$

or, in matrix form,

$$\begin{pmatrix} L_{11} & L_{12} \\ L_{21} & L_{22} \end{pmatrix} \begin{pmatrix} a \\ b \end{pmatrix} = \lambda \begin{pmatrix} g_{11} & g_{12} \\ g_{21} & g_{22} \end{pmatrix} \begin{pmatrix} a \\ b \end{pmatrix}. \qquad (6.27)$$

This remark leads to the following fundamental proposition.

Proposition 6.4.4. *The maximum and minimum values κ_1 and κ_2 of $II_p(\vec{w})$ restricted to the unit circle are the negatives of the eigenvalues of dn_p. Furthermore, there exists an orthonormal basis $\{\vec{e}_1, \vec{e}_2\}$ of T_pS such that $dn_p(\vec{e}_1) = -\kappa_1\vec{e}_1$ and $dn_p(\vec{e}_2) = -\kappa_2\vec{e}_2$.*

Proof: Set $\begin{pmatrix} a' \\ b' \end{pmatrix} = g\begin{pmatrix} a \\ b \end{pmatrix}$, where $g = (g_{ij})$ is the metric tensor. Then Equation (6.27) becomes

$$Lg^{-1}\begin{pmatrix} a' \\ b' \end{pmatrix} = \lambda \begin{pmatrix} a' \\ b' \end{pmatrix},$$

and since the matrix $[dn_p] = -Lg^{-1}$, the first part of the proposition follows.

Since the first fundamental form $I_p(\cdot, \cdot)$ is positive definite and since, by Proposition 6.3.1, dn_p is self-adjoint with respect to this form, then by the Spectral Theorem, dn_p is diagonalizable. (Many linear algebra textbooks discuss the Spectral Theorem exclusively using the dot product as the positive definite form. See [19, Section 7, Chapter XV] for a more general presentation of the Spectral Theorem, which we use here.)

Now, since dn_p is self-adjoint with respect to the first fundamental form,

$$I_p(dn_p(\vec{e}_1), \vec{e}_2) = I_p(-\kappa_1\vec{e}_1, \vec{e}_2) = -\kappa_1 I_p(\vec{e}_1, \vec{e}_2),$$

$$\|$$

$$I_p(\vec{e}_1, dn_p(\vec{e}_2)) = I_p(\vec{e}_1, -\kappa_2\vec{e}_2) = -\kappa_2 I_p(\vec{e}_1, \vec{e}_2),$$

and thus,

$$(\kappa_1 - \kappa_2)I_p(\vec{e}_1, \vec{e}_2) = 0.$$

Hence, if $\kappa_1 \neq \kappa_2$, then $I_p(\vec{e}_1, \vec{e}_2) = \vec{e}_1 \cdot \vec{e}_2 = 0$, where in the latter dot product we view $\vec{e}_1$ and $\vec{e}_2$ as vectors in $\mathbb{R}^3$, and if $\kappa_1 = \kappa_2$, then by the Spectral Theorem, any orthonormal basis satisfies the claim of the proposition.   □

Definition 6.4.5. Let S be a regular surface, and let p be a point on S. The maximum and minimum normal curvatures κ_1 and κ_2 at p are called the *principal curvatures* of S at p. The corresponding directions, i.e., unit eigenvectors $\vec{e}_1$ and $\vec{e}_2$ with $dn_p(\vec{e}_i) = -\kappa_i \vec{e}_i$, are called *principal directions* at p.

In a plane, the second fundamental form is identically 0, all normal curvatures including the principal curvatures are 0, and hence, all directions at all points are principal directions. Similarly, it is not hard to show that at every point of the sphere the normal curvature in every direction is the same, and hence, all directions are principal.

Example 6.4.6 (Ellipsoids). As a contrast to the plane or sphere, consider the ellipsoid parametrized by

$$\vec{X}(u, v) = (a \cos u \sin v, b \sin u \sin v, c \cos v).$$

It turns out that the formulas for κ_1 and κ_2 as functions of (u, v) are quite long so we shall calculate the principal curvatures at the point corresponding to $(u, v) = \left(\frac{\pi}{3}, \frac{\pi}{6}\right)$. It is not too hard to calculate the coefficients of the first and second fundamental forms

$$g_{11} = a^2 \sin^2 u \sin^2 v + b^2 \cos^2 u \sin^2 v, \qquad L_{11} = \frac{abc \sin^2 v}{\sqrt{\det g}},$$

$$g_{12} = (b^2 - a^2) \sin u \cos u \sin v \cos v, \qquad L_{12} = 0,$$

$$g_{22} = a^2 \cos^2 u \cos^2 v + b^2 \sin^2 u \cos^2 v + c^2 \sin^2 v, \qquad L_{22} = \frac{abc}{\sqrt{\det g}}.$$

At $(u, v) = \left(\frac{\pi}{3}, \frac{\pi}{6}\right)$, this leads to

$$(a_i^j) = \frac{8abc}{(12a^2 b^2 + 3a^2 c^2 + b^2 c^2)^{3/2}} \begin{pmatrix} -3a^2 - 9b^2 - 4c^2 & -3a^2 + 3b^2 \\ -12a^2 + 12b^2 & -12a^2 - 4b^2 \end{pmatrix},$$

and the eigenvalues of this matrix are

$$\lambda = \frac{4abc\left(-15a^2 - 13b^2 - 4c^2 \pm \sqrt{225a^4 - 378a^2b^2 - 72a^2c^2 + 169b^2 + 40b^2c^2 + 16c^4}\right)}{(12a^2b^2 + 3a^2c^2 + b^2c^2)^{3/2}}.$$

Though there are negative signs under the square root, one knows that this is a positive number for all a, b, and c since the expression under the square root is

$$(a_1^1 + a_2^2)^2 - 4(a_1^1a_2^2 - a_1^2a_2^1) = (a_1^1 - a_2^2)^2 + 4a_1^2a_2^1,$$

and in this case both a_1^2 and a_2^1 are positive. Testing various values for a, b, and c shows that the eigenvalues are often distinct.

If a curve on the surface given by $\gamma(t) = \vec{X}(u(t), v(t))$ is such that at $\gamma(t_0)$ its direction $\gamma'(t_0)$ is a principal direction with principal curvature κ_i, then by Equation (6.27), with $\lambda = -\kappa_i$, one has

$$\begin{cases} L_{11}u' + L_{12}v' = -\kappa_i(g_{11}u' + g_{12}v'), \\ L_{21}u' + L_{22}v' = -\kappa_i(g_{21}u' + g_{22}v'). \end{cases}$$

Eliminating κ_i from these two equations leads to the relationship

$$(L_{11}g_{21}-L_{21}g_{11})(u')^2+(L_{11}g_{22}-L_{22}g_{11})u'v'+(L_{12}g_{22}-L_{22}g_{12})(v')^2 = 0,$$
(6.28)

or equivalently,

$$\begin{vmatrix} (v')^2 & -u'v' & (u')^2 \\ g_{11} & g_{12} & g_{22} \\ L_{11} & L_{12} & L_{22} \end{vmatrix} = 0. \tag{6.29}$$

Also, in light of the Weingarten equations in Equation (6.26), one can summarize the above two formulas by

$$\begin{pmatrix} u' \\ v' \end{pmatrix} \cdot dn_p \begin{pmatrix} v' \\ -u' \end{pmatrix} = 0. \tag{6.30}$$

Definition 6.4.7. Any curve on a regular oriented surface given by $\gamma(t) = \vec{X}(u(t), v(t))$ in a coordinate neighborhood parametrized by $\vec{X} : U \to \mathbb{R}^3$ that satisfies Equation (6.28) for all t is called a *line of curvature*.

The lines of curvature on a surface form an orthogonal family of curves on the surface, that is, two sets of curves intersecting at right angles. With an appropriate change of variables and in an interval of t where $u'(t) \neq 0$, using the chain rule $\frac{dv}{du} = \frac{v'(t)}{u'(t)}$, (6.28) can be rewritten as

$$(L_{12}g_{22}-L_{22}g_{12})\left(\frac{dv}{du}\right)^2 + (L_{11}g_{22}-L_{22}g_{11})\frac{dv}{du} + (L_{11}g_{21}-L_{21}g_{11}) = 0.$$

$$(6.31)$$

If in the neighborhood we are studying $u'(t) = 0$ for some t, then since we assume the curve is regular, $v'(t)$ cannot also be 0, and hence, we can rewrite Equation (6.28) with u as a function of v. In general, solving the above differential equation is an intractable problem. Of course, as long as the coefficients of the first and second fundamental form are not proportional to each other, one can easily solve algebraically for $\frac{dv}{du}$ in Equation (6.31) and obtain two distinct solutions of $\frac{dv}{du}$ as a function of u and v. Then according to the theory of differential equations (see [2, Section 9.2] for a reference), given any point (u_0, v_0) and for each of the two solutions of $\frac{dv}{du}$ as a function of u and v, there exists a unique function $v = f(u)$ solving Equation (6.31).

Consequently, a regular oriented surface of class C^2 can be covered by lines of curvature wherever the coefficients of the first and second fundamental form are continuous and are not proportional to each other. Such points have a special name.

Definition 6.4.8. Let S be a regular oriented surface of class C^2. An *umbilical point* is a point p on S such that, given a parametrization $\vec{X}$ of a neighborhood of p, the corresponding first and second fundamental forms have coefficients that are proportional, namely,

$$L_{11}g_{12} - L_{12}g_{11} = 0 \quad \text{and} \quad L_{22}g_{12} - L_{12}g_{22} = 0.$$

Proposition 6.4.9. *Let S be a regular oriented surface. A point p on S is an umbilical point if and only if the eigenvalues of dn_p are equal.*

Proof: Let $\vec{X}$ be a parametrization of a neighborhood of p. With respect to this parametrization and the associated basis $\{\vec{X}_u, \vec{X}_v\}$ on T_pS, the matrix of dn_p is $[dn_p] = (a_i^j)$. Proposition 6.4.4 implies that (a_i^j) is diagonalizable.

The matrix (a_i^j) has equal eigenvalues if and only if there exists an invertible matrix B such that

$$(a_i^j) = B \begin{pmatrix} \lambda & 0 \\ 0 & \lambda \end{pmatrix} B^{-1},$$

and then

$$(a_i^j) = B(\lambda I)B^{-1} = \lambda B B^{-1} = \lambda I.$$

Consequently (a_i^j) has equal eigenvalues if and only if it is already diagonal, with elements on the diagonal being equal. Then $\lambda I = (a_i^j) = -Lg^{-1}$, and therefore, $L = -\lambda g$, which is tantamount to saying that the coefficients of the first and second fundamental forms are proportional. The proposition follows. $\qquad\square$

The proof of Proposition 6.4.4 shows that if $\kappa_1 \neq \kappa_2$, then the orthonormal basis of principal directions is unique up to signs, while if $\kappa_1 = \kappa_2$, any orthogonal basis is principal. Therefore, in light of Proposition 6.4.9, at umbilical points, there is no preferred basis of principal directions, and thus it makes sense that the only solutions to Equation (6.28) at umbilical points have $u'(t) = v'(t) = 0$, that is, $(u(t), v(t)) = (u_0, v_0)$.

Solving Equation (6.28) gives the lines of curvature on a surface. At every point p on the surface that is not umbilical, there are two lines of curvature through p, and they intersect at right angles. This simple remark leads to the following nice characterization.

Proposition 6.4.10. *The coordinate lines of a parametrization $\vec{X}$ of a surface are curvature lines if and only if $g_{12} = L_{12} = 0$.*

Proof: Suppose that coordinate lines are curvature lines. Since any two lines of curvature intersect at a point at a right angle, we know that $g_{12} = \vec{X}_u \cdot \vec{X}_v = 0$. Furthermore, coordinate lines are given by $\vec{\gamma}(t) = \vec{X}(t, v_0)$ or $\vec{\gamma}(t) = \vec{X}(u_0, t)$. If $\vec{\gamma}(t) = \vec{X}(t, v_0)$, then $u'(t) = 1$ and $v'(t) = 0$, and if $\vec{\gamma}(t) = \vec{X}(u_0, t)$, then $u'(t) = 0$ and $v'(t) = 1$. Then Equation (6.28) implies that both of the following hold:

$$L_{11}g_{21} - L_{21}g_{11} = 0 \quad \text{and} \quad L_{12}g_{22} - L_{22}g_{12} = 0.$$

Since $g_{12} = 0$, we have $L_{21}g_{11} = 0$ or $L_{12}g_{22} = 0$. However, since $\det(g) = g_{11}g_{22} - g_{12}^2 > 0$, the functions g_{11} and g_{22} cannot both be 0. Since $L_{12} = L_{21}$, we deduce that $L_{12} = 0$.

Conversely, if $g_{12} = L_{12} = 0$, then (g_{ij}) and (L_{ij}) are both diagonal matrices, making (a^i_j) a diagonal matrix, and hence, $\vec{X}_u$ and $\vec{X}_v$ are eigenvectors of dn_p. Hence, coordinate lines are curvature lines.   □

As a linear transformation from T_pS to itself, dn_p is independent of a parametrization of a neighborhood of p. Therefore, if p is not an umbilical point, the principal directions $\{\vec{e}_1, \vec{e}_2\}$ as eigenvalues of dn_p provide a basis of T_pS that possesses more geometric meaning than the coordinate basis of any particular parametrization of a neighborhood of p.

In the orthonormal basis $\{\vec{e}_1, \vec{e}_2\}$, a unit vector $\vec{w} \in T_pS$ is written as $\vec{w} = \cos\theta\vec{e}_1 + \sin\theta\vec{e}_2$ for some angle θ. Using these coordinates, the normal curvature of S at p in the direction of $\vec{w}$ is

$$II_p(\vec{w}) = -\vec{w} \cdot dn_p(\vec{w})$$
$$= -(\cos\theta\vec{e}_1 + \sin\theta\vec{e}_2) \cdot dn_p(\cos\theta\vec{e}_1 + \sin\theta\vec{e}_2)$$
$$= -(\cos\theta\vec{e}_1 + \sin\theta\vec{e}_2) \cdot (\cos\theta dn_p(\vec{e}_1) + \sin\theta dn_p(\vec{e}_2))$$
$$= -(\cos\theta\vec{e}_1 + \sin\theta\vec{e}_2) \cdot (-\cos\theta\kappa_1\vec{e}_1 - \sin\theta\kappa_2\vec{e}_2)$$
$$II_p(\vec{w}) = (\cos^2\theta)\kappa_1 + (\sin^2\theta)\kappa_2. \tag{6.32}$$

Equation (6.32) is called *Euler's curvature formula*.

Another useful geometric characterization of the behavior of S near p is called the *Dupin indicatrix* (see Figure 6.9), which consists of all vectors $\vec{u} \in T_pS$ such that $II_p(\vec{u}) = \pm 1$. If $\vec{u} = (u_1, u_2) = (\rho\cos\theta, \rho\sin\theta)$ are expressions of $\vec{u}$ in Cartesian and polar coordinates referenced in terms of the orthonormal frame $\{\vec{e}_1, \vec{e}_2\}$, then Euler's curvature formula gives

$$\pm 1 = II_p(\vec{u}) = \rho^2 II_p(\vec{u}) = \kappa_1\rho^2\cos^2\theta + \kappa_2\rho^2\sin^2\theta.$$

Thus, the Dupin indicatrix satisfies the equation

$$\kappa_1 u_1^2 + \kappa_2 u_2^2 = \pm 1. \tag{6.33}$$

Since the principal curvatures at a point p are the negatives of the eigenvalues of dn_p, Proposition 6.3.8 provides a characterization of whether a point is elliptic, hyperbolic, parabolic, or planar in terms of the principal curvatures. More precisely, a point $p \in S$ is

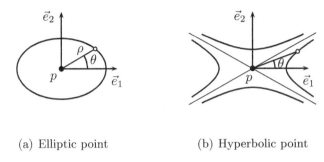

(a) Elliptic point   (b) Hyperbolic point

Figure 6.9. The Dupin indicatrix.

1. elliptic if κ_1 and κ_2 have the same sign;

2. hyperbolic if κ_1 and κ_2 have opposite signs;

3. parabolic if exactly one of κ_1 and κ_2 is 0;

4. planar if $\kappa_1 = \kappa_2 = 0$.

The Dupin indicatrix justifies this terminology because p is elliptic or hyperbolic if and only if Equation(6.33) is the equation for a single ellipse or two hyperbolas, respectively. Furthermore, the half-axes for the corresponding ellipse of the hyperbola are

$$\sqrt{\frac{1}{|\kappa_1|}} \quad \text{and} \quad \sqrt{\frac{1}{|\kappa_2|}}.$$

If a point is parabolic, then the Dupin indicatrix is simply a pair of parallel lines equidistant from one of the principal direction lines, and if a point is planar, the Dupin indicatrix is the empty set.

When a point is hyperbolic, it is possible to find the asymptotes of the Dupin indicatrix without referring directly to the principal curvatures.

Definition 6.4.11. Let p be a point on a regular oriented surface S of class C^2. An *asymptotic direction* of S at p is a unit vector $\vec{w}$ in T_pS such that the normal curvature is 0. An *asymptotic curve* on S is a regular curve C such that at every point $p \in C$, the unit tangent vector at p is an asymptotic direction.

Since the principal curvatures at a point p are the maximum and minimum normal curvatures, at an elliptic point, there exist no asymptotic directions, which one can see from the Dupin indicatrix. If, on the other hand, p is hyperbolic, then the principal curvatures have opposite signs. If a unit vector $\vec{w}$ is an asymptotic direction at p and makes an angle θ with $\vec{e}_1$, then by Euler's formula we deduce that

$$\cos^2\theta = -\frac{\kappa_2}{\kappa_1 - \kappa_2}.$$

The right-hand side is always positive and less than 1, so this equation leads to four solutions for $\vec{w}$, i.e., two mutually negative pairs, each representing one of the asymptotes of the Dupin indicatrix hyperbola. As we see in Problem 6.4.9, the asymptotic curves in fact provide a more subtle description of the behavior of a surface near a point than the Dupin indicatrix does since it describes more than second-order phenomena.

As with the lines of curvature, it is not difficult to find a differential equation that characterizes asymptotic curves on a surface. Suppose that $\vec{X} : U \to \mathbb{R}^3$ parametrizes a neighborhood of $p \in S$ and that $\vec{\gamma} : I \to S$ is a curve on S defined by $\vec{\gamma} = \vec{X} \circ \vec{\alpha}$, with $\vec{\alpha} : I \to U$ and $\vec{\alpha}(t) = (u(t), v(t))$. By Proposition 6.4.2, if we call $\kappa_n(t)$ the normal curvature of S at $\vec{\gamma}(t)$ in the direction of $\vec{\gamma}'(t)$, then

$$\kappa_n(t) = II_{\vec{\gamma}(t)}\left(\vec{T}(t)\right) = \frac{1}{(s'(t))^2} II_{\vec{\gamma}(t)}\left(\begin{pmatrix} u' \\ v' \end{pmatrix} \vec{T}(t)\right).$$

Thus, since an asymptotic curve must satisfy $\kappa_n(t) = 0$ for all t, then any asymptotic curve must satisfy the differential equation

$$L_{11}(u')^2 + 2L_{12}u'v' + L_{22}(v')^2 = 0. \tag{6.34}$$

Example 6.4.12 (Surfaces of Revolution). Consider a surface of revolution parametrized by

$$\vec{X}(u, v) = (f(v)\cos u, f(v)\sin u, h(v)),$$

with $f(v) > 0$, $u \in (0, 2\pi)$, and $v \in (a, b)$. It is not hard to show (see Problem 6.5.8) that the coefficients of the first fundamental form are

$$g_{11} = f(v)^2, \quad g_{12} = g_{21} = 0, \quad g_{22} = (f'(v))^2 + (h'(v))^2,$$

and the coefficients of the second fundamental form are

$$L_{11} = -fh, \quad L_{12} = L_{21} = 0, \quad L_{22} = f''h' - f'h''.$$

By Problem 6.4.4, we deduce that the meridians (where $u = $ const.) and the parallels (where $v = $ const.) are lines of curvature. We can also conclude that the principal curvatures are

$$\kappa_1(u, v) = -\frac{h'(v)}{f(v)}, \quad \kappa_2(u, v) = \frac{f''(v)h'(v) - f'(v)h''(v)}{(f'(v))^2 + (h'(v))^2}, \quad (6.35)$$

though, as written, one can make no assumption that $\kappa_1 > \kappa_2$. Obviously, the first and second fundamental forms depend only on the coordinate v, and hence, all properties of points, such as whether they are elliptic, hyperbolic, parabolic, planar, or umbilical, depend only on v. Setting $\kappa_1 = \kappa_2$ in Equation (6.35) produces an equation that determines for what v the points on the surface are umbilical points.

Problems

6.4.1. Provide the details for Example 6.4.6.

6.4.2. Consider the ellipsoid with half-axes a, b, and c.

    (a) Prove the ellipsoid has four umbilical points when a, b, and c are distinct.

    (b) Calculate the coordinates of the umbilical points when all half-axes have different length.

    (c) What happens when two of the half-axes are equal?

6.4.3. (ODE) Determine the asymptotic curves and the lines of curvature of the helicoid

$$\vec{X}(u, v) = (v \cos u, v \sin u, cu).$$

6.4.4. Let $\vec{X}$ be the parametrization for a neighborhood of S. Prove that if (g_{ij}) and (L_{ij}) are diagonal matrices, then the lines of curvature are the coordinate curves (curves on S where $u = $ const. or $v = $ const.).

6.4.5. Let $\vec{X}$ be the parametrization for a neighborhood of S. Prove that the coordinate curves are asymptotic curves if and only if $L_{11} = L_{22} = 0$.

6.4.6. (ODE) Determine the asymptotic curves of the catenoid

$$\vec{X}(u, v) = (\cosh v \cos u, \cosh v \sin u, v).$$

6.4.7. Consider Enneper's surface. Using the results of Problem 6.3.3, show the following:

(a) The principal curvatures are

$$\kappa_1 = \frac{2}{(1 + u^2 + v^2)^2}, \qquad \kappa_2 = -\frac{2}{(1 + u^2 + v^2)^2}.$$

(b) The lines of curvature are the coordinate curves.

(c) The asymptotic curves are $u + v =$const. and $u - v =$const.

6.4.8. Find equations for the lines of curvature when the surface is given by $z = f(x, y)$.

6.4.9. Consider the monkey saddle given by the graph of the function $z = x^3 - 3xy^2$. Prove that the set of asymptotic curves that possesses $(0, 0, 0)$ as a limit point consists of three straight lines through $(0, 0, 0)$ with equal angles between them.

6.4.10. Let S_1 and S_2 be two regular surfaces that intersect along a regular curve C. Let p be a point on C, and call λ_1 and λ_2 the normal curvatures of S at p in the direction of C. Prove that the curvature κ of C at p satisfies

$$\kappa^2 \sin^2 \theta = \lambda_1^2 + \lambda_2^2 - 2\lambda_1\lambda_2 \cos^2 \theta,$$

where θ is the angle between S_1 and S_2 at p (calculated using the normals to S_1 and S_2 at p).

6.4.11. Let S be a regular oriented surface and p a point on S. Two nonzero vectors $\vec{u}_1, \vec{u}_2 \in T_p S$ are called *conjugate* if

$$I_p(dn_p(\vec{u}_1), \vec{u}_2) = I_p(\vec{u}_1, dn_p(\vec{u}_2)) = 0. \qquad (6.36)$$

Prove the following:

(a) A curve C on S parametrized by $\vec{\gamma} : I \to S$ is a line of curvature if and only if the unit tangent vector $\vec{T}$ and any normal vector to $\vec{T}$ in $T_{\vec{\gamma}(t)}(S)$ are conjugate to each other.

(b) Let $\vec{\gamma}_1(t)$ and $\vec{\gamma}_2(t)$ be regular space curves; define a surface S by the parametrization $\vec{X}(u, v) = \vec{\gamma}_1(u) + \vec{\gamma}_2(v)$. Surfaces constructed in this manner are called *translation surfaces*. Show that the coordinate lines of S are conjugate lines.

6.5   Gaussian and Mean Curvature

We now arrive at two fundamental geometric invariants that encapsulate a considerable amount of useful information about the local shape of a surface. Again, in this section, we must assume that S is a regular surface of class C^2.

Definition 6.5.1. Let κ_1 and κ_2 be the principal curvatures of a regular surface S at a point p. Define

1. the *Gaussian curvature* of S at p as the product $K = \kappa_1\kappa_2$;

2. the *mean curvature* of S at p as the average $H = \frac{\kappa_1+\kappa_2}{2}$ of the principal curvatures.

By Problem 6.3.6, the matrix for the Gauss map in terms of any particular coordinate system on a neighborhood of p on S is conjugate to the corresponding matrix for a different coordinate system via the change of basis matrix $\left(\frac{\partial x_i}{\partial \bar{x}_j}\right)$. Therefore, since $\det(BAB^{-1}) = \det(A)$ and $\mathrm{Tr}(BAB^{-1}) = \mathrm{Tr}(A)$, which is a standard result in linear algebra, the eigenvalues of the Gauss map, the principal curvatures, the Gaussian curvature, and the mean curvature are invariant under coordinate changes. Furthermore, since $\det(-A) = \det(A)$ when A is a square matrix with an even number of rows, then

$$K = \kappa_1\kappa_2 = \det(a_i^j) = \det(dn_p),$$

$$H = \frac{\kappa_1 + \kappa_2}{2} = -\frac{1}{2}\mathrm{Tr}(a_i^j) = -\frac{1}{2}\mathrm{Tr}(dn_p).$$

Consequently, as claimed above, the Gaussian curvature and the mean curvature of S at p are geometric invariants. This means that they do not depend on the orientation or position of S in space, and they do not depend on any particular coordinate system on S in a neighborhood of the point p.

In order to calculate the mean curvature $H(u,v)$, one has no choice but to calculate the matrix of the differential of the Gauss map $(a_j^i) = [dn_p]$. However, from a computational perspective, Equation (6.26) leads to a much simpler formula for the Gaussian curvature function $K(u,v)$. Recall that $\det(AB) = \det(A)\det(B)$ for all square matrices of the same size and also that $\det(A^{-1}) = 1/\det(A)$. Then Equation (6.26) implies that

$$K = \frac{\det(L_{ij})}{\det(g_{ij})} = \frac{L_{11}L_{22} - L_{12}^2}{g_{11}g_{22} - g_{12}^2}. \tag{6.37}$$

This equation lends itself readily to calculations since one does not need to fully compute the matrix of the Gauss map, let alone find its eigenvalues. However, we can also obtain an alternate characterization of the Gaussian curvature.

Proposition 6.5.2. *Let S be a regular surface of class C^2, and let V be a neighborhood parametrized by $\vec{X} : U \subset \mathbb{R}^2 \to \mathbb{R}^3$. Define the unit normal vector $\vec{N}$ as*

$$\vec{N} = \frac{\vec{X}_u \times \vec{X}_v}{\|\vec{X}_u \times \vec{X}_v\|}.$$

Then over the domain U, the Gaussian curvature is the unique function $K(u, v)$ satisfying

$$\vec{N}_u \times \vec{N}_v = K(u, v)\vec{X}_u \times \vec{X}_v$$

Proof: From the definition of the (a_j^i) functions in Equation (6.22), we have

$$\vec{N}_u \times \vec{N}_v = (a_1^1 \vec{X}_u + a_1^2 \vec{X}_v) \times (a_2^1 \vec{X}_u + a_2^2 \vec{X}_v).$$

It is then easy to see that

$$\vec{N}_u \times \vec{N}_v = \det(a_j^i)\vec{X}_u \times \vec{X}_v.$$

The result follows since $K = \det(a_j^i)$. $\square$

Example 6.5.3 (Spheres). Consider the sphere parametrized by the vector function

$$\vec{X}(u, v) = (R\cos u \sin v, R\sin u \sin v, R\cos v),$$

with $(u, v) \in (0, 2\pi) \times (0, \pi)$. Example 6.1.5 and Example 6.3.3 gave us

$$(g_{ij}) = \begin{pmatrix} R^2\sin^2 v & 0 \\ 0 & R^2 \end{pmatrix} \quad \text{and} \quad (L_{ij}) = \begin{pmatrix} -R\sin^2 v & 0 \\ 0 & -R \end{pmatrix}.$$

Using Equation (6.37), it is then easy to see that the Gaussian curvature function on the sphere is

$$K(u, v) = \frac{R^2 \sin^2 v}{R^4 \sin^2 v} = \frac{1}{R^2},$$

which is a constant function. (Note that we assumed that $v \neq 0, \pi$, which correspond to the north and south poles of the parametrization. However, another parametrization that includes the north and south poles would show that at these points as well we have $K = 1/R^2$.)

The matrix of the Gauss map is then

$$[dn_p] = -Lg^{-1} = \begin{pmatrix} -\frac{1}{R} & 0 \\ 0 & -\frac{1}{R} \end{pmatrix},$$

from which one immediately deduces that the principal curvatures are $\kappa_1(u, v) = \kappa_2(u, v) = \frac{1}{R}$. This shows that all points on the sphere are umbilical points. In addition, the mean curvature is also a constant function $H(u, v) = \frac{1}{R}$.

Example 6.5.4 (Function Graphs). Consider the graph of a function $f : U \subset \mathbb{R}^2 \to \mathbb{R}$. The graph can be parametrized by $\vec{X} : U \to \mathbb{R}^3$, with

$$\vec{X}(u, v) = (u, v, f(u, v)).$$

Problem 6.3.7 asks the reader to calculate the matrix (L_{ij}). It is not hard to show that

$$(g_{ij}) = \begin{pmatrix} 1 + (f_u)^2 & f_u f_v \\ f_v f_u & 1 + (f_v)^2 \end{pmatrix} \quad \text{and} \quad (L_{ij}) = \frac{1}{\sqrt{1 + f_u^2 + f_v^2}} \begin{pmatrix} f_{uu} & f_{uv} \\ f_{vu} & f_{vv} \end{pmatrix},$$

where f_u is the typical shorthand to mean $\frac{\partial f}{\partial u}$. But then $\det(g) = 1 + f_u^2 + f_v^2$, and we find that the Gaussian curvature function on a function graph is

$$K(u, v) = \frac{1}{(1 + f_u^2 + f_v^2)^2} \begin{vmatrix} f_{uu} & f_{uv} \\ f_{vu} & f_{vv} \end{vmatrix} = \frac{f_{uu} f_{vv} - f_{uv}^2}{(1 + f_u^2 + f_v^2)^2}.$$

This result allows us to rephrase the second derivative test in the calculus of a function f from $\mathbb{R}^2$ to $\mathbb{R}$ as follows: if f has continuous second partial derivatives and (u_0, v_0) is a critical point (i.e., $f_u(u_0, v_0) = f_v(u_0, v_0) = 0$), then

1. (u_0, v_0) is a local maximum if $K(u_0, v_0) > 0$ and $f_{uu}(u_0, v_0) < 0$;

2. (u_0, v_0) is a local minimum if $K(u_0, v_0) > 0$ and $f_{uu}(u_0, v_0) > 0$;

3. (u_0, v_0) is a saddle point if $K(u_0, v_0) < 0$;

4. the test is inconclusive if $K(u_0, v_0) = 0$.

In the language we have introduced in this chapter, local minima and local maxima of the function $z = f(u, v)$ are elliptic points, saddle points are hyperbolic points, and points where the second derivative test is inconclusive are either parabolic or planar points.

Example 6.5.5 (Pseudosphere). A *tractrix* is a curve in the plane with parametric equations

$$\vec{x}(t) = (t - \tanh t, \operatorname{sech} t),$$

and it has the x-axis as an asymptote. The *pseudosphere* is defined as half of the surface of revolution of a tractrix about its asymptote. More precisely, we can parametrize the pseudosphere by

$$\vec{X}(u, v) = (\operatorname{sech} v \cos u, \operatorname{sech} v \sin u, v - \tanh v),$$

with $u \in [0, 2\pi)$ and $v \in [0, \infty)$. (See Figure 6.10 for a picture of the tractrix and the pseudosphere.) We leave it as an exercise for the reader to prove that

$$(g_{ij}) = \begin{pmatrix} \operatorname{sech}^2 v & 0 \\ 0 & \tanh^2 v \end{pmatrix} \quad \text{and} \quad (L_{ij}) = \begin{pmatrix} \operatorname{sech} v \tanh v & 0 \\ 0 & -\operatorname{sech} v \tanh v \end{pmatrix}.$$

We find that the Gaussian curvature for the pseudosphere is $K = -1$, which motivates the name "pseudosphere" since it is analogous to the unit sphere but with a constant Gaussian curvature of -1 instead of 1.

Proposition 6.5.6. *Let S be a regular oriented surface of class C^2 with Gauss map $n : S \to \mathbb{S}^2$, and let p be a point on S. Call $K(p)$ the Gaussian curvature of S at p, and suppose that $K(p) \neq 0$. Let B_ε be the ball of radius ε around p and define $V_\varepsilon = B_\varepsilon \cap S$. Calling $K(p)$ the Gaussian curvature of S at p, we have*

$$|K(p)| = \lim_{\varepsilon \to 0} \frac{\operatorname{Area}(n(V_\varepsilon))}{\operatorname{Area}(V_\varepsilon)}.$$

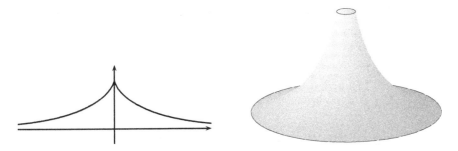

Figure 6.10. Tractrix and pseudosphere.

In other words, the absolute value of the Gaussian curvature at a point p is the limit around p of the ratio of the surface area on $\mathbb{S}^2$ mapped under the Gauss map to the corresponding surface area on S.

Proof: Let $\vec{X} : U \to \mathbb{R}^3$ be a regular parametrization of a neighborhood of $p = \vec{X}(u_0, v_0)$, where U is an open subset of $\mathbb{R}^2$. Since $\vec{X}(U)$ is an open set, $V_\varepsilon \subset \vec{X}(U)$ for all ε small enough. Furthermore, since we assumed that $K(p) \neq 0$, by the continuity of the Gaussian curvature function $K : U \to \mathbb{R}$, we deduce that K does not change sign for ε chosen small enough. Therefore, from now on, we assume that K does not change sign.

Define $U_\varepsilon = \vec{X}^{-1}(V_\varepsilon)$, the preimage of the neighborhood V_ε under $\vec{X}$. Since $\vec{X}$ is bijective as a regular parametrization, $\{(u_0, v_0)\}$ is the unique point in all U_ε for all $\varepsilon > 0$. Then by the formula for surface area,

$$\text{Area}(V_\varepsilon) = \iint_{V_\varepsilon} dA = \iint_{U_\varepsilon} \|\vec{X}_u \times \vec{X}_v\| \, du \, dv.$$

Similarly, on the unit sphere,

$$\text{Area}(n(V_\varepsilon)) = \iint_{U_\varepsilon} \|\vec{N}_u \times \vec{N}_v\| \, du \, dv.$$

However, by Proposition 6.5.2, we also have

$$\text{Area}(n(V_\varepsilon)) = \iint_{U_\varepsilon} |K(u, v)| \|\vec{X}_u \times \vec{X}_v\| \, du \, dv.$$

By the Mean Value Theorem for double integrals, for every $\varepsilon > 0$, there exist points $(u_\varepsilon, v_\varepsilon)$ and $(u'_\varepsilon, v'_\varepsilon)$ in the open set U_ε such that

$$\iint_{U_\varepsilon} \|\vec{X}_u \times \vec{X}_v\| \, du \, dv = \|\vec{X}_u(u_\varepsilon, v_\varepsilon) \times \vec{X}_v(u_\varepsilon, v_\varepsilon)\|$$

and

$$\iint_{U_\varepsilon} \|\vec{N}_u \times \vec{N}_v\| \, du \, dv = |K(u'_\varepsilon, v'_\varepsilon)| \, \|\vec{X}_u(u'_\varepsilon, v'_\varepsilon) \times \vec{X}_v(u'_\varepsilon, v'_\varepsilon)\|.$$

Thus, since $\lim_{\varepsilon \to 0}(u_\varepsilon, v_\varepsilon) = \lim_{\varepsilon \to 0}(u'_\varepsilon, v'_\varepsilon) = (u_0, v_0)$, we have

$$\lim_{\varepsilon \to 0} \frac{\mathrm{Area}(n(V_\varepsilon))}{\mathrm{Area}(V_\varepsilon)} = \lim_{\varepsilon \to 0} \frac{|K(u'_\varepsilon, v'_\varepsilon)| \, \|\vec{X}_u(u'_\varepsilon, v'_\varepsilon) \times \vec{X}_v(u'_\varepsilon, v'_\varepsilon)\|}{\|\vec{X}_u(u_\varepsilon, v_\varepsilon) \times \vec{X}_v(u_\varepsilon, v_\varepsilon)\|}$$

$$= \frac{|K(u_0, v_0)| \, \|\vec{X}_u(u_0, v_0) \times \vec{X}_v(u_0, v_0)\|}{\|\vec{X}_u(u_0, v_0) \times \vec{X}_v(u_0, v_0)\|}$$

$$= |K(u_0, v_0)| = |K(p)|. \qquad \square$$

Problems

6.5.1. Calculate the Gaussian curvature of the torus parametrized by

$$\vec{X}(u, v) = ((a + b\cos v)\cos u, (a + b\cos v)\sin u, b\sin v)$$

where $a > b$ are constants and $(u, v) \in (0, 2\pi) \times (0, 2\pi)$.

6.5.2. Consider a regular space curve $\vec{\gamma} : I \to \mathbb{R}^3$ and the tangential surface defined by

$$\vec{X}(t, u) = \vec{\gamma}(t) + u\vec{\gamma}'(t)$$

for $(t, u) \in I \times \mathbb{R}$. Calculate the mean and Gaussian curvature functions.

6.5.3. Let $\vec{\alpha}(t)$ and $\vec{\beta}(t)$ be differentiable vector functions with common domain I. Define the *secant surface* between the two resulting curves by

$$\vec{X}(t, u) = (1 - u)\vec{\alpha}(t) + u\vec{\beta}(t)$$

for $(t, u) \in I \times \mathbb{R}$. Assume that the corresponding surface is regular.

(a) Prove that $K(u,v) = 0$ for any point with $u = \frac{1}{2}$.

(b) Prove that $K(u,v) = 0$ if and only if $u = \frac{1}{2}$ or $(\vec{\beta}(t) - \vec{\alpha}(t))$ is in the plane spanned by $\vec{\alpha}'(t)$ and $\vec{\beta}'(t)$.

6.5.4. If M is a nonorientable surface, one can define the Gaussian curvature on a coordinate patch of M by using $\vec{N} = \vec{X}_u \times \vec{X}_v / \|\vec{X}_u \times \vec{X}_v\|$ and Equation (6.37) without reference to dn_p, which is not well defined since no Gauss map $n : M \to \mathbb{S}^2$ exists. Consider the Möbius strip M depicted in Figure 5.14. Show that, using the parametrization of the Möbius strip in Example 5.5.3, the Gaussian curvature is

$$K = -\frac{1}{\left(\frac{1}{4}v^2 + (2 - v\sin\frac{u}{2})^2\right)^2}.$$

6.5.5. *Tubes.* Let $\vec{\gamma} : I \to \mathbb{R}^3$ be a regular space curve and let r be a positive real number. Consider the tube of radius r around $\vec{\gamma}(t)$ parametrized by

$$\vec{X}(u,v) = \vec{\gamma}(u) + (r\cos v)\vec{P}(u) + (r\sin v)\vec{B}(u).$$

(a) Calculate the second fundamental form (L_{ij}) and the matrix for dn_p.

(b) Calculate the Gaussian curvature function $K(u,v)$ on the tube.

(c) Prove that all the points with $K = 0$ are either points on curves $\vec{\gamma}(t) \pm r\vec{B}(t)$ for all $t \in I$ or points on circles $\vec{\gamma}(u_0) + (r\cos v)\vec{P}(u_0) + (r\sin v)\vec{B}(u_0)$, where u_0 is a value where $\kappa(u_0) = 0$.

6.5.6. Consider the pseudosphere and the parametrization provided in Example 6.5.5.

(a) Prove the statements about the pseudosphere in Example 6.5.5.

(b) Find the mean curvature function on the pseudosphere.

(c) Modify the given parametric equations to find a parametrization of a surface with constant Gaussian curvature $K = -\frac{1}{R^2}$.

(d) Determine the lines of curvature on the pseudosphere.

6.5.7. Consider the monkey saddle $\vec{X}(u,v) = (u,v,u^3 - 3uv^2)$. Calculate the Gaussian curvature function of $\vec{X}$, and indicate the points where $K > 0$, $K < 0$, or $K = 0$.

6.5.8. *Surfaces of revolution.* Consider the surface of revolution about the z-axis given by the parametrization $\vec{X}(u,v) = (f(v)\cos u, f(v)\sin u, g(v))$ for $u \in [0, 2\pi)$ and $v \in I$, where I is some interval.

(a) Calculate the second fundamental form (L_{ij}).

(b) Calculate the coefficients of the matrix for dn_p.

(c) Calculate the Gaussian curvature function.

(d) Determine which points are elliptic, hyperbolic, parabolic, or planar.

(e) Prove that the lines of curvature are the coordinate lines.

6.5.9. *Normal Variations.* Let $\vec{X}(u, v)$ be the parametrization for a coordinate patch V of a regular surface S. Consider the normal variation of S over V that is parametrized by

$$\vec{Y_r}(u, v) = \vec{X}(u, v) + r\vec{N}(u, v)$$

for some constant $r \in \mathbb{R}$ and where $\vec{N}$ is the unit normal vector associated to $\vec{X}$.

(a) Prove that the unit normal $\vec{N}_Y$ to $\vec{Y}$ is everywhere equal to $\vec{N}$ (except perhaps up to a sign).

(b) Call K_Y the Gaussian curvature for $\vec{Y_r}$ and K the Gaussian curvature of $\vec{X}$. Prove that

$$K(\vec{X}_u \times \vec{X}_v) = K_Y \frac{\partial \vec{Y_r}}{\partial u} \times \frac{\partial \vec{Y_r}}{\partial v}.$$

(c) Call V_Y the corresponding coordinate patches on the normal variation. Conclude that

$$\iint_V K \, dA = \iint_{V_Y} K_Y \, dA.$$

6.5.10. Consider two plane curves $\vec{\alpha}(t) = (\alpha_1(s), \alpha_2(s))$ and $\vec{\beta}(t) = (\beta_1(t), \beta_2(t))$, both parametrized by arc length. Assume we are in $\mathbb{R}^3$ with standard basis $\{\vec{i}, \vec{j}, \vec{k}\}$. Consider now the parametrized surface S given by

$$\vec{X}(s, t) = \vec{\alpha}(s) + \beta_1(t)\vec{N}(s) + \beta_2(t)\vec{k},$$

where $\vec{N}(s)$ is the usual unit normal vector for plane curves.

(a) Prove that if S is regular, then $1 - \beta_1(t)\kappa_\alpha(s) \neq 0$ for all (s, t), where $\kappa_\alpha(s)$ is the curvature of the plane curve $\vec{\alpha}$.

(b) Calculate the Gaussian curvature of S.

(c) Prove that $\kappa_\alpha(s) = 0$ or $\kappa_\beta(t) = 0$ imply that $K(s, t) = 0$, but explain why the converse is not true.

6.5.11. Let S be a regular oriented surface and p a point of S. Let $\vec{u}$ be any fixed unit vector in T_pS. Show that the mean curvature H at p is given by

$$H = \frac{1}{\pi} \int_0^\pi \kappa_n(\theta)\, d\theta,$$

where $\kappa_n(\theta)$ is the normal curvature of S along a direction making and angle θ with $\vec{u}$.

6.5.12. (*) *Theorem of Beltrami-Enneper.* Prove that the absolute value of the torsion at any point on an asymptotic curve with nonzero curvature is given by

$$|\tau| = \sqrt{-K},$$

where K is the Gaussian curvature of the surface at that point.

6.5.13. Let S be a regular surface in $\mathbb{R}^3$ parametrized by $\vec{X}$, and consider the linear transformation $T : \mathbb{R}^3 \to \mathbb{R}^3$ given by $T(\vec{x}) = A\vec{x}$ with respect to the standard basis, where A is an invertible matrix. It is usually an intractable problem to determine the Gaussian or mean curvature of the image surface $S' = T(S)$ from those of S. Nonetheless, it is possible to answer the following question: prove that T preserves the sign of the Gaussian curvature of any surface in $\mathbb{R}^3$, more precisely, if $p \in S$ and $q = T(p)$ is the corresponding point on S', then $K(p) = 0 \Leftrightarrow K(q) = 0$ and $\operatorname{sign} K(p) = \operatorname{sign} K(q)$. [Hint: Use Equation (6.17).]

6.5.14. (*) Consider a regular surface $S \in \mathbb{R}^3$ defined by $F(x, y, z) = 0$ and a point p on S. Use implicit differentiation to find a formula for Gaussian curvature K at the point p. (The formula is not particularly pretty but can be written concisely using the notation that will be introduced in Section 7.1.)

6.6 Ruled Surfaces and Minimal Surfaces

When studying plane curves, we showed in Proposition 1.3.4 that if the curvature $\kappa_g(t)$ of a regular plane curve is always 0, then the curve is a line segment. In this section, we wish to study properties of surfaces with either Gaussian curvature everywhere 0 or mean curvature everywhere 0. Since there exist formulas for the mean curvature and Gaussian curvature in terms of a particular parametrization, the equations $K = 0$ and $H = 0$ are partial differential equations. However, since they are nonlinear differential equations that a priori involve three unknown functions in two variables, they are intractable in general. Nonetheless, we shall study various classes of surfaces that satisfy $K = 0$ or $H = 0$.

Planes satisfy $K = 0$, but other surfaces do as well, for example, cylinders and cones. In a geometric sense, cylinders and cones in fact resemble a plane because, as any elementary school student knows, one can create a cylinder or a cone out of a flat paper without folding, stretching, or crumpling. In the first half of this section, we introduce ruled surfaces and determine the conditions for these ruled surfaces to have Gaussian curvature everywhere 0.

As of yet, we have neither seen a particularly intuitive interpretation of the mean curvature nor have presented surfaces that satisfy $H = 0$. However, as we shall see, surfaces that satisfy $H = 0$ have minimal surface area in the following sense. Given a curve C in $\mathbb{R}^3$, among surfaces S that have $C = \partial S$ as a boundary, a surface with minimal surface area will have mean curvature H everywhere 0. Consequently, a surface that satisfies $H = 0$ is called a *minimal surface*. Many articles and books are devoted to the study of minimal surfaces (see [20], [26], or [25] to name a few; an internet search will reveal many more), so in the interest of space, the second half of this section gives only a brief introduction to minimal surfaces.

6.6.1   Ruled Surfaces

Geometrically, we define a *ruled surface* as the union of a (differentiable) one-parameter family of straight lines in $\mathbb{R}^3$. One can specify each line in the family by a point $\vec{\alpha}(t)$ and by a direction given by another vector $\vec{w}(t)$. The adjective "differentiable" means that both $\vec{\alpha}$ and $\vec{w}$ are differentiable vector functions over some interval $I \subset \mathbb{R}$. We parametrize a ruled surface by

$$\vec{X}(t, u) = \vec{\alpha}(t) + u\,\vec{w}(t), \qquad \text{with } (t, u) \in I \times \mathbb{R}. \qquad (6.38)$$

We call the lines L_t passing through $\vec{\alpha}(t)$ with direction $\vec{w}(t)$ the *rulings*, and the curve $\vec{\alpha}(t)$ is called the *directrix* of the surface. We should note that this definition does not insist that ruled surfaces be regular; in particular, we allow singular points, that is, points where $\vec{X}_t \times \vec{X}_u = \vec{0}$.

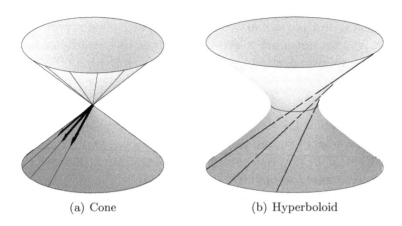

(a) Cone   (b) Hyperboloid

Figure 6.11. Ruled surfaces.

Example 6.6.1 (Cylinders). The simplest example of a ruled surface is a cylinder. The general definition of a cylinder is a ruled surface that can be given as a one-parameter family of lines where the directrix $\vec{\alpha}(t)$ is planar and $\vec{w}(t)$ is a constant vector in $\mathbb{R}^3$.

Example 6.6.2 (Cones). The general definition of a cone is a surface that can be given as a one-parameter family of lines where the directrix $\vec{\alpha}(t)$ lies in a plane P and the rulings L_t all pass through some common point $p \notin P$. Therefore, the cone over $\vec{\alpha}(t)$ through p can be parametrized by

$$\vec{X}(t, u) = \vec{\alpha}(t) + u(p - \vec{\alpha}(t)).$$

Example 6.6.3. As a perhaps initially surprising example, the hyperboloid of one sheet is also a ruled surface. The standard parametrization for the hyperboloid of one sheet is

$$\vec{X}(u, v) = (\cosh v \cos u, \cosh v \sin u, \sinh v).$$

Consider alternatively the ruled surface given by the directrix $\vec{\alpha}(t) = (\cos t, \sin t, 0)$, with ruling directions $\vec{w}(t) = \vec{\alpha}'(t) + \vec{k} = (-\sin t, \cos t, 1)$ (see Figure 6.11(b)). We obtain the parametrized surface

$$\vec{Y}(t, u) = (\cos t - u \sin t, \sin t + u \cos t, u).$$

Though it is not particularly easy to express $\vec{Y}$ as a reparametrization of $\vec{X}$, it is not hard to see that both of these parametrizations satisfy the following usual equation that gives the hyperboloid as a conic surface:

$$x^2 + y^2 - z^2 = 1.$$

We will determine the Gaussian curvature for ruled surfaces, but, for the purposes of subsequent calculations, we make two simplifications. First, note that one can impose the condition that $\|\vec{w}(t)\| = 1$ without changing the definition or the image of the parametrization in Equation (6.38). In defining particular ruled surfaces, it is usually easier not to make this assumption, but it does simplify calculations since $\|\vec{w}(t)\| = 1$ for all $t \in I$ implies that $\vec{w}(t) \cdot \vec{w}(t)' = 0$. Secondly, note that different curves $\vec{\alpha}(t)$ can serve as the directrix for the same ruled surface, so we wish to employ a curve $\vec{\beta}(t)$ as a directrix, which will simplify the algebra in our calculations. We choose a curve $\vec{\beta}(t)$ satisfying $\vec{\beta}'(t) \cdot \vec{w}'(t) = 0$.

Since $\vec{\beta}(t)$ lies on $\vec{X}$, we can write

$$\vec{\beta}(t) = \vec{\alpha}(t) + u(t)\vec{w}(t). \tag{6.39}$$

Then

$$\vec{\beta}'(t) = \vec{\alpha}'(t) + u'(t)\vec{w}(t) + u(t)\vec{w}'(t),$$

and since $\vec{\beta}' \cdot \vec{w}' = 0$, we have

$$\vec{\alpha}'(t) \cdot \vec{w}'(t) + u(t)\vec{w}'(t) \cdot \vec{w}'(t) = 0.$$

Thus, we determine that

$$u(t) = -\frac{\vec{\alpha}'(t) \cdot \vec{w}'(t)}{\|\vec{w}'(t)\|^2}. \tag{6.40}$$

Furthermore, it is easy to prove (see Problem 6.6.3) that, with our present assumptions, $\vec{\beta}(t)$ is unique. We call the curve $\vec{\beta}(t)$ the *line of stricture* of the ruled surface.

We now use the parametrization for the ruled surface $\vec{X} = \vec{\beta}(t) + u\vec{w}(t)$ and proceed to calculate the first and second fundamental forms. Understanding that $\vec{\beta}$ and $\vec{w}$ are functions of t, we find that

$$\vec{X}_t = \vec{\beta}' + u\,\vec{w}', \qquad\qquad \vec{X}_{tt} = \vec{\beta}'' + u\,\vec{w}'',$$
$$\vec{X}_u = \vec{w}, \qquad \text{and} \qquad \vec{X}_{tu} = \vec{w}',$$
$$\vec{X}_t \times \vec{X}_u = \vec{\beta}' \times \vec{w} + u\,\vec{w}' \times \vec{w}, \qquad\qquad \vec{X}_{uu} = \vec{0}.$$

One can already notice that if $\vec{w}(t)$ is constant, then $L_{12} = L_{21} = L_{22} = 0$, which leads to $K = 0$ and proves that all cylinders have Gaussian curvature $K = 0$. We will assume now that $\vec{w}'(t)$ is not identically 0.

Because of the conditions $\vec{w} \cdot \vec{w}' = 0$ and $\vec{\beta}' \cdot \vec{w}' = 0$, we can conclude that $\vec{\beta}' \times \vec{w}$ is parallel to $\vec{w}'$. Thus, $\vec{\beta}' \times \vec{w}$ is perpendicular to $\vec{w} \times \vec{w}'$, so

$$\|\vec{X}_t \times \vec{X}_u\|^2 = \|\vec{\beta}' \times \vec{w}\|^2 + u^2\|\vec{w}'\|^2.$$

Furthermore, $\vec{\beta}' \times \vec{w} = (\vec{\beta}' \times \vec{w}) \cdot \vec{w}'/\|\vec{w}'\|$ and hence,

$$\|\vec{X}_t \times \vec{X}_u\|^2 = (\vec{\beta}'\vec{w}\vec{w}')^2/\|\vec{w}'\|^2 + u^2\|\vec{w}'\|^2 \tag{6.41}$$

$$= \frac{1}{\|\vec{w}'\|^2}\left((\vec{\beta}'\vec{w}\vec{w}')^2 + u^2\|\vec{w}'\|^4\right), \tag{6.42}$$

where we use the notation $(\vec{\beta}'\vec{w}\vec{w}')$ for $(\vec{\beta}' \times \vec{w}) \cdot \vec{w}'$, the triple-vector product in $\mathbb{R}^3$. Consequently, we can write the unit normal vector $\vec{N}$ as

$$\vec{N}(t,u) = \frac{\|\vec{w}'\|}{\sqrt{(\vec{\beta}'\vec{w}\vec{w}')^2 + u^2\|\vec{w}'\|^4}}\left(\vec{\beta}' \times \vec{w} + u\,\vec{w}' \times \vec{w}\right), \tag{6.43}$$

and therefore, again using the conditions that $\vec{w}\cdot\vec{w}' = 0$ and $\vec{\beta}'\cdot\vec{w}' = 0$, we get

$$(g_{ij}) = \begin{pmatrix} \|\vec{\beta}'\|^2 + u^2\|\vec{w}'\|^2 & 0 \\ 0 & 1 \end{pmatrix} \tag{6.44}$$

and

$$(L_{ij}) = \frac{\|\vec{w}'\|}{\sqrt{(\vec{\beta}'\vec{w}\vec{w}')^2 + u^2\|\vec{w}'\|^4}}\begin{pmatrix} (\vec{\beta}'' + u\vec{w}'') \cdot (\vec{\beta}' \times \vec{w} + u\vec{w}' \times \vec{w}) & (\vec{\beta}'\vec{w}\vec{w}') \\ (\vec{\beta}'\vec{w}\vec{w}') & 0 \end{pmatrix}. \tag{6.45}$$

We cannot say much for the entry L_{11}, but thanks to Equation (6.37) for the Gaussian curvature and the fact that $\|\vec{X}_u \times \vec{X}_v\|^2 = \det(g_{ij})$ (see Equation (6.6)), one can calculate the Gaussian curvature for a ruled surface as

$$K = \frac{\det(L_{ij})}{\det(g_{ij})} = -\frac{\|\vec{w}'\|^4(\vec{\beta}'\vec{w}\vec{w}')^2}{\left((\vec{\beta}'\vec{w}\vec{w}')^2 + u^2\|\vec{w}'\|^4\right)^2}. \tag{6.46}$$

This formula for the Gaussian curvature of a ruled surface makes a few facts readily apparent. First, $K \leq 0$ for all points of a ruled surface. Secondly, by Equation (6.41), all of the singular points of a ruled surface, i.e., points where $\vec{X}_t \times \vec{X}_u = \vec{0}$, must have $u = 0$ and therefore occur on the line of stricture. Finally, we see that a noncylindrical ruled surface satisfies $K(t,u) = 0$ if and only if $(\vec{\beta}'\vec{w}\vec{w}') = 0$ for all $t \in I$.

Let us return now to the general definition of a ruled surface from Equation (6.38) with regular curves $\vec{\alpha}(t)$ and $\vec{w}(t)$ with no conditions. Since both $\vec{w}$ and $\vec{w}'$ are perpendicular to $\vec{w} \times \vec{w}'$,

$$(\vec{\alpha}'\vec{w}\vec{w}') = \vec{\alpha}' \cdot (\vec{w} \times \vec{w}') = \vec{\beta}' \cdot (\vec{w} \times \vec{w}') = (\vec{\beta}'\vec{w}\vec{w}').$$

Furthermore, if we call $\hat{w} = \vec{w}/\|\vec{w}\|$, we remark that

$$(\vec{\alpha}'\vec{w}\vec{w}') = \|w\|^2 (\vec{\alpha}'\hat{w}\hat{w}') = \|w\|^2 (\vec{\beta}'\hat{w}\hat{w}').$$

Consequently, $(\vec{\beta}'\hat{w}\hat{w}') = 0$ if and only if $(\vec{\alpha}'\vec{w}\vec{w}') = 0$.

The ruled surface $\vec{X}(t,u)$ is called a *developable surface* if $(\vec{\alpha}'\vec{w}\vec{w}') = 0$. Equation (6.46) shows that all developable surfaces have Gaussian curvature identically 0. This result provides us with a class of surfaces that satisfy $K = 0$, though it by no means proves whether $K = 0$ implies that a surface is developable.

It is interesting to remark that a ruled surface is a cone if and only if $\vec{\alpha}'(t) = 0$ and is a cylinder if and only if $\vec{w}'(t) = 0$, showing again that both cones and cylinders are developable and have $K = 0$. The exercises present examples of developable surfaces that are neither cones nor cylinders.

6.6.2   Minimal Surfaces

Definition 6.6.4. A *minimal surface* is a parametrized surface of class C^2 that satisfies the regularity condition and for which the mean curvature is identically 0.

We first wish to justify the name "minimal." Let $\vec{X} : U \to \mathbb{R}^3$ be a coordinate neighborhood of a regular parametrized surface of class C^2. Let D' be a connected compact set in U, and let $D = \vec{X}(D')$. Let $h : U \to \mathbb{R}$ be a differentiable function. A normal variation of

$\vec{X}$ over D' determined by h is the family of surfaces with $t \in (-\varepsilon, \varepsilon)$ defined by

$$\vec{X}^t : D' \longrightarrow \mathbb{R}^3$$

$$(u, v) \longmapsto \vec{X}(u, v) + th(u, v)\vec{N}(u, v). \qquad (6.47)$$

Proposition 6.6.5. *Let $\vec{X}^t$ be a normal variation of $\vec{X}$ over a compact region D' and determined by some function h. For ε small enough, $\vec{X}^t$ satisfies the regularity condition for all $t \in (-\varepsilon, \varepsilon)$. In this case, the area of $\vec{X}^t$ is*

$$A(t) = \iint_{D'} \sqrt{1 - 4thH + t^2 R} \sqrt{\det(g_{ij})} \, du \, dv \qquad (6.48)$$

for some function $R(u, v, t)$ that is polynomial in t.

Proof: Denote g_{ij}^t as the coefficients of the metric tensor. The surfaces of the normal variation have

$$\vec{X}_u^t = \vec{X}_u + th_u\vec{N} + th\vec{N}_u,$$
$$\vec{X}_v^t = \vec{X}_v + th_v\vec{N} + th\vec{N}_v.$$

Thus, we calculate,

$$g_{11}^t = g_{11} + 2th\vec{X}_u \cdot \vec{N}_u + t^2 h^2 \vec{N}_u \cdot \vec{N}_u + t^2 h_u^2,$$
$$g_{12}^t = g_{12} + th(\vec{X}_u \cdot \vec{N}_v + \vec{X}_v \cdot \vec{N}_u) + t^2 h^2 \vec{N}_u \cdot \vec{N}_v + t^2 h_u h_v,$$
$$g_{22}^t = g_{22} + 2th\vec{X}_v \cdot \vec{N}_v + t^2 h^2 \vec{N}_v \cdot \vec{N}_v + t^2 h_v^2.$$

However, by Equation (6.16), one can summarize the above equations as

$$g_{ij}^t = g_{ij} - 2thL_{ij} + t^2 h^2 \vec{N}_i \cdot \vec{N}_j + t^2 h_i h_j,$$

where we use the notation h_1 (respectively h_2) to indicate the partial derivative h_u (respectively h_v). Therefore, we calculate that

$$\det(g_{ij}^t) = \det(g_{ij}) - 2th(g_{11}L_{22} - 2g_{12}L_{12} + g_{22}L_{11}) + t^2 \bar{R},$$

where $\bar{R}(u, v, t)$ is a function of the form $A_0(u, v) + tA_1(u, v) + t^2 A_2(u, v)$ for continuous functions A_i defined over D'. However, the mean curvature is

$$H = \frac{1}{2} \frac{g_{11}L_{22} - 2g_{12}L_{12} + g_{22}L_{11}}{g_{11}g_{22} - g_{12}^2},$$

which leads to

$$\det(g_{ij}^t) = \det(g_{ij})(1 - 4thH) + t^2\bar{R} = \det(g_{ij})(1 - 4thH + t^2R),$$
(6.49)

where $R = \bar{R}/\det(g_{ij})$. Since D' is compact, the functions h, H, and R are bounded over D', which shows that

$$\lim_{t \to 0} \det(g_{ij}^t) = \det(g_{ij}) \quad \text{for all } (u,v) \in D'.$$

Hence, if $\varepsilon > 0$ is small enough, then $\det(g_{ij}^t) \neq 0$ for all $t \in (-\varepsilon, \varepsilon)$, and thus, all normal variations satisfy the regularity condition. The rest of the proposition follows from Equation (6.49) since

$$A(t) = \iint\limits_{D'} \sqrt{\det(g_{ij}^t)} \, du \, dv.$$

$\square$

Proposition 6.6.6. *Let $\vec{X} : U \to \mathbb{R}^3$ be a parametrized surface of class C^2, and let $D' \in U$ be a compact set. Let $A(t)$ be the area function defined in Equation (6.48). Then $\vec{X}$ parametrizes a minimal surface if and only if $A'(0) = 0$ for all D' and all normal variations of $\vec{X}$ over D'.*

Proof: We calculate

$$A'(t) = \iint\limits_{D'} \frac{-4hH + 2tR + t^2 R_t}{2\sqrt{1 - 4thH + t^2R}} \sqrt{\det(g_{ij})} \, du \, dv,$$

which implies that

$$A'(0) = -2 \iint\limits_{D'} hH \sqrt{\det(g_{ij})} \, du \, dv.$$

Obviously, if $\vec{X}$ parametrizes a minimal surface, which means that $H = 0$ for all $(u,v) \in U$, then $A'(0) = 0$ regardless of the function $h(u,v)$ or the compact set D'.

To prove the converse, suppose that $A'(0) = 0$ for all continuous functions $h(u,v)$ and all compact $D' \subset U$. Choosing $h(u,v) = H(u,v)$, since $\det(g_{ij}) > 0$, we have

$$A'(0) = -2 \iint\limits_{D'} H^2 \sqrt{\det(g_{ij})} \, du \, dv,$$

so $A'(0) \leq 0$ for all choices of D'. Since $H(u,v)$ is continuous, if $H(u_0, v_0) \neq 0$ for any point (u_0, v_0), there is a compact set D' containing (u_0, v_0) such that $H(u,v)^2 > 0$ over D'. Thus, if H is anywhere nonzero, there exists a compact subset D' such that $A'(0) < 0$. This is a contradiction, so we conclude that $A'(0) = 0$ for all h and all D' implies that $H(u,v) = 0$ for all $(u,v) \in U$. $\square$

Proposition 6.6.6 shows that a parametrized surface $\vec{X}$ that satisfies the regularity condition $\|\vec{X}_u \times \vec{X}_u\| \neq 0$ is minimal (has $H = 0$ everywhere) if it is a surface such that over every patch of surface there is no way to deform it along normal vectors to obtain a surface of lesser surface area.

Minimal surfaces have enjoyed considerable popular attention, especially in museums of science, as soap films on a wire frame. When a wire frame is dipped into a soapy liquid and pulled out, the surface tension on the soap film pulls the film into the state of least potential energy, which turns out to be the surface such that no normal variation can lead to a surface with less surface area. Of course, when experimenting with soap films, one can experiment with wire frames that are nonregular curves or not even curves at all (the skeleton of a cube, for example) and investigate what kinds of soap film surfaces result.

The problem of determining a minimal surface with a given closed regular space curve C as a boundary was raised by Lagrange in 1760. However, the mathematical problem became known as Plateau's problem, named after Joseph Plateau who specifically studied soap film surfaces. The nineteenth century saw a few specialized solutions to the problem, but it was not solved until 1930. (See [25] for a more complete history of Plateau's problem.)

Though one can easily construct a minimal surface with a soap film on a wire frame, either checking that a surface is minimal or finding a parametrization for a minimal surface is quite difficult. In fact, the study of minimal surfaces continues to provide new areas of research and connections with other branches of analysis. Perhaps among the most interesting results is a connection between complex analytic functions and minimal surfaces [10, p. 206] or the use of elliptic integrals to parametrize special minimal surfaces [25, Chapter 1], topics that go beyond the scope of this book. However, it is quite likely that having both the simple experimentative aspect and connections to advanced mathematics has fueled interest in minimal surfaces.

Problems

6.6.1. Prove that the surface given by $z = kxy$, where k is a constant, is a ruled surface and give it as a one-parameter family of lines.

6.6.2. The tangential surface to a space curve $\vec{\alpha}(t)$ was presented in Problem 6.5.2. Show that the curve $\vec{\alpha}$ is the line of stricture for the tangential surface. Show that the tangential surface to a regular space curve is a developable surface.

6.6.3. Suppose that $\{\vec{\alpha}_1(t), \vec{w}(t)\}$ and $\{\vec{\alpha}_2(t), \vec{w}(t)\}$, where $\|\vec{w}(t)\| = 1$, are two one-parameter families of lines that trace out the same ruled surface. Prove that Equation (6.39) with, (6.40), using either $\vec{\alpha}_1$ or $\vec{\alpha}_2$, produce the same line of stricture.

6.6.4. Let S be an orientable surface, and let $\vec{\alpha}(s)$ be a curve on S parametrized by arc length. Assume that $\vec{\alpha}$ is nowhere tangent to the asymptotic direction on S, and define $\vec{N}(s)$ as the unit normal vector to S along $\vec{\alpha}(s)$. Consider the ruled surface

$$\vec{X}(s, u) = \vec{\alpha}(s) + u \frac{\vec{N}(s) \times \vec{N}'(s)}{\|\vec{N}'(s)\|}.$$

The assumption that $\vec{\alpha}'(s)$ is not an asymptotic direction ensures that $\vec{N}'(s) \neq \vec{0}$. Prove that $\vec{X}(s, v)$ is a developable surface. (This kind of surface is called the *envelope* of a family of tangent planes along a curve of a surface.)

6.6.5. Show that a surface $F(x, y, z) = 0$ is developable if

$$\begin{vmatrix} F_{xx} & F_{xy} & F_{xz} & F_x \\ F_{yx} & F_{yy} & F_{yz} & F_y \\ F_{zx} & F_{zy} & F_{zz} & F_z \\ F_x & F_y & F_z & 0 \end{vmatrix} = 0.$$

6.6.6. Let $\vec{X}(t, v) = \vec{\alpha}(t) + v\vec{w}(t)$ be a developable surface. Prove that at a regular point

$$\vec{N}_v \cdot \vec{X}_t = 0 \qquad \text{and} \qquad \vec{N}_v \cdot \vec{X}_v = 0.$$

Use this result to prove that the tangent plane of a developable surface is constant along a line of ruling.

6.6.7. Consider the helicoid parametrized by $\vec{X}(u, v) = (v\cos u, v\sin u, cu)$, where c is a fixed number.

    (a) Determine the asymptotic curves.

    (b) Determine the lines of curvature.

    (c) Show that the mean curvature of the helicoid is 0.

6.6.8. Prove that the catenoid parametrized by $\vec{X}(u, v) = (a\cosh v\cos u, a\cosh v\sin u, av)$ is a minimal surface.

6.6.9. Prove that Enneper's minimal surface $\vec{X}(u, v) = (u - \frac{u^3}{3} + uv^2, v - \frac{v^3}{3} + vu^2, u^2 - v^2)$ deserves its name.

6.6.10. Show that the minimal surface with the space curve $\vec{\alpha}(t) = (\cos t, \sin t, \sin 2t)$ as its boundary is the surface given by $z = 2xy$.

6.6.11. Let $U \subset \mathbb{R}^2$, and suppose that $\vec{X} : U \to \mathbb{R}^3$ and $\vec{Y} : U \to \mathbb{R}^3$ parametrize two minimal surfaces defined over the same domain. Prove that for all $t \in [0, 1]$, the surfaces parametrized by $\vec{Z}_t(u, v) = (1 - t)\vec{X}(u, v) + t\vec{Y}(U, v)$ are also minimal surfaces.

6.6.12. (ODE) Prove that the *only* surface of revolution that is a minimal surface is a catenoid (see Problem 6.6.8).

6.6.13. Suppose that S is a minimal regular surface with no planar points. Prove that at all points $p \in S$, the Gauss map $n : S \to \mathbb{S}^2$ satisfies

$$dn_p(\vec{w}_1) \cdot dn_p(\vec{w}_2) = -K(p)\vec{w}_1 \cdot \vec{w}_2$$

for all $\vec{w}_1, \vec{w}_2 \in T_pS$, where $K(p)$ is the Gaussian curvature of S at p. Use this result to show that on a minimal surface the angle between two intersecting curves on S is the angle between their images on $\mathbb{S}^2$ under the Gauss map n.

6.6.14. Let $\vec{X}(u, v)$ be a parametrization of a regular, orientable surface, and let $\vec{N}(u, v) = \vec{X}_u \times \vec{X}_v / \|\vec{X}_u \times \vec{X}_v\|$ be the orientation. A *parallel surface* to $\vec{X}$ is a surface parametrized by

$$\vec{Y}(u, v) = \vec{X}(u, v) + a\vec{N}(u, v).$$

    (a) If K and H are the Gaussian and mean curvatures of $\vec{X}$, prove that

$$\vec{Y}_u \times \vec{Y}_v = (1 - 2Ha + Ka^2)\vec{X}_u \times \vec{X}_v.$$

(b) Prove that at regular points, the Gaussian curvature of $\vec{Y}$ is

$$\frac{K}{1 - 2Ha + Ka^2},$$

and the mean curvature is

$$\frac{H - Ka}{1 - 2Ha + Ka^2}.$$

(c) Use the above to prove that if $\vec{X}$ is a surface with constant mean curvature $H = c \neq 0$, then there is a parallel surface to $\vec{X}$ that has constant Gaussian curvature.

CHAPTER 7

The Fundamental
Equations of Surfaces

In the previous chapter, we began to study the local geometry of a surface and saw how many computations depend on the coefficients of the first and second fundamental forms. In fact, in Section 6.1 we saw that many metric calculations (angles between curves, arc length, area, etc.) depend only on the first fundamental form. Of course, we first approached such concepts from the perspective of objects in $\mathbb{R}^3$, but with the first fundamental form, one can perform all the calculations by using only the coordinates that parametrize the surface, without referring to the ambient space. In fact, it is not at all difficult to imagine a regular surface as a subset of $\mathbb{R}^n$, with $n > 3$, and the formulas that depend on the first fundamental form would remain unchanged. Concepts that rely only on the first fundamental form are called *intrinsic properties* of a surface.

On the other hand, concepts such as the second fundamental form, the Gauss map, Gaussian curvature, and principal curvatures were defined in reference to the unit normal vector, and these are not necessarily intrinsic properties. To illustrate this point, consider a parametrized surface S that is a subset of $\mathbb{R}^4$, and let p be a point of S. One can still define the tangent plane as the span of two linearly independent tangent vectors, but there no longer exists a unique (up to sign) unit normal vector to S at p. In this situation, one cannot define the Gauss map. Properties that we presented as depending on the Gauss map either cannot be defined in this situation or need to be defined in an alternate way.

This chapter studies what kind of information about a surface one can know from just the first fundamental form or from knowing both the first and the second fundamental forms.

Section 7.2 introduces the Christoffel symbols and studies relations between the coefficients of the first and second fundamental forms. In Section 7.3, we present the famous Theorema Egregium, which proves that the Gaussian curvature of a surface is in fact an intrinsic property. More precisely, the theorem expresses the Gaussian curvature in terms of the g_{ij} functions. In Section 7.4, we present another landmark result for surfaces in $\mathbb{R}^3$, the Fundamental Theorem of Surface Theory, which proves that under appropriate conditions, the coefficients of the first and second fundamental forms determine the surface up to position and orientation in space. However, in Section 7.1, we introduce tensorial notation. As we do not need tensors in their full generality in the rest of this book (as one does in the theory of manifolds), the reader should feel free to treat this section as optional.

7.1   Tensor Notation

In Chapter 6, we gave to the first fundamental form the alternate name of metric tensor and delayed the explanation of what a tensor is. Mathematicians and physicists often present tensors and the tensor product in very different ways, sometimes making it difficult for a reader to see that authors in different fields are talking about the same thing. In this section, we introduce tensor notation in what one might call the "physics style," which emphasizes how components of objects change under a coordinate transformation. Readers of mathematics who are well aquainted with tensor algebras on vector spaces might find this approach unsatisfactory, but physicists should recognize it. (The reader who wishes to understand the full modern mathematical formulation of tensors and see how the physics style meshes with the mathematical style should consult Appendix C in [22].)

The description of tensors we introduce below relies heavily on transformations between coordinate systems. Though we discuss coordinates and transformations between them generally, one can keep in mind as running examples Cartesian or polar coordinates for regions of the plane or Cartesian, cylindrical, and spherical coordinates for regions of $\mathbb{R}^3$. Ultimately, our discussion will apply to changes of coordinates between overlapping coordinate patches on regular surfaces. (The reader should be aware that [22], which follows the present text, provides a rigorous introduction to these concepts, whereas our presentation here is a little more based on intuition.)

7.1.1 Curvilinear Coordinate Systems

Let S be an open set in $\mathbb{R}^n$. A continuous surjective function $f : U \to S$, where U is an open set in $\mathbb{R}^n$, defines a coordinate system on S by associating to every point $P \in S$ an n-tuple $x(P) = (x^1(P), x^2(P), \ldots, x^n(P))$, such that $f(x(P)) = P$. In this notation, the superscripts do not indicate powers of a variable x but the ith coordinate for that point in the given coordinate system. Though a possible source of confusion at the beginning, differential geometry literature uses superscripts instead of the usual subscripts in order to mesh properly with subsequent tensor notation. One also sometimes says that $f : U \to S$ parametrizes S. As with polar coordinates, where (r_0, θ_0) and $(r_0, \theta_0 + 2\pi)$ correspond to the same point in the plane, the n-tuple need not be uniquely associated to the point P.

Let S be an open set in $\mathbb{R}^n$, and consider two coordinate systems on S relative to which the coordinates of a point P are denoted by $(x^1, x^2, \ldots, x^n)$ and $(\bar{x}^1, \bar{x}^2, \ldots, \bar{x}^n)$.

Suppose that the open set $U \subset \mathbb{R}^n$ parametrizes S in the system $(x^1, x^2, \ldots, x^n)$ and that the open set $V \subset \mathbb{R}^n$ parametrizes S in the system $(\bar{x}^1, \bar{x}^2, \ldots, \bar{x}^n)$. We assume that there exists a bijective change-of-coordinates function $F : U \to V$ so that we can write

$$(\bar{x}^1, \bar{x}^2, \ldots, \bar{x}^n) = F(x^1, x^2, \ldots, x^n). \tag{7.1}$$

Again using the superscript notation, we might write explicitly

$$\begin{cases} \bar{x}^1 & = F^1(x^1, x^2, \ldots, x^n), \\ \ \ \vdots \\ \bar{x}^n & = F^n(x^1, x^2, \ldots, x^n), \end{cases}$$

where F^i are functions from U to $\mathbb{R}$. We will assume from now on that the change of variables function F is always of class C^2, i.e., that all the second partial derivatives are continuous. Unless it becomes necessary for clarity, one often abbreviates the notation and writes

$$\bar{x}^i = \bar{x}^i(x^j),$$

by which one understands that the coordinates $(\bar{x}^1, \bar{x}^2, \ldots, \bar{x}^n)$ functionally depend on the coordinates $(x^1, x^2, \ldots, x^n)$. Therefore, one writes $\frac{\partial \bar{x}^i}{\partial x^j}$ for $\frac{\partial F^i}{\partial x^j}$, and the matrix of the differential dF_P (see Equation (5.2)) is given by

$$(dF_P)_{ij} = \left(\frac{\partial \bar{x}^i}{\partial x^j} \right).$$

Just as the functions $\bar{x}^i = \bar{x}^i(x^j)$ represent the change of variables F, we write $x^j = x^j(\bar{x}^k)$ to indicate the component functions of the inverse $F^{-1} : V \to U$. In this notation, Proposition 5.3.3 states that, as matrices,

$$\left(\frac{\partial x^j}{\partial \bar{x}^i} \right) = \left(\frac{\partial \bar{x}^i}{\partial x^j} \right)^{-1}, \tag{7.2}$$

where we assume that the functions in the first matrix are evaluated at p in the $(\bar{x}^1, \ldots, \bar{x}^n)$-coordinates, while the functions in the second matrix are evaluated at p in the $(x^1, \ldots, x^n)$-coordinates. One can express the same relationship in an alternate way by writing $x^i = x^i(\bar{x}^j(x^k))$ and applying the chain rule when differentiating with respect to x^k as follows:

$$\frac{\partial x^i}{\partial x^k} = \frac{\partial x^i}{\partial \bar{x}^1} \frac{\partial \bar{x}^1}{\partial x^k} + \frac{\partial x^i}{\partial \bar{x}^2} \frac{\partial \bar{x}^2}{\partial x^k} + \cdots + \frac{\partial x^i}{\partial \bar{x}^n} \frac{\partial \bar{x}^n}{\partial x^k} = \sum_{j=1}^{n} \frac{\partial x^i}{\partial \bar{x}^j} \frac{\partial \bar{x}^j}{\partial x^k}.$$

However, by definition of a coordinate system in $\mathbb{R}^n$, there must be no function dependence of one variable on another, so

$$\frac{\partial x^i}{\partial x^k} = \delta^i_k,$$

where δ^i_k is the Kronecker delta symbol defined by

$$\delta^i_j = \begin{cases} 1, & \text{if } i = j, \\ 0, & \text{if } i \neq j. \end{cases} \tag{7.3}$$

Therefore, since δ^i_j are essentially the entries of the identity matrix, we conclude that

$$\sum_{j=1}^{n} \frac{\partial x^i}{\partial \bar{x}^j} \frac{\partial \bar{x}^j}{\partial x^k} = \delta^i_k \tag{7.4}$$

and hence recover Equation (7.2).

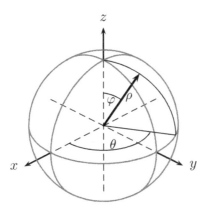

Figure 7.1. Spherical coordinates.

The space $\mathbb{R}^n$ is a vector space, so to each point P, we can associate the position vector $\vec{r} = \overrightarrow{OP}$. Let $(x^1, x^2, \ldots, x^n)$ be a coordinate system of an open set containing P. The natural basis of $\mathbb{R}^n$ at a point P associated to this coordinate system is the set of vectors

$$\left\{ \frac{\partial \vec{r}}{\partial x^1}, \frac{\partial \vec{r}}{\partial x^2}, \ldots, \frac{\partial \vec{r}}{\partial x^n} \right\},$$

where all the derivatives are evaluated at P. Note that one often expresses the vector $\vec{r}$ in terms of the Cartesian coordinate system, and this expression is precisely the transformation functions between the given coordinate system and the Cartesian coordinate system.

Example 7.1.1 (Spherical Coordinates). Spherical coordinates for points in $\mathbb{R}^3$ consist of a triple (ρ, θ, φ), where ρ is the distance of P to the origin, θ is the angle between the xz-half-plane and the vertical half-plane containing the ray $[OP)$, and φ is the angle between the positive z-axis and the ray $[OP)$ (see Figure 7.1).

In Cartesian coordinates, the position vector $\vec{r}$ for a point with spherical coordinates (ρ, θ, φ) is

$$\vec{r} = (\rho \cos \theta \sin \varphi, \rho \sin \theta \sin \varphi, \rho \cos \varphi). \tag{7.5}$$

This corresponds to the coordinate transformation from spherical to Cartesian coordinates. Then

$$\frac{\partial \vec{r}}{\partial \rho} = (\cos\theta\sin\varphi, \sin\theta\sin\varphi, \cos\varphi),$$

$$\frac{\partial \vec{r}}{\partial \theta} = (-\rho\sin\theta\sin\varphi, \rho\cos\theta\sin\varphi, 0),$$

$$\frac{\partial \vec{r}}{\partial \varphi} = (\rho\cos\theta\cos\varphi, \rho\sin\theta\cos\varphi, -\rho\sin\varphi).$$

It is interesting to note that these three vectors are orthogonal for all triples (ρ, θ, φ). When this is the case, we say that the coordinate system is *orthogonal*. In this case, one obtains a natural orthonormal basis at P associated to this coordinate system by simply dividing by the length of each vector (or its negative). For spherical coordinates, the associated orthonormal basis at any point consists of the three vectors

$$\vec{e}_\rho = (\cos\theta\sin\varphi, \sin\theta\sin\varphi, \cos\varphi),$$

$$\vec{e}_\theta = (-\sin\theta, \cos\theta, 0),$$

$$\vec{e}_\varphi = (\cos\theta\cos\varphi, \sin\theta\cos\varphi, -\sin\varphi).$$

The factors 1, $\rho\sin\varphi$, and ρ between $\partial\vec{r}/\partial x^i$ and its normalized vector are sometimes called the *scaling factors* for each coordinate. Again, it is important to note that unlike the usual bases in linear algebra, the basis $\{\vec{e}_\rho, \vec{e}_\theta, \vec{e}_\varphi\}$ depends on the coordinates (ρ, θ, φ) of a point in $\mathbb{R}^n$.

We are now in a position to discuss how components of various quantities defined locally, namely in a neighborhood U of a point $p \in \mathbb{R}^3$, change under a coordinate transformation on U. As mentioned before, our definitions do not possess the usual mathematical flavor, and the supporting discussion might feel like a game of symbols. However, it is important to understand the transformational properties of tensor components even before becoming familiar with the machinery of linear algebra of tensors. We begin with the simplest situation.

Definition 7.1.2. Let $p \in \mathbb{R}^n$, and let U be a neighborhood of p. Suppose that $(x^1, x^2, \ldots, x^n)$ and $(\bar{x}^1, \bar{x}^2, \ldots, \bar{x}^n)$ are two systems of coordinates on U. A function $f(x^1, \ldots, x^n)$ given in the $(x^1, x^2, \ldots, x^n)$-coordinates is said to be a *scalar* if its expression $\bar{f}(\bar{x}^1, \ldots, \bar{x}^n)$ in the $(\bar{x}^1, \ldots, \bar{x}^n)$-coordinates has the same numerical value. In other words, if

$$\bar{f}(\bar{x}^1, \ldots, \bar{x}^n) = f(x^1, \ldots, x^n). \tag{7.6}$$

In Definition 7.1.2, it is understood that both of the coordinates $(x^1, x^2, \ldots, x^n)$ and $(\bar{x}^1, \bar{x}^2, \ldots, \bar{x}^n)$ refer to the same point p in $\mathbb{R}^n$.

This definition might appear at first glance not to hold much content in that every function defined in reference to some coordinate system should possess this property, but that is not true. Suppose that f is a scalar. The quantity that gives the derivative of f in the first coordinate is not a scalar, for though $\bar{f}(\bar{x}^1, \ldots, \bar{x}^n) = f(x^1, \ldots, x^n)$, we have

$$\frac{\partial \bar{f}}{\partial \bar{x}^1} = \frac{\partial f}{\partial x^1} \frac{\partial x^1}{\partial \bar{x}^1} + \frac{\partial f}{\partial x^2} \frac{\partial x^2}{\partial \bar{x}^1} + \cdots + \frac{\partial f}{\partial x^n} \frac{\partial x^n}{\partial \bar{x}^1}.$$

As a second example of how quantities change under coordinate transformations, we consider the gradient $\vec{\nabla} f$ of a differentiable scalar function f. Recall that

$$\vec{\nabla} f = \left(\frac{\partial f}{\partial x^1}, \frac{\partial f}{\partial x^2}, \ldots, \frac{\partial f}{\partial x^n} \right)$$

in usual Cartesian coordinates. The gradient is a vector field, or we may simply consider the gradient of f at P, namely $\vec{\nabla} f_P$, which is a vector. We highlight the transformational properties of the gradient. The chain rule gives

$$\frac{\partial \bar{f}}{\partial \bar{x}^j} = \sum_{i=1}^{n} \frac{\partial x^i}{\partial \bar{x}^j} \frac{\partial f}{\partial x^i}. \tag{7.7}$$

However, it turns out that this is not the only way components of what we usually call a "vector" can change under a coordinate transformation.

Again, consider on an open set of $\mathbb{R}^n$ two coordinate systems $(x^1, x^2, \ldots, x^n)$ and $(\bar{x}^1, \bar{x}^2, \ldots, \bar{x}^n)$. Let $\vec{A}$ be a vector in $\mathbb{R}^n$, which we consider based at P. The components of the vector $\vec{A}$ in the respective coordinate systems is $(A^1, \ldots, A^n)$ and $(\bar{A}^1, \ldots, \bar{A}^n)$, where

$$\vec{A} = \sum_{i=1}^{n} A^i \frac{\partial \vec{r}}{\partial x^i} = \sum_{j=1}^{n} \bar{A}^j \frac{\partial \vec{r}}{\partial \bar{x}^j}.$$

Since $\partial \vec{r}/\partial x^i = \sum_j (\partial \vec{r}/\partial \bar{x}^j)(\partial \bar{x}^j/\partial x^i)$, we find that the components of $\vec{A}$ in the two systems of coordinates are related by

$$\bar{A}^j = \sum_{i=1}^{n} \frac{\partial \bar{x}^j}{\partial x^i} A^i. \tag{7.8}$$

Example 7.1.3 (Velocity in Spherical Coordinates). Consider a space curve that is parametrized by $\vec{r}(t)$ for $t \in I$. The chain rule allows us to write the velocity vector in spherical or Cartesian coordinates as

$$\vec{r}'(t) = \rho'(t)\frac{\partial \vec{r}}{\partial \rho} + \theta'(t)\frac{\partial \vec{r}}{\partial \theta} + \varphi'(t)\frac{\partial \vec{r}}{\partial \varphi}$$
$$= x'(t)\vec{i} + y'(t)\vec{j} + z'(t)\vec{k}.$$

However, differentiating Equation (7.5), assuming ρ, θ, and φ are functions of t, allows us to identify the Cartesian coordinates of the velocity vector as:

$$\begin{pmatrix} x' \\ y' \\ z' \end{pmatrix} = \begin{pmatrix} \cos\theta\sin\varphi & -\rho\sin\theta\sin\varphi & \rho\cos\theta\cos\varphi \\ \sin\theta\sin\varphi & -\rho\cos\theta\sin\varphi & \rho\sin\theta\cos\varphi \\ \cos\varphi & 0 & -\rho\sin\varphi \end{pmatrix} \begin{pmatrix} \rho' \\ \theta' \\ \varphi' \end{pmatrix}. \tag{7.9}$$

If we label the spherical coordinates as $(\bar{x}^1, \bar{x}^2, \bar{x}^3)$ and the Cartesian coordinates as (x^1, x^2, x^3), then the transition matrix between components of the velocity vector from spherical coordinates to Cartesian coordinates is precisely $(\frac{\partial x^i}{\partial \bar{x}^j})$. Thus, the velocity vector of any curve $\vec{\gamma}(t)$ through a point $P = \vec{\gamma}(t_0)$ does not change according to Equation (7.7) but according to Equation (7.8).

Though we have presented the notion of curvilinear coordinates in general, one should keep in mind the linear coordinate changes

$$\begin{pmatrix} \bar{x}^1 \\ \bar{x}^2 \\ \vdots \\ \bar{x}^n \end{pmatrix} = M \begin{pmatrix} x^1 \\ x^2 \\ \vdots \\ x^n \end{pmatrix}$$

where M is an $n \times n$ matrix. If either of these coordinate systems is given as coordinates in a basis of $\mathbb{R}^n$, then the other system simply corresponds to a change of basis in $\mathbb{R}^n$. It is easy to show that the transition matrix is then

$$\left(\frac{\partial \bar{x}^i}{\partial x^j}\right) = M,$$

the usual basis transition matrix. Furthermore, $\left(\frac{\partial \bar{x}^i}{\partial x^j}\right)$ is constant over all $\mathbb{R}^n$.

7.1.2 Tensors: Definitions and Notation

The relations established in Equations (7.7) and (7.8) show that *there are two different kinds of vectors*, each following different transformational properties under a coordinate change. This distinction is not emphasized in most linear algebra courses but essentially corresponds to the difference between a column vector and a row vector, which in turn corresponds to vectors in $\mathbb{R}^n$ and its dual $(\mathbb{R}^n)^*$. (The $*$ notation denotes the dual of a vector space. See Appendix C.3 in [22] for background.) We summarize this dichotomy in the following two definitions.

Definition 7.1.4. Let $(x^1, \ldots, x^n)$ and $(\bar{x}^1, \ldots, \bar{x}^n)$ be two coordinate systems in a neighborhood of a point $p \in \mathbb{R}^n$. An n-tuple $(A^1, A^2, \ldots, A^n)$ is said to constitute the components of a *contravariant vector* at a point p if these components transform according to the relation

$$\bar{A}^j = \sum_{i=1}^{n} \frac{\partial \bar{x}^j}{\partial x^i} A^i,$$

where we assume the partial derivatives are evaluated at p.

Definition 7.1.5. Under the same conditions as above, an n-tuple $(B_1, B_2, \ldots, B_n)$ is said to constitute the components of a *covariant vector* at a point p if these components transform according to the relation

$$\bar{B}_j = \sum_{i=1}^{n} \frac{\partial x^i}{\partial \bar{x}^j} B_i,$$

where we assume the partial derivatives are evaluated at p.

A few comments are in order at this point. Though the above two definitions are unsatisfactory from the modern perspective of set theory, these are precisely what one is likely to find in a classical mathematics text or a physics text presenting differential geometry. Nonetheless, we will content ourselves with these definitions and with the more general Definition 7.1.6. We defer until Chapter 4 in [22] what is considered the proper modern definition of a tensor on a manifold.

Next, we point out that the quantities $(A^1, A^2, \ldots, A^n)$ in Definition 7.1.4 or $(B_1, B_2, \ldots, B_n)$ in Definition 7.1.5 can either be constant or be functions of the coordinates in a neighborhood of p. If the quantities are constant, one says they form the components of an affine vector. If the quantities are functions in the coordinates, then the components define a different vector for every point in an open set, and thus, one views these quantities as the components of a vector field over a neighborhood of p.

Finally, in terms of notation, we distinguish between the two types of vectors by using subscripts for covariant vectors and superscripts for contravariant vectors. This convention of notation is consistent throughout the literature and forms a central part of tensor calculus. This convention also explains the use of superscripts for the coordinates since (x^i) represents the components of a contravariant vector, namely, the position vector of a point.

As a further example to motivate the definition of a tensor, recall the transformational properties of the components of the first fundamental form given in Equation (6.7):

$$\bar{g}_{ij} = \sum_{k=1}^{2} \sum_{l=1}^{2} \frac{\partial x_k}{\partial \bar{x}_i} \frac{\partial x_l}{\partial \bar{x}_j} g_{kl},$$

where g_{kl} (respectively $\bar{g}_{ij}$) represents the coefficients of the first fundamental form in the (x^1, x^2)- (respectively $(\bar{x}^1, \bar{x}^2)$-) coordinates. This formula mimics but generalizes the transformational properties in Definition 7.1.4 and Definition 7.1.5. Many objects of interest that arise in differential geometry possess similar properties and lead to the following definition of a tensor.

Definition 7.1.6. Let $(x^1, \ldots, x^n)$ and $(\bar{x}^1, \ldots, \bar{x}^n)$ be two coordinate systems in a neighborhood of a point $p \in \mathbb{R}^n$. A set of n^{r+s} quantities

$T^{i_1 i_2 \cdots i_r}_{j_1 j_2 \cdots j_s}$ is said to constitute the components of a *tensor* of type (r, s) if under a coordinate transformation these quantities transform according to

$$\bar{T}^{k_1 k_2 \cdots k_r}_{l_1 l_2 \cdots l_s} = \sum_{i_1=1}^{n} \sum_{i_2=1}^{n} \cdots \sum_{i_r=1}^{n} \sum_{j_1=1}^{n} \sum_{j_2=1}^{n} \cdots \sum_{j_s=1}^{n} \frac{\partial \bar{x}^{k_1}}{\partial x^{i_1}} \frac{\partial \bar{x}^{k_2}}{\partial x^{i_2}} \cdots \frac{\partial \bar{x}^{k_r}}{\partial x^{i_r}} \frac{\partial x^{j_1}}{\partial \bar{x}^{l_1}} \frac{\partial x^{j_2}}{\partial \bar{x}^{l_2}} \cdots \frac{\partial x^{j_s}}{\partial \bar{x}^{l_s}} T^{i_1 i_2 \cdots i_r}_{j_1 j_2 \cdots j_s},$$

$$(7.10)$$

where we assume the partial derivatives are evaluated at p. The *rank* of the tensor is the integer $r + s$.

From the above definition, one could rightly surmise that basic calculations with tensors involve numerous repeated summations. In order to alleviate this notational burden, mathematicians and physicists who use the tensor notation as presented above utilize the *Einstein summation convention*. In this convention of notation, one assumes that one takes a sum from 1 to n (the dimension of $\mathbb{R}^n$ or the number of coordinates) over any index that appears both in a superscript and a subscript of a product. Furthermore, for this convention, in a partial derivative $\frac{\partial \bar{x}^i}{\partial x^j}$, the index i is considered a superscript, and the index j is considered a subscript.

For example, if A_{ij} form the components of a $(0, 2)$-tensor and B^k constitutes the components of a contravariant vector, with the Einstein summation convention, the expression $A_{ij} B^j$ means

$$\sum_{j=1}^{n} A_{ij} B^j.$$

As another example, with the Einstein summation convention, the transformational property in Equation (7.10) of a tensor is written as

$$\bar{T}^{k_1 k_2 \cdots k_r}_{l_1 l_2 \cdots l_s} = \frac{\partial \bar{x}^{k_1}}{\partial x^{i_1}} \frac{\partial \bar{x}^{k_2}}{\partial x^{i_2}} \cdots \frac{\partial \bar{x}^{k_r}}{\partial x^{i_r}} \frac{\partial x^{j_1}}{\partial \bar{x}^{l_1}} \frac{\partial x^{j_2}}{\partial \bar{x}^{l_2}} \cdots \frac{\partial x^{j_s}}{\partial \bar{x}^{l_s}} T^{i_1 i_2 \cdots i_r}_{j_1 j_2 \cdots j_s}, \quad (7.11)$$

where the summations from 1 to n over the indices i_1, i_2, ... i_r, j_1, j_2, ... j_s is understood. As a third example, if C^{ij}_{kl} is a tensor of type $(2, 2)$, then C^{ij}_{kj} means

$$\sum_{j=1}^{n} C^{ij}_{kj}.$$

On the other hand, with this convention, we do not sum over the index i in the expression $A^i + B^i$ or even in $A^i + B_i$. In fact, as we shall see, though the former expression has an interpretation, the latter does not.

In the rest of this book, we will use the Einstein summation convention when working with components of tensors.

7.1.3   Operations on Tensors

It is possible to construct new tensors from old ones. (Again, the reader is encouraged to consult Appendix C.4 in [22] to see the underlying algebraic meaning of the following operations.)

First of all, if $S^{i_1 i_2 \cdots i_r}_{j_1 j_2 \cdots j_s}$ and $T^{i_1 i_2 \cdots i_r}_{j_1 j_2 \cdots j_s}$ are both components of tensors of type (r, s), then the quantities

$$W^{i_1 i_2 \cdots i_r}_{j_1 j_2 \cdots j_s} = S^{i_1 i_2 \cdots i_r}_{j_1 j_2 \cdots j_s} + T^{i_1 i_2 \cdots i_r}_{j_1 j_2 \cdots j_s}$$

form the components of another (r, s)-tensor. In other words, tensors of the same type can be added to obtain another tensor of the same type. The proof is very easy and follows immediately from the transformational properties and distributivity.

Secondly, if $S^{i_1 i_2 \cdots i_r}_{j_1 j_2 \cdots j_s}$ and $T^{k_1 k_2 \cdots k_t}_{l_1 l_2 \cdots l_u}$ are components of tensors of type (r, s) and (t, u), respectively, then the quantities obtained by multiplying these components as in,

$$W^{i_1 i_2 \cdots i_r k_1 k_2 \cdots k_t}_{j_1 j_2 \cdots j_s l_1 l_2 \cdots l_u} = S^{i_1 i_2 \cdots i_r}_{j_1 j_2 \cdots j_s} T^{k_1 k_2 \cdots k_t}_{l_1 l_2 \cdots l_u},$$

form the components of another tensor but of type $(r + t, s + u)$. Again, the proof is very easy, but one must be careful with the plethora of indices. One should note that this operation of tensor product works also for multiplying a tensor by a scalar since a scalar is a tensor of rank 0.

Finally, another common operation on tensors is the *contraction* between two indices. We illustrate the contraction with an example. Let A^{ijk}_{rs} be the components of a $(3, 2)$-tensor, and define the quantities

$$B^{ij}_r = A^{ijk}_{rk} = \sum_{k=1}^{n} A^{ijk}_{rk} \qquad \text{by Einstein summation convention.}$$

It is not hard to show (left as an exercise for the reader) that B_r^{ij} constitute the components of a tensor of type $(2, 1)$. More generally, starting with a tensor of type (r, s), if one sums over an index that appears both in the superscript and in the subscript, one obtains the components of a $(r-1, s-1)$-tensor. This is the contraction of a tensor over the stated indices.

7.1.4   Examples

Example 7.1.7. Following the terminology of Definition 7.1.6, a covariant vector is often called a $(0, 1)$-tensor, and similarly, a contravariant vector is called a $(1, 0)$-tensor.

Example 7.1.8. In Problem 6.3.4, one showed that the coefficients L_{ij} of the second fundamental form constitute the components of a $(0, 2)$-tensor, just as the metric tensor does.

Example 7.1.9 (Inverse of a $(0, 2)$-tensor). As a more involved example, consider the components A_{ij} of a $(0, 2)$-tensor in $\mathbb{R}^n$. Denote by A^{ij} the quantities given as the coefficients of the inverse matrix of (A_{ij}). We prove that A^{ij} form the components of a $(2, 0)$-tensor.

Suppose that the coefficients A^{ij} given in a coordinate system with variables $(x^1, \ldots, x^n)$ and $\bar{A}^{rs}$ are given in the $(\bar{x}^1, \ldots, \bar{x}^n)$-coordinate system. That they are the inverse to the matrices (A_{ij}) and $(\bar{A}_{rs})$ means that A^{ij} and $\bar{A}^{rs}$ are the unique quantities such that

$$A^{ij} A_{jk} = \delta_k^i, \qquad \text{and} \qquad (7.12)$$
$$\bar{A}^{rs} \bar{A}_{st} = \delta_t^r, \qquad (7.13)$$

where the reader must remember that we are using the Einstein summation convention. Combining Equation (7.13) and the transformational properties of A_{jk}, we get

$$\bar{A}^{rs} \frac{\partial x^i}{\partial \bar{x}^s} \frac{\partial x^j}{\partial \bar{x}^t} A_{ij} = \delta_t^r.$$

Multiplying both sides by $\frac{\partial \bar{x}^t}{\partial x^\alpha}$ and summing over t, we obtain

$$\bar{A}^{rs}\frac{\partial x^i}{\partial \bar{x}^s}\frac{\partial x^j}{\partial \bar{x}^t}A_{ij}\frac{\partial \bar{x}^t}{\partial x^\alpha} = \delta^r_t\frac{\partial \bar{x}^t}{\partial x^\alpha}$$

$$\Longleftrightarrow \qquad \bar{A}^{rs}\frac{\partial x^i}{\partial \bar{x}^s}\delta^j_\alpha A_{ij} = \frac{\partial \bar{x}^r}{\partial x^\alpha}$$

$$\Longleftrightarrow \qquad \bar{A}^{rs}\frac{\partial x^i}{\partial \bar{x}^s}A_{i\alpha} = \frac{\partial \bar{x}^r}{\partial x^\alpha}.$$

Multiplying both sides by $A^{\alpha\beta}$ and then summing over α, we get

$$\bar{A}^{rs}\frac{\partial x^i}{\partial \bar{x}^s}\delta^\beta_i = \bar{A}^{rs}\frac{\partial x^\beta}{\partial \bar{x}^s} = \frac{\partial \bar{x}^r}{\partial x^\alpha}A^{\alpha\beta}.$$

Finally, multiplying the rightmost equality by $\frac{\partial \bar{x}^s}{\partial x^\beta}$ and summing over β, one concludes that

$$\bar{A}^{rs} = \frac{\partial \bar{x}^r}{\partial x^\alpha}\frac{\partial \bar{x}^s}{\partial x^\beta}A^{\alpha\beta}.$$

This shows that the quantities A^{ij} satisfy Definition 7.1.6 and form the components of a $(2,0)$-tensor.

By a similar manipulation, one can show that if B^{ij} are the components of a $(2,0)$-tensor, then the quantities B_{ij} corresponding to the inverse of the matrix of B^{ij} form the components of a $(0,2)$-tensor.

Example 7.1.10 (Gauss Map Coefficients). In differential geometry, one denotes by g^{ij} the coefficients of the inverse of the matrix associated to the first fundamental form. Example 7.1.9 shows that g^{ij} is a $(2,0)$-tensor. Furthermore, recall that the Weingarten equations in Equation (6.26) give the components (associated to the standard basis on $T_p(S)$ given by a particular parametrization) of the Gauss map as

$$a^i_j = -L_{jk}g^{ki}.$$

By tensor product and contraction, we see that the functions a^i_j form the components of a $(1,1)$-tensor.

Example 7.1.11 (Metric Tensors). It is important to understand some standard operations of vectors in the context of tensor notation. Consider two vectors in a vector space V of dimension n. Using tensor notation, one refers to these vectors as affine contravariant vectors with components A^i and B^j, with $i, j = 1, 2, \ldots, n$. We have seen that addition of the vectors or scalar multiplication are the usual operations from linear algebra. Another operation between vectors in V is the dot product, which was originally defined as

$$\sum_{i=1}^{n} A^i B^i,$$

but this is not the correct way to understand the dot product in the context of tensor algebra. The very fact that one cannot use the Einstein summation convention is a hint that we must adjust our notation. The use of the usual dot product for its intended geometric purpose makes an assumption of the given basis of V, namely, that the basis is orthonormal. When using tensor algebra, one makes no such assumption. Instead, one associates a $(0, 2)$-tensor g_{ij}, called the metric tensor, to the basis of V with respect to which coordinates are defined. Then the first fundamental form (or *scalar product*) between A^i and B^j is

$$g_{ij} A^i B^j \qquad \text{(Einstein summation)}.$$

One immediately notices that because of tensor multiplication and contraction, the result is a scalar quantity, and hence, will remain unchanged under a coordinate transformation. In this formulation, the assumption that a basis is orthonormal is equivalent to having

$$g_{ij} = \begin{cases} 1, & \text{if } i = j, \\ 0, & \text{if } i \neq j. \end{cases}$$

7.1.5 Symmetries

The usual operations of tensor addition and scalar multiplication were explained above. We should point out that, using distributivity and associativity, one notices that the set of affine tensors of type

(r, s) in $\mathbb{R}^n$ form a vector space. The $(r + s)$-tuple of all the indices can take on n^{r+s} values, so this vector space has dimension n^{r+s}.

However, it is not uncommon that there exist symmetries within the components of a tensor. For example, as we saw for the metric tensor, we always have $g_{ij} = g_{ji}$. In the context of matrices, we said that the matrix (g_{ij}) is a symmetric matrix, but in the context of tensor notation, we say that the components g_{ij} are symmetric in the indices i and j. More generally, if $T^{i_1 i_2 \cdots i_r}_{j_1 j_2 \cdots j_s}$ are the components of a tensor of type (r, s), we say that the components are *symmetric* in a set $\mathcal{S}$ of indices if the components remain equal when we interchange any two indices from among the indices in $\mathcal{S}$.

For example, let A^{ijk}_{rs} be the components of a $(3, 2)$-tensor. To say that the components are symmetric in $\{i, j, k\}$ affirms the equalities

$$A^{ijk}_{rs} = A^{ikj}_{rs} = A^{jik}_{rs} = A^{jki}_{rs} = A^{kij}_{rs} = A^{kji}_{rs}$$

for all $i, j, k \in \{1, 2, \ldots, n\}$. Note that, because of the additional conditions, the dimension of the space of all $(3, 2)$-tensors that are symmetric in their contravariant indices is smaller than n^5 but not simply $n^5/6$ either. We can find the dimension of this vector space by determining the cardinality of

$$\mathcal{I} = \left\{ (i, j, k) \in \{1, 2, \ldots, n\}^3 \, | \, 1 \le i \le j \le k \le n \right\}.$$

We will see shortly that $|\mathcal{I}| = \binom{n+2}{3}$, and therefore, the dimension of the vector space of $(3, 2)$-tensors that are symmetric in their contravariant indices is $\binom{n+2}{3} n^2$. We provide the following proposition for completeness.

Proposition 7.1.12. *Let $A^{j_1 \cdots j_r}_{k_1 \cdots k_s}$ be the components of a tensor over $\mathbb{R}^n$ that is symmetric in a set $\mathcal{S}$ of its indices. Assuming that all the indices are fixed except for the indices of $\mathcal{S}$, the number of independent components of the tensor is equal to the cardinality of*

$$\mathcal{I} = \left\{ (i_1, \ldots, i_m) \in \{1, 2, \ldots, n\}^m \, | \, 1 \le i_1 \le i_2 \le \cdots \le i_m \le n \right\}.$$

This cardinality is

$$\binom{n - 1 + m}{m} = \frac{(n - 1 + m)!}{(n - 1)! m!}.$$

Proof: Since the components are symmetric in the set $\mathcal{S}$ of indices, one gets a unique representative of equivalent components by imposing that the indices in question be listed in nondecreasing order. This remark proves the first part of the proposition. To prove the second part, consider the set of integers $\{1, 2, \ldots, n + m\}$ and pick m distinct integers $\{l_1, \ldots, l_m\}$ that are greater than 1 from among this set. We know from the definition of combinations that there are $\binom{n-1+m}{m}$ ways to do this. Assuming that $l_1 < l_2 < \cdots < l_m$, define $i_t = l_t - t$. It is easy to see that the resulting m-tuple $(i_1, \ldots, i_m)$ is in the set $\mathcal{I}$. Furthermore, since one can reverse the process by defining $l_t = i_t + t$ for $1 \le t \le m$, there exists a bijection between $\mathcal{I}$ and the m-tuples $(l_1, \ldots, l_m)$ described above. This establishes that

$$|\mathcal{I}| = \binom{n - 1 + m}{m}.$$

$\square$

Another common situation with relationships between the components of a tensor is when components are antisymmetric in a set of indices. We say that components are *antisymmetric* in a set $\mathcal{S}$ of indices if the components are negated when we interchange any two indices from among the indices in $\mathcal{S}$. This condition imposes a number of immediate consequences. Consider, for example, the components of a $(0, 3)$-tensor A_{ijk} that are antisymmetric in all its indices. If k is any value but $i = j$, then

$$A_{ijk} = A_{iik} = A_{jik} = -A_{ijk},$$

and so $A_{iik} = 0$. Given any triple (i, j, k) in which at least two of the indices are equal, the corresponding component is equal to 0. As another consequence of the antisymmetric condition, consider the component A_{231}. One obtains the triple $(2, 3, 1)$ from $(1, 2, 3)$ by first interchanging 1 and 2 to get $(2, 1, 3)$ and then interchanging the last two to get $(2, 3, 1)$. Therefore, we see that

$$A_{123} = -A_{213} = A_{231}.$$

In modern algebra, a permutation (a bijection on a finite set) that interchanges two inputs and leaves the rest fixed is called a *transposition*. We say that we used two transpositions to go from $(1, 2, 3)$ to $(2, 3, 1)$.

The above example illustrates that the value of the component involving a particular m-tuple $(i_1, \ldots, i_m)$ of distinct indices determines the value of any component involving a permutation $(j_1, \ldots, j_m)$ of $(i_1, \ldots, i_m)$, as

$$A_{j_1 \ldots j_m} = \pm A_{i_1 \ldots i_m},$$

where the sign $\pm$ is $+$ (respectively $-$) if it takes an even (respectively odd) number of interchanges to get from $(i_1, \ldots, i_m)$ to $(j_1, \ldots, j_m)$. A priori, if one could get from $(i_1, \ldots, i_m)$ to $(j_1, \ldots, j_m)$ with both an odd and an even number of transpositions, then $A_{i_1 \ldots i_m}$ and all components indexed by a permutation of $(i_1, \ldots, i_m)$ would be 0. However, a fundamental fact in modern algebra (see Theorem 5.5 in [13]) states that given a permutation σ on $\{1, 2, \ldots, m\}$, if we have two ways to write σ as a composition of transpositions,

$$\sigma = \tau_1 \circ \tau_2 \circ \cdots \circ \tau_a = \tau_1' \circ \tau_2' \circ \cdots \circ \tau_b',$$

then a and b have the same parity.

Definition 7.1.13. We call a permutation *even* (respectively *odd*) if this common parity is even (respectively odd) and the *sign* of σ is

$$\mathrm{sign}(\sigma) = \begin{cases} 1, & \text{if } \sigma \text{ is even,} \\ -1, & \text{if } \sigma \text{ is odd.} \end{cases}$$

The above discussion leads to the following proposition about the components of an antisymmetric tensor.

Proposition 7.1.14. *Let $A^{j_1 \cdots j_r}_{k_1 \cdots k_s}$ be the components of a tensor over $\mathbb{R}^n$ that is antisymmetric in a set $\mathcal{S}$ of its indices. If any of the indices in $\mathcal{S}$ are equal, then*

$$A^{j_1 \cdots j_r}_{k_1 \cdots k_s} = 0.$$

If $|\mathcal{S}| = m$, then fixing all but the indices in $\mathcal{S}$, the number of independent components of the tensor is equal to

$$\binom{n}{m} = \frac{n!}{m!(n-m)!}.$$

Finally, if the indices of $A^{i_1 \cdots i_r}_{j_1 \cdots j_s}$ differ from $A^{k_1 \cdots k_r}_{l_1 \cdots l_s}$ only by a permutation σ on the indices in $\mathcal{S}$, then

$$A^{k_1 \cdots k_r}_{l_1 \cdots l_s} = \mathrm{sign}(\sigma) \, A^{i_1 \cdots i_r}_{j_1 \cdots j_s}.$$

7.1.6 Numerical Tensors

As a motivating example of what are called numerical tensors, note that the quantities δ_j^i form the components of a $(1,1)$-tensor. To see this, suppose that the quantities δ_j^i are expressed in a system of coordinates $(x^1, \ldots, x^n)$ and suppose that $\bar{\delta}_l^k$ are its transformed coefficients in another system of coordinates $(\bar{x}^1, \ldots, \bar{x}^n)$. Obviously, for all fixed i and j, the values of δ_j^i are constant, and therefore

$$\bar{\delta}_l^k = \begin{cases} 1, & \text{if } k = l, \\ 0, & \text{if } k \neq l. \end{cases}$$

But using the properties of the δ_j^i coefficients and the chain rule,

$$\frac{\partial \bar{x}^k}{\partial x^i} \frac{\partial x^j}{\partial \bar{x}^l} \delta_j^i = \frac{\partial \bar{x}^k}{\partial x^i} \frac{\partial x^i}{\partial \bar{x}^l} = \frac{\partial \bar{x}^k}{\partial \bar{x}^l} = \bar{\delta}_l^k.$$

Therefore, δ_j^i is a $(1,1)$-tensor in a tautological way.

A *numerical tensor* is a tensor of rank greater than 0 whose components are constant in the variables $(x^1, \ldots, x^n)$ and hence also $(\bar{x}^1, \ldots, \bar{x}^n)$. The Kronecker delta is just one example of a numerical tensor and we have already seen that it plays an important role in many complicated calculations. The Kronecker delta is the simplest case of the most important numerical tensor, the generalized Kronecker delta. The *generalized Kronecker delta* of order r is a tensor of type (r, r), with components denoted by $\delta_{j_1 \cdots j_r}^{i_1 \cdots i_r}$ defined as the following determinant:

$$\delta_{j_1 \cdots j_r}^{i_1 \cdots i_r} = \begin{vmatrix} \delta_{j_1}^{i_1} & \delta_{j_2}^{i_1} & \cdots & \delta_{j_r}^{i_1} \\ \delta_{j_1}^{i_2} & \delta_{j_2}^{i_2} & \cdots & \delta_{j_r}^{i_2} \\ \vdots & \vdots & \ddots & \vdots \\ \delta_{j_1}^{i_r} & \delta_{j_2}^{i_r} & \cdots & \delta_{j_r}^{i_r} \end{vmatrix}. \tag{7.14}$$

It is not obvious from Equation (7.14) that the quantities $\delta_{j_1 \cdots j_r}^{i_1 \cdots i_r}$ form the components of a tensor. However, one can write the components of the generalized Kronecker delta of order 2 as

$$\delta_{kl}^{ij} = \delta_k^i \delta_l^j - \delta_l^i \delta_k^j,$$

which presents δ^{ij}_{kl} as the difference of two $(2,2)$-tensors, which shows that the coefficients δ^{ij}_{kl} indeed constitute a tensor. More generally, expanding out Equation (7.14) gives the generalized Kronecker delta of order r as a sum of $r!$ components of tensors of type (r,r), proving that $\delta^{i_1\cdots i_r}_{j_1\cdots j_r}$ are the components of an (r,r)-tensor.

Properties of the determinant imply that $\delta^{i_1\cdots i_r}_{j_1\cdots j_r}$ is antisymmetric in both the superscript indices and the subscript indices. That is to say, $\delta^{i_1\cdots i_r}_{j_1\cdots j_r}=0$ if any of the superscript indices are equal or if any of the subscript indices are equal. Hence, the value of a component is negated if any two superscript indices are interchanged and similarly for subscript indices. We also note that if $r>n$ where we assume $\delta^{i_1\cdots i_r}_{j_1\cdots j_r}$ is a tensor in $\mathbb{R}^n$, then $\delta^{i_1\cdots i_r}_{j_1\cdots j_r}=0$ for all choices of indices since at least two superscript (and at least two subscript) indices would be equal.

We introduce one more symbol related to the generalized Kronecker delta, namely the permutation symbol. Define

$$\varepsilon^{i_1\cdots i_n}=\delta^{i_1\cdots i_n}_{1\cdots n},$$
$$\varepsilon_{j_1\cdots j_n}=\delta^{1\cdots n}_{j_1\cdots j_n},$$
(7.15)

Note that the use of the maximal index n in Equation (7.15) as opposed to r is intentional. Because of the properties of the determinant, it is not hard to see that one has the values

$$\varepsilon^{i_1\cdots i_n}=\varepsilon_{i_1\cdots i_n}=\begin{cases}1, & \text{if }(i_1,\ldots,i_n)\text{ is an even permutation of }(1,2,\ldots,n),\\ -1, & \text{if }(i_1,\ldots,i_n)\text{ is an odd permutation of }(1,2,\ldots,n),\\ 0, & \text{if }(i_1,\ldots,i_n)\text{ is not a permutation of }(1,2,\ldots,n).\end{cases}$$

We are careful, despite the notation, not to call the permutation symbols the components of a tensor, for they are not. Instead, we have the following proposition.

Proposition 7.1.15. *Let* $(x^1,\ldots,x^n)$ *and* $(\bar{x}^1,\ldots,\bar{x}^n)$ *be two coordinate systems. The permutation symbols transform according to*

$$\bar{\varepsilon}^{j_1\cdots j_n}=J\frac{\partial\bar{x}^{j_1}}{\partial x^{i_1}}\cdots\frac{\partial\bar{x}^{j_n}}{\partial x^{i_n}}\varepsilon^{i_1\cdots i_n},$$
$$\bar{\varepsilon}_{k_1\cdots k_n}=J^{-1}\frac{\partial x^{h_1}}{\partial\bar{x}^{k_1}}\cdots\frac{\partial x^{h_n}}{\partial\bar{x}^{k_n}}\varepsilon_{h_1\cdots h_n},$$

where $J = \det\left(\frac{\partial \bar{x}^i}{\partial x^j}\right)$ is the Jacobian of the transformation of coordinates function.

Proof: (Left as an exercise for the reader.) □

Example 7.1.16 (Cross Product). As an example of the permutation symbol, consider two contravariant vectors A^i and B^j in $\mathbb{R}^3$. If we define $C_k = \varepsilon_{ijk}A^iB^j$, we easily find that

$$C_1 = A^2B^3 - A^3B^2, \quad C_2 = A^3B^1 - A^1B^3, \quad C_3 = A^1B^2 - A^2B^1.$$

The values C_k are precisely the terms of the cross product of the vectors A^i and B^j. However, a quick check shows that the quantities C_k do *not* form the components of a covariant tensor.

One explanation in relation to standard linear algebra for the fact that C_k does not give a contravariant vector is that if $\vec{a}$ and $\vec{b}$ are vectors in $\mathbb{R}^3$ given with coordinates in a certain basis and if M is a coordinate-change matrix, then

$$(M\vec{a}) \times (M\vec{b}) \neq M(\vec{a} \times \vec{b}).$$

In many physics textbooks, when one assumes that we use the usual metric, (g_{ij}) being the identity matrix, one is not always careful with the superscript and subscript indices. This is because one can obtain a contravariant vector B^j from a covariant vector A_i simply by defining $B^j = g^{ij}A_i$, and the components (B^1, B^2, B^3) are numerically equal to (A_1, A_2, A_3). Therefore, in this context, one can define the cross product as the vector with components

$$C^l = g^{kl}\varepsilon_{ijk}A^iB^j, \tag{7.16}$$

However, one must remember that this is not a contravariant vector since it does not satisfy the transformational properties of a tensor.

The generalized Kronecker delta has a close connection to determinants, which we will elucidate here. Note that if the superscript indices are exactly equal to the subscript indices, then $\delta^{i_1 \cdots i_r}_{j_1 \cdots j_r}$ is the determinant of the identity matrix. Thus, the contraction over all indices $\delta^{j_1 \cdots j_r}_{j_1 \cdots j_r}$ counts the number of permutations of r indices taken from the set $\{1, 2, \ldots, n\}$. Thus,

$$\delta^{j_1 \cdots j_r}_{j_1 \cdots j_r} = \frac{n!}{(n-r)!}. \tag{7.17}$$

Another property of the generalized Kronecker delta is that $\varepsilon^{j_1 \cdots j_n}$ $\cdot \varepsilon_{i_1 \cdots i_n} = \delta^{j_1 \cdots j_n}_{i_1 \cdots i_n}$, the proof of which is left as an exercise for the reader (Problem 7.1.10). Now let a^i_j be the components of a $(1,1)$-tensor, which we can view as the matrix of a linear transformation from $\mathbb{R}^n$ to $\mathbb{R}^n$. By definition of the determinant,

$$\det(a^i_j) = \varepsilon^{j_1 \cdots j_n} a^1_{j_1} \cdots a^n_{j_n}.$$

Then, by properties of the determinant related to rearranging rows or columns, we have

$$\varepsilon^{i_1 \cdots i_n} \det(a^i_j) = \varepsilon^{j_1 \cdots j_n} a^{i_1}_{j_1} \cdots a^{i_n}_{j_n}.$$

Multiplying by $\varepsilon_{i_1 \cdots i_n}$ and summing over all the indices $i_1, \ldots, i_n$, we have

$$\varepsilon_{i_1 \cdots i_n} \varepsilon^{i_1 \cdots i_n} \det(a^i_j) = \delta^{j_1 \cdots j_n}_{i_1 \cdots i_n} a^{i_1}_{j_1} \cdots a^{i_n}_{j_n},$$

and since $\varepsilon_{i_1 \cdots i_n} \varepsilon^{i_1 \cdots i_n}$ counts the number of permutations of $\{1, \ldots, n\}$, we have

$$n! \det(a^i_j) = \delta^{j_1 \cdots j_n}_{i_1 \cdots i_n} a^{i_1}_{j_1} \cdots a^{i_n}_{j_n}. \tag{7.18}$$

Problems

7.1.1. Prove that

(a) $\delta^i_j \delta^j_k \delta^k_l = \delta^i_l$,

(b) $\delta^i_j \delta^j_k \delta^k_i = n$.

7.1.2. Let B_i be the components of a covariant vector. Prove that the quantities

$$C_{jk} = \frac{\partial B_j}{\partial x^k} - \frac{\partial B_k}{\partial x^j}$$

form the components of a $(0,2)$-tensor.

7.1.3. Let $T^{i_1 i_2 \cdots i_r}_{j_1 j_2 \cdots j_s}$ be the components of a tensor of type (r, s). Prove that the quantities $T^{i i_2 \cdots i_r}_{i j_2 \cdots j_s}$, obtained by contracting over the first two indices, form the components of a tensor of type $(r-1, s-1)$. Explain why one still obtains a tensor when one contracts over any superscript and subscript index.

7.1.4. Let S_{ijk} be the components of a tensor, and suppose they are antisymmetric in $\{i, j\}$. Find a tensor with components T_{ijk} that is antisymmetric in $\{j, k\}$ satisfying $-T_{ijk} + T_{jik} = S_{ijk}$.

7.1.5. If A^{jk} is antisymmetric in its indices and B_{jk} is symmetric in its indices, show that the scalar $A^{jk} B_{jk}$ is 0.

7.1.6. Consider A^i, B^j, C^k the components of three contravariant vectors. Prove that $\varepsilon_{ijk} A^i B^j C^k$ is the value triple product $(\vec{A}\vec{B}\vec{C}) = \vec{A} \cdot (\vec{B} \times \vec{C})$, which is the volume of the parallelopiped spanned by these three vectors.

7.1.7. Prove that $\varepsilon_{ijk} \varepsilon^{rsk} = \delta^{rs}_{ij}$. Assume that we use a metric g_{ij} that is the identity matrix, and define the cross product $\vec{C}$ of two contravariant vectors $\vec{A} = (A^i)$ and $\vec{B} = (B^j)$ as $C^k = g^{kl} \varepsilon_{ijl} A^i B^j$. Use what you just proved to show that

$$\vec{A} \times (\vec{B} \times \vec{C}) = (\vec{A} \cdot \vec{C})\vec{B} - (\vec{A} \cdot \vec{B})\vec{C}.$$

7.1.8. Let $\vec{A}$, $\vec{B}$, $\vec{C}$, and $\vec{D}$ be vectors in $\mathbb{R}^3$. Use the ε_{ijk} symbols to prove that

$$(\vec{A} \times \vec{B}) \times (\vec{C} \times \vec{D}) = (\vec{A}\vec{B}\vec{D})\vec{C} - (\vec{A}\vec{B}\vec{C})\vec{D}.$$

7.1.9. Prove Proposition 7.1.15.

7.1.10. Prove that $\varepsilon^{i_1 \cdots i_n} \varepsilon_{j_1 \cdots j_n} = \delta^{i_1 \cdots i_n}_{j_1 \cdots j_n}$.

7.1.11. Let A_{ij} be the components of an antisymmetric tensor of type $(0, 2)$, and define the quantities

$$B_{rst} = \frac{\partial A_{st}}{\partial x^r} + \frac{\partial A_{tr}}{\partial x^s} + \frac{\partial A_{rs}}{\partial x^t}.$$

(a) Prove that B_{rst} are the components of a tensor of type $(0, 3)$.

(b) Prove that the components B_{rst} are antisymmetric in all the indices.

(c) Determine the number of independent components of antisymmetric tensors of type $(0, 3)$ over $\mathbb{R}^n$.

(d) Would the quantities B_{rst} still be the components of a tensor if A_{ij} were symmetric?

7.1.12. Let A be an $n \times n$ matrix with coefficients $A = (A_i^j)$, and consider the coordinate transformation

$$\bar{x}^j = \sum_{i=1}^{n} A_i^j x^i.$$

Recall that this transformation is called orthogonal if $AA^T = I$, where A^T is the transpose of A and I is the identity matrix. The orthogonality condition implies that $\det(A) = \pm 1$. An orthogonal transformation is called *special* or *proper* if, in addition, $\det(A) = 1$. A set of quantities $T_{j_1 \cdots j_s}^{i_1 \cdots i_r}$ is called a *proper tensor* of type (r, s) if it satisfies the tensor transformation property from Equation (7.10) for all proper orthogonal transformations.

(a) Prove that the orthogonality condition is equivalent to requiring that

$$\eta_{ij} = \eta_{hk} A_i^h A_j^k,$$

where

$$\eta_{ij} = \begin{cases} 1, & \text{if } i = j, \\ 0, & \text{if } i \neq j. \end{cases}$$

(b) Prove that the orthogonality condition is also equivalent to saying that orthogonal transformations are the invertible linear transformations that preserve the quantity $(x^1)^2 + (x^2)^2 + \cdots + (x^n)^2$.

(c) Prove that (1) the space of proper tensors of type (r, s) form a vector space over $\mathbb{R}$, (2) the product of a proper tensor of type (r_1, s_1) and a proper tensor of type (r_2, s_2) is a proper tensor of type $(r_1 + r_2, s_1 + s_2)$, and (3) contraction over two indices of a proper tensor of type (r, s) produces a proper tensor of type $(r - 1, s - 1)$.

(d) Prove that the permutation symbols are proper tensors of type $(n, 0)$ or $(0, n)$, as appropriate.

(e) Use this to prove that the cross product of two contravariant vectors in $\mathbb{R}^3$ as defined in Equation (7.16) is a proper tensor of type $(1, 0)$. [Hint: This explains that the cross product of two vectors transforms correctly only if we restrict ourselves to proper orthogonal transformations on $\mathbb{R}^3$.]

(f) Suppose that we are in $\mathbb{R}^3$. Prove that the rotation with matrix

$$A = \begin{pmatrix} \cos\alpha & -\sin\alpha & 0 \\ \sin\alpha & \cos\alpha & 0 \\ 0 & 0 & 1 \end{pmatrix}$$

is a proper orthogonal transformation.

(g) Again, suppose that we are in $\mathbb{R}^3$. Prove that the linear transformation with matrix given with respect to the standard basis

$$B = \begin{pmatrix} \cos\beta & \sin\beta & 0 \\ \sin\beta & -\cos\beta & 0 \\ 0 & 0 & 1 \end{pmatrix}$$

is an orthogonal transformation that is not proper.

7.1.13. Consider the vector space $\mathbb{R}^{n+1}$ with coordinates $(x^0, x^1, \ldots, x^n)$, where $(x^1, \ldots, x^n)$ are called the space coordinates and x^0 is the time coordinate. The usual connection between x^0 and time t is $x^0 = ct$, where c is the speed of light. We equip this space with the metric $\eta_{\mu\nu}$ where $\eta_{00} = -1$, $\eta_{ii} = 1$ for $1 \leq i \leq n$, and $\eta_{ij} = 0$ if $i \neq j$. (The quantities $\eta_{\mu\nu}$ do not give a metric in the sense we have presented in this text so far because it is not positive definite. Though we do not provide the details here, this unusual metric gives a mathematical justification for why it is impossible to travel faster than the speed of light c.) This vector space equipped with the metric $\eta_{\mu\nu}$ is called the n-dimensional *Minkowski spacetime*, and $\eta_{\mu\nu}$ is called the *Minkowski metric*.

Let L be an $(n+1)\times(n+1)$ matrix with coefficients L^α_β, and consider the linear transformation

$$\bar{x}^j = \sum_{i=0}^{n} L^j_i x^i.$$

A *Lorentz transformation* is an invertible linear transformation on Minkowski spacetime with matrix L such that

$$\eta_{\alpha\beta} = \eta_{\mu\nu} L^\mu_\alpha L^\nu_\beta.$$

Finally, a set of quantities $T^{i_1\cdots i_r}_{j_1\cdots j_s}$, with indices ranging in $\{0, 1, \ldots, n\}$, is called a *Lorentz tensor* of type (r, s) if it satisfies the tensor transformation property from Equation (7.10) for all Lorentz transformations.

(a) Prove that a transformation of Minkowski spacetime is a Lorentz transformation if and only if it preserves the quantity

$$-(x^0)^2 + (x^1)^2 + (x^2)^2 + \cdots + (x^n)^2.$$

(b) Suppose we are working in three-dimensional Minkowski space-time. Prove that the rotation matrix

$$A = \begin{pmatrix} 1 & 0 & 0 & 0 \\ 0 & \cos\alpha & -\sin\alpha & 0 \\ 0 & \sin\alpha & \cos\alpha & 0 \\ 0 & 0 & 0 & 1 \end{pmatrix}$$

represents a Lorentz transformation.

(c) Again, suppose we are working in three-dimensional Minkowski spacetime. Consider the matrix

$$L = \begin{pmatrix} \gamma & -\beta\gamma & 0 & 0 \\ -\beta\gamma & \gamma & 0 & 0 \\ 0 & 0 & 1 & 0 \\ 0 & 0 & 0 & 1 \end{pmatrix},$$

where β is a positive real number satisfying $-1 < \beta < 1$ and $\gamma = 1/\sqrt{1-\beta^2}$. Prove that L represents a Lorentz transformation.

(d) Prove that (1) the space of Lorentz tensors of type (r,s) form a vector space over $\mathbb{R}$, (2) the product of a Lorentz tensor of type (r_1, s_1) and a Lorentz tensor of type (r_2, s_2) is a Lorentz tensor of type $(r_1 + r_2, s_1 + s_2)$, and (3) contraction over two indices of a Lorentz tensor of type (r,s) produces a Lorentz tensor of type $(r-1, s-1)$.

7.1.14. Let a^i_j be the components of a $(1,1)$-tensor, or in other words, the matrix of a linear transformation from $\mathbb{R}^n$ to $\mathbb{R}^n$ given with respect to some basis. Recall that the characteristic equation for the matrix is

$$\det(a^i_j - \lambda\delta^i_j) = 0. \tag{7.19}$$

[Hint: The solutions to this equation are the eigenvalues of the matrix.] Prove that Equation (7.19) is equivalent to

$$\lambda^n + \sum_{r=1}^{n}(-1)^r a_{(r)}\lambda^{n-r} = 0$$

where

$$a_{(r)} = \frac{1}{r!}\delta^{i_1\cdots i_r}_{j_1\cdots j_r} a^{i_1}_{j_1}\cdots a^{i_r}_{j_r}.$$

7.1.15. *Moment of Inertia Tensor.* Suppose that $\mathbb{R}^3$ is given a basis that is not necessarily orthonormal. Let g_{ij} be the metric tensor corresponding to this basis, which means that the scalar product between two (contravariant) vectors A^i and B^j is given by

$$\langle \vec{A}, \vec{B} \rangle = g_{ij} A^i B^j.$$

In the rest of the problem, call (x^1, x^2, x^3) the coordinates of the position vector $\vec{r}$.

Let S be a solid in space with a density function $\rho(\vec{r})$, and suppose that it rotates about an axis ℓ through the origin. The angular velocity vector $\vec{\omega}$ is defined as the vector along the axis ℓ, pointing in the direction that makes the rotation a right-hand corkscrew motion, and with magnitude ω that is equal to the radians per second swept out by the motion of rotation. Let $(\omega^1, \omega^2, \omega^3)$ be the components of $\vec{\omega}$ in the given basis. The *moment of inertia* of the solid S about the direction $\vec{\omega}$ is defined as the quantity

$$I_\ell = \iiint_S \rho(\vec{r}) r_\perp^2 \, dV,$$

where $r_\perp$ is the distance from a point $\vec{r}$ with coordinate (x^1, x^2, x^3) to the axis ℓ.

The *moment of inertia tensor* of a solid is often presented using cross products, but we define it here using a characterization that is equivalent to the usual definition but avoids cross products. We define the moment of inertia tensor as the unique $(0, 2)$-tensor I_{ij} such that

$$I_{ij} \omega^i \omega^j = I_\ell \omega^2, \tag{7.20}$$

where $\omega = \|\vec{\omega}\| = \sqrt{\langle \vec{\omega}, \vec{\omega} \rangle}$.

(a) Prove that

$$r_\perp^2 = g_{ij} x^i x^j - \frac{(g_{kl} \omega^k x^l)^2}{g_{rs} \omega^r \omega^s}.$$

(b) Prove that, using the metric g_{ij}, the moment of inertia tensor is given by

$$I_{ij} = \iiint_S \rho(x^1, x^2, x^3)(g_{ij} g_{kl} - g_{ik} g_{jl}) x^k x^l \, dV. \tag{7.21}$$

(c) Prove that I_{ij} is symmetric in its indices.

(d) Prove that if the basis of $\mathbb{R}^3$ is orthonormal (which means that (g_{ij}) is the identity matrix), one recovers the following usual formulas one finds in physics texts:

$$I_{11} = \iiint\limits_S \rho((x^2)^2 + (x^3)^2)\, dV, \quad I_{12} = -\iiint\limits_S \rho x^1 x^2\, dV, \quad (7.22)$$

$$I_{22} = \iiint\limits_S \rho((x^1)^2 + (x^3)^2)\, dV, \quad I_{13} = -\iiint\limits_S \rho x^1 x^3\, dV, \quad (7.23)$$

$$I_{33} = \iiint\limits_S \rho((x^1)^2 + (x^2)^2)\, dV, \quad I_{23} = -\iiint\limits_S \rho x^2 x^3\, dV. \quad (7.24)$$

(We took the relation in Equation (7.20) as the defining property of the moment of inertia tensor because of the theorem that $I_\ell \omega$ is the component of the angular moment vector along the axis of rotation that is given by $(I_{ij}\omega^i)\frac{\omega^j}{\omega}$. See [12] p. 221-222, and in particular, Equation (9.7) for an explanation.

The interesting point about this approach is that it avoids the use of an orthonormal basis and provides a formula for the moment of inertia tensor when one has an affine metric tensor that is not the identity. Furthermore, since it avoids the cross product, the above definitions for the moment of inertia tensor of a solid about an axis are generalizable to solids in $\mathbb{R}^n$.)

7.1.16. This problem considers formulas for curvatures of curves or surfaces defined implicitly by one equation.

(a) In Problem 1.3.13, the reader was asked to prove a formula for the geodesic curvature κ_g at a point p on a curve given implicitly by the equation $F(x, y) = 0$. Show that the formula found there can be written as

$$\kappa_g = \frac{1}{(F_x^2 + F_y^2)^{3/2}} \varepsilon^{i_1 i_2} \varepsilon^{j_1 j_2} F_{i_1} F_{j_1} F_{i_2 j_2}, \quad (7.25)$$

where, in the Einstein summation convention, F_1 means $\frac{\partial F}{\partial x^1} = \frac{\partial F}{\partial x}$ and F_2 means $\frac{\partial F}{\partial x^2} = \frac{\partial F}{\partial y}$, and where all the functions are evaluated at the point p.

(b) Prove that Equation (7.25) can be written as

$$\kappa_g = -\frac{1}{(F_x^2 + F_y^2)^{3/2}} \begin{vmatrix} F_{xx} & F_{xy} & F_x \\ F_{yx} & F_{yy} & F_y \\ F_x & F_y & 0 \end{vmatrix}. \quad (7.26)$$

(c) Problem 6.5.14 asked the reader to do the same exercise but to find a formula for the Gaussian curvature at a point on a surface given implicitly by the equation $F(x, y, z) = 0$. Show that the Gaussian curvature K at a point p on a curve given implicitly by the equation $F(x, y, z) = 0$ can be written as

$$K = \frac{1}{(F_x^2 + F_y^2 + F_z^2)^2} \varepsilon^{i_1 i_2 i_3} \varepsilon^{j_1 j_2 j_3} F_{i_1} F_{j_1} F_{i_2 j_2} F_{i_3 j_3} \quad (7.27)$$

with the same conventions as used for a curve defined implicitly.

(d) Prove that Equation (7.27) can be written as

$$K = -\frac{1}{(F_x^2 + F_y^2 + F_z^2)^2} \begin{vmatrix} F_{xx} & F_{xy} & F_{xz} & F_x \\ F_{yx} & F_{yy} & F_{yz} & F_y \\ F_{zx} & F_{zy} & F_{zz} & F_z \\ F_x & F_y & F_z & 0 \end{vmatrix}. \quad (7.28)$$

7.2   Gauss's Equations and the Christoffel Symbols

We now return to the study of surfaces. As we shall see, the results of this section assume that one can take the third derivative of a coordinate parametrization. Therefore, in this section and in the remainder of the chapter, we consider only regular surfaces of class C^3.

If one compares the theory of surfaces developed so far to the theory of curves, one will point out one major gap in the presentation of the former. In the theory of space curves, we discussed natural equations, namely the curvature $\kappa(s)$ and torsion $\tau(s)$ with respect to arc length, and we proved that these two functions locally define a unique curve up to its position in space. Implicit in the proof of this result for natural equations of curves was the fact that in general there do not exist algebraic relations between the functions $\kappa(s)$ and $\tau(s)$.

In the theory of surfaces, the problem of finding natural equations cannot be quite so simple. For example, even when restricting our attention to the first fundamental form, we know that given any three functions $E(u, v)$, $F(u, v)$, and $G(u, v)$, there does not necessarily exist a surface with

$$\begin{pmatrix} g_{11} & g_{12} \\ g_{21} & g_{22} \end{pmatrix} = \begin{pmatrix} E & F \\ F & G \end{pmatrix}$$

because we need the nontrivial requirement that $EG - F^2 > 0$. Furthermore, one might suspect that, because of the smoothness conditions that imply that $\vec{X}_{uuv} = \vec{X}_{uvu} = \vec{X}_{vuu}$, the (g_{ij}) and (L_{ij}) coefficients may satisfy some inherent relations.

Also, when discussing natural equations for space curves, one often uses the Frenet frame as a basis for $\mathbb{R}^3$, with origin a point p of the curve. In particular, one must use the Frenet frame to obtain a parametrization of a curve in the neighborhood of p from the knowledge of the curvature $\kappa(s)$ and torsion $\tau(s)$. When performing calculations in the neighborhood of a point p on a surface S parametrized by $\vec{X} : U \to \mathbb{R}^3$, the basis $\{\vec{X}_1, \vec{X}_2, \vec{N}\}$ is the most natural, but, unlike the Frenet frame, this basis is not orthonormal. In addition, the basis $\{\vec{X}_1, \vec{X}_2, \vec{N}\}$ depends significantly on the parametrization $\vec{X}$, while a reparametrization of a curve can at most change the sign of $\vec{T}$ and $\vec{B}$. One might propose $\{\vec{e}_1, \vec{e}_2, \vec{N}\}$, where $\vec{e}_1$ and $\vec{e}_2$ are the principal directions, as a basis more related to the geometry of a surface. The orthonormality has advantages, but calculations using this basis quickly become intractable, and it does not lend itself well to subsequent generalizations. Therefore, we use the basis $\{\vec{X}_1, \vec{X}_2, \vec{N}\}$ for $\mathbb{R}^3$ in a neighborhood of p and study the relations that arise between the (g_{ij}) and (L_{ij}) coefficients.

Recapping earlier definitions, we have

$$\begin{pmatrix} \vec{X}_1 \cdot \vec{X}_1 & \vec{X}_1 \cdot \vec{X}_2 & \vec{X}_1 \cdot \vec{N} \\ \vec{X}_2 \cdot \vec{X}_1 & \vec{X}_2 \cdot \vec{X}_2 & \vec{X}_2 \cdot \vec{N} \\ \vec{N} \cdot \vec{X}_1 & \vec{N} \cdot \vec{X}_2 & \vec{N} \cdot \vec{N} \end{pmatrix} = \begin{pmatrix} g_{11} & g_{12} & 0 \\ g_{21} & g_{22} & 0 \\ 0 & 0 & 1 \end{pmatrix}.$$

Every vector in $\mathbb{R}^3$ can be expressed as a linear combination in this basis. In particular, one would like to express the second derivatives $\vec{X}_{11}$, $\vec{X}_{12}$, $\vec{X}_{21}$, and $\vec{X}_{22}$ as linear combinations of this basis. From Equation (6.17), we know that $L_{ij} = \vec{X}_{ij} \cdot \vec{N}$, and since $\vec{N} \cdot \vec{X}_i = 0$, we deduce that L_{ij} is the coordinate of $\vec{X}_{ij}$ along basis vector $\vec{N}$. However, we do not know the coordinates of $\vec{X}_{ij}$ along the other two basis vectors $\vec{X}_1$ and $\vec{X}_2$. We define eight functions $\Gamma^i_{jk} : U \to \mathbb{R}$ with indices $1 \le i, j, k \le 2$ by the unique formula

$$\vec{X}_{jk} = L_{jk}\vec{N} + \Gamma^1_{jk}\vec{X}_1 + \Gamma^2_{jk}\vec{X}_2. \tag{7.29}$$

This uniquely determines the functions Γ^i_{jk}, but we now proceed to find formulas for them in terms of the metric coefficients.

Note that though we have eight combinations for three indices each ranging between 1 and 2, we do not in fact have eight distinct functions because $\vec{X}_{jk} = \vec{X}_{kj}$. Thus, we already know that

$$\Gamma^i_{jk} = \Gamma^i_{kj}.$$

Before determining formulas for the Γ^i_{jk} functions, we will first establish expressions for $\vec{X}_{ij} \cdot \vec{X}_k$ that are easier to find. Since these quantities occur frequently, there is a common shorthand symbol for them, namely,

$$[ij, k] = \vec{X}_{ij} \cdot \vec{X}_k. \tag{7.30}$$

Begin by fixing $j = k = 1$. Using the formulas

$$\frac{\partial g_{11}}{\partial x^1} = \frac{\partial}{\partial x^1}(\vec{X}_1 \cdot \vec{X}_1) = 2\vec{X}_{11} \cdot \vec{X}_1,$$

$$\frac{\partial g_{12}}{\partial x^1} = \frac{\partial}{\partial x^1}(\vec{X}_1 \cdot \vec{X}_2) = \vec{X}_{11} \cdot \vec{X}_2 + \vec{X}_{21} \cdot \vec{X}_1,$$

$$\frac{\partial g_{11}}{\partial x^2} = \frac{\partial}{\partial x^2}(\vec{X}_1 \cdot \vec{X}_1) = 2\vec{X}_{12} \cdot \vec{X}_1,$$

one deduces that

$$[11, 1] = \frac{1}{2}\frac{\partial g_{11}}{\partial x^1} \quad \text{and} \quad [11, 2] = \frac{\partial g_{12}}{\partial x^1} - \frac{1}{2}\frac{\partial g_{11}}{\partial x^2}.$$

Pursuing the calculations for the remaining cases, using the fact that (g_{ij}) is symmetric, and rewriting the results in a convenient manner, one can prove the following lemma.

Lemma 7.2.1. *For all indices $1 \leq i, j, k \leq 2$, we have*

$$[ij, k] = \vec{X}_{ij} \cdot \vec{X}_k = \frac{1}{2}\left(\frac{\partial g_{jk}}{\partial x^i} + \frac{\partial g_{ki}}{\partial x^j} - \frac{\partial g_{ij}}{\partial x^k}\right). \tag{7.31}$$

Using this lemma, one can easily establish a formula for the functions Γ^i_{jk}. We remind the reader that we shall often use the Einstein summation convention when an index is repeated in one superscript and one subscript position. (See Section 7.1 for a more accurate explanation of the convention.)

Proposition 7.2.2. *Let* $\vec{X} : U \to \mathbb{R}^3$ *be the parametrization of a regular surface of class* C^2 *in the neighborhood of some point* $p = \vec{X}(q)$. *Then the coefficients* Γ^i_{jk} *satisfy*

$$\Gamma^i_{jk} = \sum_{l=1}^{2} g^{il}[jk,l] = \sum_{l=1}^{2} g^{il} \frac{1}{2} \left(\frac{\partial g_{kl}}{\partial x^j} + \frac{\partial g_{lj}}{\partial x^k} - \frac{\partial g_{jk}}{\partial x^l}, \right) \quad (7.32)$$

where (g^{ij}), *with the indices in superscript, is the inverse matrix* $(g_{ij})^{-1}$.

Proof: We have already determined formulas for $[ij,k] = \vec{X}_{ij} \cdot \vec{X}_k$. However, from Equation (7.29), taking a dot product with respect to the derivative vectors $\vec{X}_i$, we obtain

$$[ij,k] = \Gamma^l_{ij} g_{lk}.$$

Multiplying the matrix of the first fundamental form (g_{lk}) by its inverse, we can write

$$g_{lk} g^{k\alpha} = \delta^\alpha_l$$

where δ^α_l is the Kronecker delta (defined in Equation (7.3)). Then we get

$$\sum_{k=1}^{2} g^{k\alpha}[ij,k] = \Gamma^l_{ij} g_{lk} g^{k\alpha} = \Gamma^l_{ij} \delta^\alpha_l = \Gamma^\alpha_{ij}.$$

The proposition follows from the symmetry of the (g^{ij}) matrix. $\square$

Definition 7.2.3. The symbols $[ij,k]$ are called the Christoffel symbols of the first kind, while the functions Γ^i_{jk} are called the Christoffel symbols of the second kind or, more simply, just the Christoffel symbols. With the Christoffel symbols understood, the the equations in Equation (7.29) are usually referred to collectively as *Gauss's equations* for a surface.

One of the first values of Gauss's equations is that knowing the functions L_{jk} and Γ^i_{jk} allows one to write not just the second partial derivatives of the parametrization $\vec{X}$ in terms of the basis $\{\vec{X}_1, \vec{X}_2, \vec{N}\}$, but also all higher derivatives of $\vec{X}(x^1, x^2)$. We point out that knowing g_{ij} in the neighborhood of a point allows us to determine the lengths of and angle between $\vec{X}_1$ and $\vec{X}_2$. Furthermore, as we shall see later, knowing the three distinct functions L_{jk} and the six distinct functions Γ^i_{jk} for a particular surface S in the neighborhood of p allows one to use a Taylor series expansion in the two variables x^1 and x^2 to write an infinite sum that provides a parametrization of S in a neighborhood of p once p, $\vec{X}_1$, and $\vec{N}$ are given. However, we still have not identified any relations that must exist between L_{ij} and g_{ij}, so we cannot yet state an equivalent to the natural equations theorem for space curves.

We point out that though we used a superscript and subscripts for the Christoffel symbol Γ^i_{jk}, these functions do *not* form the components of a tensor but rather transform according to the following proposition.

Proposition 7.2.4. *Let (x^1, x^2) and $(\bar{x}^1, \bar{x}^2)$ be two coordinate systems for a neighborhood of a point p on a surface S. If we denote by Γ^m_{ij} and $\bar{\Gamma}^\mu_{\alpha\beta}$ the Christoffel symbols in the respective coordinate systems, then they are related by*

$$\bar{\Gamma}^\mu_{\alpha\beta} = \frac{\partial x^i}{\partial \bar{x}^\alpha} \frac{\partial x^j}{\partial \bar{x}^\beta} \frac{\partial \bar{x}^\mu}{\partial x^m} \Gamma^m_{ij} + \frac{\partial^2 x^m}{\partial \bar{x}^\alpha \partial \bar{x}^\beta} \frac{\partial \bar{x}^\mu}{\partial x^m}.$$

Proof: We leave some of the details of this proof for the reader but present an outline here.

Consider two systems of coordinates (x^1, x^2) and $(\bar{x}^1, \bar{x}^2)$ for a neighborhood of a point p on a surface S. We know that the metric coefficients change according to

$$\bar{g}_{\alpha\beta} = \frac{\partial x^i}{\partial \bar{x}^\alpha} \frac{\partial x^j}{\partial \bar{x}^\beta} g_{ij} \qquad \text{and} \qquad \bar{g}^{\alpha\beta} = \frac{\partial \bar{x}^\alpha}{\partial x^i} \frac{\partial \bar{x}^\beta}{\partial x^j} g^{ij}. \qquad (7.33)$$

In the $(\bar{x}^1, \bar{x}^2)$ coordinate system, we denote the Christoffel symbols of the first kind by

$$\overline{[\alpha\beta, \nu]} = \frac{1}{2}\left(\frac{\partial \bar{g}_{\beta\nu}}{\partial \bar{x}^\alpha} + \frac{\partial \bar{g}_{\nu\alpha}}{\partial \bar{x}^\beta} - \frac{\partial \bar{g}_{\alpha\beta}}{\partial \bar{x}^\nu} \right), \qquad (7.34)$$

and we must first relate this to the Christoffel symbols $[ij, k]$ in the (x^1, x^2) coordinate system. Note that in this proof, we make indices (i, j, k, m) in the (x^1, x^2) coordinate system correspond to indices $(\alpha, \beta, \nu, \mu)$. Using Equation (7.33), the first term in Equation (7.34) transforms according to

$$\frac{\partial \bar{g}_{\beta\nu}}{\partial \bar{x}^\alpha} = \frac{\partial^2 x^j}{\partial \bar{x}^\alpha \partial \bar{x}^\beta}\frac{\partial x^k}{\partial \bar{x}^\nu}g_{jk} + \frac{\partial x^j}{\partial \bar{x}^\beta}\frac{\partial^2 x^k}{\partial \bar{x}^\alpha \partial \bar{x}^\nu}g_{jk} + \frac{\partial x^j}{\partial \bar{x}^\beta}\frac{\partial x^k}{\partial \bar{x}^\nu}\frac{\partial x^i}{\partial \bar{x}^\alpha}\frac{\partial g_{jk}}{\partial x^i}. \quad (7.35)$$

Equation (7.35) and the corresponding results of the other two terms in the sum in Equation (7.34) produce an expression for the transformation property of the Christoffel symbol of the first kind. The expression is not pleasing, especially since it involves the coefficients g_{ij} explicitly. One must now calculate

$$\bar{\Gamma}^\mu_{\alpha\beta} = \sum_{\nu=1}^{2} \bar{g}^{\mu\nu}\overline{[\alpha\beta, \nu]} = \sum_{\nu=1}^{2} \frac{\partial \bar{x}^\mu}{\partial x^m}\frac{\partial \bar{x}^\nu}{\partial x^l}g^{ml}\overline{[\alpha\beta, \nu]},$$

where we replace $\overline{[\alpha\beta, \nu]}$ with the terms found in Equation (7.35) and similar equalities. After appropriate simplifications, we find that

$$\begin{aligned}
\bar{\Gamma}^\mu_{\alpha\beta} = {} & \frac{\partial \bar{x}^\mu}{\partial x^m}\frac{\partial x^i}{\partial \bar{x}^\alpha}\frac{\partial x^j}{\partial \bar{x}^\beta}\Gamma^m_{ij} + \frac{1}{2}\left(\frac{\partial \bar{x}^\mu}{\partial x^j}\frac{\partial^2 x^j}{\partial \bar{x}^\alpha \partial \bar{x}^\beta} + \frac{\partial \bar{x}^\mu}{\partial x^i}\frac{\partial^2 x^i}{\partial \bar{x}^\alpha \partial \bar{x}^\beta}\right) \\
& + \frac{1}{2}\frac{\partial \bar{x}^\nu}{\partial x^l}\frac{\partial \bar{x}^\mu}{\partial x^m}\left(\frac{\partial x^j}{\partial \bar{x}^\alpha}\frac{\partial^2 x^k}{\partial \bar{x}^\alpha \partial \bar{x}^\nu}g^{ml}g_{jk} - \frac{\partial x^j}{\partial \bar{x}^\alpha}\frac{\partial^2 x^i}{\partial \bar{x}^\alpha \partial \bar{x}^\nu}g^{ml}g_{ij}\right) \quad (7.36) \\
& + \frac{1}{2}\frac{\partial \bar{x}^\nu}{\partial x^l}\frac{\partial \bar{x}^\mu}{\partial x^m}\left(\frac{\partial x^i}{\partial \bar{x}^\alpha}\frac{\partial^2 x^k}{\partial \bar{x}^\beta \partial \bar{x}^\nu}g^{ml}g_{ki} - \frac{\partial x^i}{\partial \bar{x}^\alpha}\frac{\partial^2 x^j}{\partial \bar{x}^\nu \partial \bar{x}^\beta}g^{ml}g_{ji}\right).
\end{aligned}$$

However, it is important to remember that when one sums over an index, the actual name of the index does not change the result of the summation. Applying this observation to Equation (7.36) and remembering that the components g_{ij} are symmetric in their indices finishes the proof of the proposition.   □

One should not view the fact that the Christoffel symbols Γ^i_{jk} do not form the components of a tensor as just a small annoyance; it is of fundamental importance in the theory of manifolds. From an intuitive perspective, one could understand tensors as objects that are related to the tangent space to a surface at a point. However, since the Γ^i_{jk} functions explicitly involve the second derivatives of a parametrization of a neighborhood of a point on a surface, one might have been able to predict that the Γ^i_{jk} functions would not necessarily form the components of a tensor.

Example 7.2.5 (Sphere). As a simple example, consider the usual parametrization of the sphere

$$\vec{X}(x^1, x^2) = (R\cos x^1 \sin x^2, R\sin x^1 \sin x^2, R\cos x^2).$$

In spherical coordinates, the variable names we used correspond to $x^1 = \theta$ as the meridian and $x^2 = \phi$ the latitude, measured as the angle down from the positive z-axis. A simple calculation leads to

$$g_{ij} = \begin{pmatrix} R^2 \sin^2(x^2) & 0 \\ 0 & R^2 \end{pmatrix} \quad \text{and} \quad g^{ij} = \frac{1}{R^2} \begin{pmatrix} \frac{1}{\sin^2(x^2)} & 0 \\ 0 & 1 \end{pmatrix}.$$

By Equation (7.31), we find that

$$[11, 1] = 0, \qquad\qquad [11, 2] = -R^2 \sin(x^2) \cos(x^2),$$
$$[12, 1] = R^2 \sin(x^2) \cos(x^2), \qquad [12, 2] = 0,$$
$$[21, 1] = R^2 \sin(x^2) \cos(x^2), \qquad [21, 2] = 0,$$
$$[22, 1] = 0, \qquad\qquad [22, 2] = 0.$$

Then, using Equation (7.32), we get

$$\Gamma^1_{11} = 0, \qquad\qquad \Gamma^2_{11} = -\sin(x^2)\cos(x^2),$$
$$\Gamma^1_{12} = \Gamma^1_{21} = \cot(x^2), \qquad \Gamma^2_{12} = \Gamma^2_{21} = 0,$$
$$\Gamma^1_{22} = 0, \qquad\qquad \Gamma^2_{22} = 0.$$

As an example of an application of Gauss's equation, recall from Equation (6.36) that two vectors $\vec{u}_1$ and $\vec{u}_2$ in the tangent space T_pS to a surface S at p are called conjugate if

$$I_p(dn_p(\vec{u}_1), \vec{u}_2) = I_p(\vec{u}_1, dn_p(\vec{u}_2)) = 0. \tag{7.37}$$

Intuitively speaking, Equation (7.37) states that conjugate directions are such that the direction of change of the unit normal vector $\vec{N}$ along $\vec{u}_1$ is perpendicular to $\vec{u}_2$ and vice versa. Given a parametrization $\vec{X}$ of a regular surface in the neighborhood of a point p, we say that $\vec{X}$ produces a conjugate set of coordinate lines if

$$dn_p(\vec{X}_1) \cdot \vec{X}_2 = 0 = dn_p(\vec{X}_2) \cdot \vec{X}_1$$

for all (x^1, x^2) in the domain of $\vec{X}$. One can state this alternatively as

$$\vec{N}_1 \cdot \vec{X}_2 = 0 = \vec{N}_2 \cdot \vec{X}_1,$$

so $\vec{X}$ produces a set of conjugate coordinate lines if and only if $L_{12} = 0$. In this situation, Gauss's equation for $\vec{X}_{12}$ reduces to

$$\vec{X}_{12} = \Gamma^1_{12}\vec{X}_1 + \Gamma^2_{12}\vec{X}_2, \tag{7.38}$$

so the parametrization $\vec{X}$ has a conjugate set of coordinate lines if and only if Equation (7.38) holds.

In addition, Equation (7.38) encompasses three independent linear differential equations in the rectangular coordinate functions of the parametrization. Therefore, given functions $P(x^1, x^2)$ and $Q(x^1, x^2)$, any three linearly independent solutions to the equation

$$\frac{\partial^2 f}{\partial x^1 \partial x^2} - P(x^1, x^2)\frac{\partial f}{\partial x^1} - Q(x^1, x^2)\frac{\partial f}{\partial x^2} = 0, \tag{7.39}$$

when taken as the coordinate functions, give a parametrization of a surface with conjugate coordinate lines in which $\Gamma^1_{12} = P(x^1, x^2)$ and $\Gamma^2_{12} = Q(x^1, x^2)$. Equation (7.39) is, in general, not simple to solve but since $\vec{N}$ does not appear in Equation (7.39), the coordinate function solutions are independent of each other.

Problems

7.2.1. Fill in the details of the proof for Proposition 7.2.4.

7.2.2. Calculate the Christoffel symbols for the torus parametrized by

$$\vec{X}(u, v) = ((a + b\cos v)\cos u, (a + b\cos v)\sin u, b\sin v),$$

where we assume that $a < b$.

7.2.3. Calculate the Christoffel symbols for functions graphs, i.e., surfaces parametrized by $\vec{X}(u, v) = (u, v, f(u, v))$, where f is a function from $U \subset \mathbb{R}^2$ to $\mathbb{R}$.

7.2.4. Calculate the Christoffel symbols for surfaces of revolution (see Problem 5.2.7).

7.2.5. Calculate the Christoffel symbols for the pseudosphere (see Problem 6.5.5).

7.2.6. *Tubes.* Let $\vec{\alpha}(t)$ be a regular space curve and let r be small enough that the tube

$$\vec{X}(t,u) = \vec{\alpha}(t) + (r\cos u)\vec{P}(t) + (r\sin u)\vec{B}(t)$$

is a regular surface. Calculate the Christoffel symbols for the tube $\vec{X}$.

7.2.7. Prove that a parametrization $\vec{X}$ of a neighborhood on a surface S is such that all the coordinate lines $x^2 = c$ are asymptotic lines if and only if $L_{11} = 0$.

7.2.8. Defining the function $D(x^1, x^2)$ as $D^2 = \det(g_{ij})$, show that

$$\frac{\partial}{\partial x^1}(\ln D) = \Gamma^1_{11} + \Gamma^2_{12} \quad \text{and} \quad \frac{\partial}{\partial x^2}(\ln D) = \Gamma^1_{12} + \Gamma^2_{22}.$$

7.2.9. Suppose that S is a surface with a parametrization that satisfies $\Gamma^1_{12} = \Gamma^2_{12} = 0$. Prove that S is a translation surface.

7.2.10. Using D as defined in Problem 7.2.8, calling $\theta(x^1, x^2)$ the angle between the coordinate lines, prove that

$$\frac{\partial \theta}{\partial x^1} = -\frac{D}{g_{11}}\Gamma^2_{11} - \frac{D}{g_{22}}\Gamma^2_{11} \quad \text{and} \quad \frac{\partial \theta}{\partial x^2} = -\frac{D}{g_{11}}\Gamma^2_{12} - \frac{D}{g_{22}}\Gamma^1_{22}. \quad (7.40)$$

Show that if the parametrization is orthogonal, i.e., $g_{12} = 0$, Equation (7.40) becomes

$$g_{22}\Gamma^2_{11} + g_{11}\Gamma^1_{12} = 0 \quad \text{and} \quad g_{22}\Gamma^2_{12} + g_{11}\Gamma^1_{22} = 0.$$

7.3   Codazzi Equations and the Theorema Egregium

The Gauss equations in Equation (7.29) define expressions for any second derivative $\vec{X}_{ij}$ in terms of the coefficients of the first and second fundamental forms and the basis $\{\vec{X}_1, \vec{X}_2, \vec{N}\}$. We remind the reader that the symmetry of the dot product imposes $g_{ij} = g_{ji}$. Furthermore, from $\vec{X}_{ij} = \vec{X}_{ji}$, we also deduced that $L_{ij} = L_{ji}$.

Definition 5.2.9 of a regular surface imposes the condition that all the higher derivatives of any parametrization $\vec{X}$ of a coordinate patch be continuous for all higher derivatives. Therefore, for a regular parametrization, the order in which one takes derivatives with respect to given variables is irrelevant. In particular, for the mixed third derivatives, we have $\vec{X}_{112} = \vec{X}_{121}$ and $\vec{X}_{221} = \vec{X}_{122}$, which can be listed in the following slightly more suggestive manner:

$$\frac{\partial \vec{X}_{11}}{\partial x^2} = \frac{\partial \vec{X}_{12}}{\partial x^1} \quad \text{and} \quad \frac{\partial \vec{X}_{22}}{\partial x^1} = \frac{\partial \vec{X}_{12}}{\partial x^2}$$

Consequently, because of the nontrivial expressions in Gauss's equations, we obtain

$$\frac{\partial}{\partial x^2}\left(\Gamma_{11}^1 \vec{X}_1 + \Gamma_{11}^2 \vec{X}_2 + L_{11}\vec{N}\right) = \frac{\partial}{\partial x^1}\left(\Gamma_{12}^1 \vec{X}_1 + \Gamma_{12}^2 \vec{X}_2 + L_{12}\vec{N}\right),$$
$$\frac{\partial}{\partial x^2}\left(\Gamma_{12}^1 \vec{X}_1 + \Gamma_{12}^2 \vec{X}_2 + L_{12}\vec{N}\right) = \frac{\partial}{\partial x^1}\left(\Gamma_{22}^1 \vec{X}_1 + \Gamma_{22}^2 \vec{X}_2 + L_{22}\vec{N}\right).$$
$$(7.41)$$

It is natural to equate the basis components of each equation. However, when we take the partial derivatives, we will obtain expressions involving second derivatives of $\vec{X}$ and derivatives of $\vec{N}$ that we need to put back into the usual $\{\vec{X}_1, \vec{X}_2, \vec{N}\}$ basis using the Weingarten equations given in Equation (6.26) or Gauss's equations in Equation (7.29). Doing this transforms Equation (7.41) into six distinct equations from which we deduce two significant theorems in the theory of surfaces.

Theorem 7.3.1 (Codazzi Equations). *For any parametrized smooth surface, the following hold:*

$$\frac{\partial L_{11}}{\partial x^2} - \frac{\partial L_{12}}{\partial x^1} = L_{11}\Gamma_{12}^1 + L_{12}\Gamma_{12}^2 - L_{12}\Gamma_{11}^1 - L_{22}\Gamma_{11}^2,$$
$$\frac{\partial L_{12}}{\partial x^2} - \frac{\partial L_{22}}{\partial x^1} = L_{11}\Gamma_{22}^1 + L_{12}\Gamma_{22}^2 - L_{12}\Gamma_{12}^1 - L_{22}\Gamma_{12}^2.$$
$$(7.42)$$

We can summarize these two equations in one by

$$\frac{\partial L_{ij}}{\partial x^l} - \Gamma_{il}^k L_{kj} = \frac{\partial L_{il}}{\partial x^j} - \Gamma_{ij}^k L_{kl} \qquad (7.43)$$

for all $1 \leq i, j, l \leq 2$, where Einstein summation is implied.

Proof: Equate the $\vec{N}$ coefficient functions in both equations from Equation (7.41). □

The classical formulation of the above relationships in Equation (7.42) are collectively called the Codazzi equations or sometimes the Mainardi-Codazzi equations. These equations present a relationship that must hold between the coefficients of the first and second fundamental form that must hold for all regular surfaces. They are a crucial part of the Fundamental Theorem of Surface Theory that we will present in Section 7.4.

The following theorem is another result that stems from equating components along $\vec{X}_1$ and $\vec{X}_2$ in Equation (7.41).

Theorem 7.3.2 (Theorema Egregium). *The Gaussian curvature of a surface is an intrinsic property of the surface, that is, it depends only on the coefficients of the metric tensor and higher derivatives thereof.*

Proof: Equating the $\vec{X}_1$ and $\vec{X}_2$ coefficient functions in the first equation of Equation (7.41), we obtain

$$\frac{\partial \Gamma^1_{11}}{\partial x^2} + \Gamma^1_{11}\Gamma^1_{12} + \Gamma^2_{11}\Gamma^1_{22} + L_{11}a^1_2 = \frac{\partial \Gamma^1_{12}}{\partial x^1} + \Gamma^1_{12}\Gamma^1_{11} + \Gamma^2_{12}\Gamma^1_{12} + L_{12}a^1_1,$$

$$\frac{\partial \Gamma^2_{11}}{\partial x^2} + \Gamma^1_{11}\Gamma^2_{12} + \Gamma^2_{11}\Gamma^2_{22} + L_{11}a^2_2 = \frac{\partial \Gamma^2_{12}}{\partial x^1} + \Gamma^1_{12}\Gamma^2_{11} + \Gamma^2_{12}\Gamma^2_{12} + L_{12}a^2_1.$$

Since $a^i_j = -\sum^2_{l=1} L_{jl}g^{li}$, where $(g^{ij}) = (g_{ij})^{-1}$, after some simplification, we can write these two equations as

$$g_{21}\frac{L_{11}L_{22} - (L_{12})^2}{\det g} = \frac{\partial \Gamma^1_{12}}{\partial x^1} - \frac{\partial \Gamma^1_{11}}{\partial x^2} + \Gamma^2_{12}\Gamma^1_{12} - \Gamma^2_{11}\Gamma^1_{22}, \quad (7.44)$$

$$-g_{11}\frac{L_{11}L_{22} - (L_{12})^2}{\det g} = \frac{\partial \Gamma^2_{12}}{\partial x^1} - \frac{\partial \Gamma^2_{11}}{\partial x^2} + \Gamma^1_{12}\Gamma^2_{11} + \Gamma^2_{12}\Gamma^2_{12}$$

$$- \Gamma^1_{11}\Gamma^2_{12} - \Gamma^2_{11}\Gamma^2_{22}. \quad (7.45)$$

Since $K = \det(L_{ij})/\det(g_{ij})$, we can use either one of these equations to obtain a formula for K in terms of the function g_{ij} and Γ^i_{jk}. However, since the Christoffel symbols Γ^i_{jk} are themselves determined by the coefficients of the metric tensor, then these equations provide formulas for the Gaussian curvature exclusively in terms of the first fundamental form. $\square$

Gauss coined the name "Theorema Egregium," which means "an excellent theorem," and indeed this result should seem rather surprising. A priori, the Gaussian curvature depends on the components of the first fundamental form and the second fundamental form. After all, the Gaussian curvature at a point p on the surface is $K = \det(dn_p)$, where dn_p is the differential of the Gauss map. Since the coefficients of the matrix dn_p with respect to the basis $\{\vec{X}_1, \vec{X}_2\}$ involve the L_{ij} coefficients, one would naturally expect that we would need the second fundamental form to calculate K. However, the Theorema Egregium shows that only the coefficient functions of the first fundamental form are necessary.

In the study of surfaces, we called a local or a global property of a curve or surface a "geometric quantity" if it does not change under the orientation and position of the curve or surface in space. On the other hand, topology studies properties of point sets that are invariant under any homeomorphism—a bijective function that is continuous in both directions. The concept of an intrinsic property lies somewhere between these two extremes. Intrinsic geometry studies what some mathematicians classically call "bending invariants," quantities that are preserved when one bends the curve or surface but does not stretch them.

In the above proof, Equations (7.44) and (7.45) motivate the following definition of the *Riemann symbols*:

$$R^l_{ijk} = \frac{\partial \Gamma^l_{ik}}{\partial x^j} - \frac{\partial \Gamma^l_{jk}}{\partial x^i} + \Gamma^m_{ik}\Gamma^l_{mj} - \Gamma^m_{jk}\Gamma^l_{mi}. \qquad (7.46)$$

As it turns out (see Problem 7.3.2), the Riemann symbols form the components of a $(1,3)$-tensor. Then, a closely associated tensor denoted by

$$R_{ijkl} = R^m_{ijk}g_{ml} \qquad (7.47)$$

has the interesting property that $R_{1212} = \det(L_{ij})$, and hence, the Gaussian curvature of a surface at a point is given by the component function

$$K = \frac{R_{1212}}{\det(g_{ij})}. \qquad (7.48)$$

The tensor associated to the components R^l_{ijk} is called the Riemann curvature tensor and plays an important role in the analysis on manifolds.

Example 7.3.3 (Cone). As a simple example, consider the right circular cone defined by the parametrization

$$\vec{X}(u, v) = (v \cos u, v \sin u, v),$$

where $u \in [0, 2\pi]$ and $v > 0$, which we know, as a developable surface, has Gaussian curvature $K = 0$ everywhere. Simple calculations give, for the metric tensor,

$$g_{11} = v^2, \qquad g_{12} = 0, \qquad g_{22} = 2.$$

For the Christoffel symbols, one obtains

$$\Gamma_{11}^2 = -\frac{1}{2}v, \qquad \Gamma_{12}^1 = \Gamma_{21}^1 = \frac{1}{v}$$

with all the other symbols of the second kind $\Gamma_{jk}^i = 0$. With this data, an application of Equation (7.46) shows that all but two of the Riemann symbols vanish simply because all the terms vanish. However, for the two nontrivial terms, one calculates

$$R_{122}^1 = \frac{\partial \Gamma_{12}^1}{\partial v} - \frac{\partial \Gamma_{22}^1}{\partial u} + \Gamma_{12}^1\Gamma_{12}^1 + \Gamma_{12}^2\Gamma_{22}^1 - \Gamma_{22}^1\Gamma_{11}^1 - \Gamma_{22}^2\Gamma_{21}^1 = -\frac{1}{v^2} + \frac{1}{v^2} = 0,$$

$$R_{121}^2 = \frac{\partial \Gamma_{11}^2}{\partial v} - \frac{\partial \Gamma_{12}^2}{\partial u} + \Gamma_{11}^1\Gamma_{12}^2 + \Gamma_{11}^2\Gamma_{22}^2 - \Gamma_{21}^1\Gamma_{11}^2 - \Gamma_{21}^2\Gamma_{21}^2 = -\frac{1}{2} - \left(\frac{1}{v}\right)\left(-\frac{1}{2}v\right) = 0.$$

Consequently, though the Christoffel symbols are not identically 0, the Riemann curvature tensor is identically 0. One finds then that $R_{1212} = 0$ as a function of u and v, and hence, using Equation (7.48), one recovers the fact that the cone has Gaussian curvature identically 0 everywhere.

The Theorema Egregium (Theorem 7.3.2) excited the mathematics community in the middle of the nineteenth century and sparked a search for a variety of alternative formulations for the Gaussian curvature of a surface as a function of the g_{ij} metric coefficients. We leave a few such formulas for K for the exercises, but we present a few here.

Recall that $L_{ij} = \vec{X}_{ij} \cdot \vec{N}$, $\det(g_{ij}) = \|\vec{X}_u \times \vec{X}_v\|^2$, and $\vec{N} = \vec{X}_u \times \vec{X}_v/\|\vec{X}_u \times \vec{X}_v\|$. From these facts, one can easily see that

$$K = \frac{L_{11}L_{22} - L_{12}^2}{\det(g_{ij})} = \frac{(\vec{X}_{uu}\vec{X}_u\vec{X}_v)(\vec{X}_{vv}\vec{X}_u\vec{X}_v) - (\vec{X}_{uv}\vec{X}_u\vec{X}_v)^2}{\det(g_{ij})^2}, \quad (7.49)$$

where $(\vec{A}\vec{B}\vec{C})$ is the triple-vector product in $\mathbb{R}^3$. However, the triple-vector product is a determinant and we know that for all matrices M_1

and M_2, one has $\det(M_1 M_2) = \det(M_1) \det(M_2)$, and so Equation (7.49) becomes

$$K = \frac{1}{\det(g_{ij})^2} \left(\begin{vmatrix} \vec{X}_{uu} \cdot \vec{X}_{vv} & \vec{X}_{uu} \cdot \vec{X}_u & \vec{X}_{uu} \cdot \vec{X}_v \\ \vec{X}_u \cdot \vec{X}_{vv} & \vec{X}_u \cdot \vec{X}_u & \vec{X}_u \cdot \vec{X}_v \\ \vec{X}_v \cdot \vec{X}_{vv} & \vec{X}_v \cdot \vec{X}_u & \vec{X}_v \cdot \vec{X}_v \end{vmatrix} \right.$$
$$\left. - \begin{vmatrix} \vec{X}_{uv} \cdot \vec{X}_{uv} & \vec{X}_{uv} \cdot \vec{X}_u & \vec{X}_{uv} \cdot \vec{X}_v \\ \vec{X}_u \cdot \vec{X}_{uv} & \vec{X}_u \cdot \vec{X}_u & \vec{X}_u \cdot \vec{X}_v \\ \vec{X}_v \cdot \vec{X}_{uv} & \vec{X}_v \cdot \vec{X}_u & \vec{X}_v \cdot \vec{X}_v \end{vmatrix} \right). \quad (7.50)$$

This equation involves only the metric tensor coefficients and Γ^i_{jk} symbols, except in the inner products $\vec{X}_{uu} \cdot \vec{X}_{vv}$ and $\vec{X}_{uv} \cdot \vec{X}_{uv}$. At first glance, this problem seems insurmountable, but if one performs the Laplace expansion on the determinants along the first row, factors terms involving $\det(g_{ij})$, and collects into determinants again, one can show that Equation (7.50) is equivalent to

$$K = \frac{1}{\det(g_{ij})^2} \left(\begin{vmatrix} \vec{X}_{uu} \cdot \vec{X}_{vv} - \vec{X}_{uv} \cdot \vec{X}_{uv} & \vec{X}_{uu} \cdot \vec{X}_u & \vec{X}_{uu} \cdot \vec{X}_v \\ \vec{X}_u \cdot \vec{X}_{vv} & \vec{X}_u \cdot \vec{X}_u & \vec{X}_u \cdot \vec{X}_v \\ \vec{X}_v \cdot \vec{X}_{vv} & \vec{X}_v \cdot \vec{X}_u & \vec{X}_v \cdot \vec{X}_v \end{vmatrix} \right.$$
$$\left. - \begin{vmatrix} 0 & \vec{X}_{uv} \cdot \vec{X}_u & \vec{X}_{uv} \cdot \vec{X}_v \\ \vec{X}_u \cdot \vec{X}_{uv} & \vec{X}_u \cdot \vec{X}_u & \vec{X}_u \cdot \vec{X}_v \\ \vec{X}_v \cdot \vec{X}_{uv} & \vec{X}_v \cdot \vec{X}_u & \vec{X}_v \cdot \vec{X}_v \end{vmatrix} \right). \quad (7.51)$$

The value in this is that though one cannot express $\vec{X}_{uu} \cdot \vec{X}_{vv}$ or $\vec{X}_{uv} \cdot \vec{X}_{uv}$ using only the metric coefficients, it can be shown that

$$\vec{X}_{uu} \cdot \vec{X}_{vv} - \vec{X}_{uv} \cdot \vec{X}_{uv} = -\frac{1}{2}g_{11,22} + g_{12,12} - \frac{1}{2}g_{22,11},$$

where by $g_{ij,kl}$ we mean the function

$$\frac{\partial^2 g_{ij}}{\partial x^k \partial x^l}.$$

This produces the following formula for K that is appealing on the grounds that it illustrates symmetry of the metric tensor coefficients in determining K:

$$K = \frac{1}{\det(g_{ij})^2} \left(\begin{vmatrix} -\frac{1}{2}g_{11,22} + g_{12,12} - \frac{1}{2}g_{11,22} & \frac{1}{2}g_{11,1} & g_{12,1} - \frac{1}{2}g_{11,2} \\ g_{12,2} - \frac{1}{2}g_{22,1} & g_{11} & g_{12} \\ \frac{1}{2}g_{22,2} & g_{21} & g_{22} \end{vmatrix} \right.$$

$$\left. - \begin{vmatrix} 0 & \frac{1}{2}g_{11,2} & \frac{1}{2}g_{22,1} \\ \frac{1}{2}g_{11,2} & g_{11} & g_{12} \\ \frac{1}{2}g_{22,1} & g_{21} & g_{22} \end{vmatrix} \right). \quad (7.52)$$

In the situation where one considers an orthogonal parametrization of a neighborhood of a surface, namely, when $g_{12} = 0$, Equation (7.52) takes on the particularly interesting form of

$$K = -\frac{1}{\sqrt{g_{11}g_{22}}} \left(\frac{\partial}{\partial x^1} \left(\frac{1}{\sqrt{g_{11}}} \frac{\partial \sqrt{g_{22}}}{\partial x^1} \right) + \frac{\partial}{\partial x^2} \left(\frac{1}{\sqrt{g_{22}}} \frac{\partial \sqrt{g_{11}}}{\partial x^2} \right) \right).$$
$$(7.53)$$

By historical habit, Equations (7.52) and (7.53) are also referred to as *Gauss's equations* and one usually refers to the collection of Equations (7.53) and (7.42) as the *Gauss-Codazzi equations*.

We now prove a theorem that we could have introduced much earlier, but it arises in part as an application of the Codazzi equations.

Definition 7.3.4. A set $S \subset \mathbb{R}^n$ is called *path-connected* if for any pair of points $p, q \in S$, there exists a continuous path (curve) $\alpha : [0, 1] \to \mathbb{R}^n$, with $\alpha(0) = p$, $\alpha(1) = q$, and $\alpha([0, 1]) \subset S$.

Proposition 7.3.5. *If S is a path-connected regular surface in which all its points are umbilical, then S is either contained in a plane or in a sphere.*

Proof: Let $U \subset \mathbb{R}^2$ be an open set, and let $\vec{X} : U \to \mathbb{R}^3$ be the parametrization of a neighborhood $V = \vec{X}(U)$ of S. Since all points of S are umbilical, then the eigenvalues of dn_p are equal to a value λ that a priori is a function of the coordinates (u, v) of the patch V. We first show that $\lambda(u, v)$ is a constant over U.

By Proposition 6.4.4, there always exist two linearly independent eigenvectors to dn_p, so dn_p is always diagonalizable. Thus, at any point p of S, the whole tangent plane T_pS is the eigenspace of the eigenvalue λ. Thus, in particular we have

$$\vec{N}_u = dn_p(\vec{X}_u) = \lambda\vec{X}_u,$$
$$\vec{N}_v = dn_p(\vec{X}_v) = \lambda\vec{X}_v.$$

Since $\vec{N}_{uv} = \vec{N}_{vu}$, a fact that is equivalent to the Codazzi equations (see Problem 7.3.1), we have

$$\lambda_v \vec{X}_u + \lambda \vec{X}_{uv} = \lambda_u \vec{X}_v + \lambda \vec{X}_{vu}.$$

Since $\vec{X}_u$ and $\vec{X}_v$ are linearly independent, we deduce that $\lambda_u = \lambda_v = 0$, and therefore λ is a constant function.

If $\lambda = 0$, then $\vec{N}_u = \vec{N}_v = \vec{0}$, so the unit normal vector $\vec{N}$ is a constant vector $\vec{N}_0$ over the domain U. Then

$$\frac{\partial}{\partial u}(\vec{X} \cdot \vec{N}_0) = \vec{X}_u \cdot \vec{N}_0 + \vec{0} = \vec{0}$$

and similarly for $\frac{\partial}{\partial v}(\vec{X} \cdot \vec{N}_0) = \vec{0}$, which shows that $\vec{X} \cdot \vec{N}_0 = \vec{0}$, which further proves that $V = \vec{X}(U)$ lies in the plane through p perpendicular to $\vec{N}_0$.

On the other hand, if $\lambda \neq 0$, then consider the vector function

$$\vec{Y}(u, v) = \vec{X}(u, v) - \frac{1}{\lambda}\vec{N}(u, v).$$

By a similar calculation as above, we see that the function $\vec{Y}(u, v)$ is a constant vector. Then we deduce that

$$\|\vec{X}(u, v) - \vec{Y}\| = \left\|\frac{1}{\lambda}\vec{N}\right\| = \frac{1}{|\lambda|},$$

which shows that $V = \vec{X}(U)$ lies on a sphere of center $\vec{Y}$ and of radius $1/|\lambda|$.

So far, the above proof has established that any parametrization of a neighborhood V of S either lies on a sphere or lies in a plane but not that all of S necessarily lies on a sphere or in a plane. The assumption that S is path-connected extends the result to the whole surface as follows.

By Example 7.1.10, if V_1 and V_1 are two coordinate patches of S with respective coordinate systems (x^1, x^2) and $(\bar{x}^1, \bar{x}^2)$ such that $V_1 \cap V_2 \neq \emptyset$, then on $V_1 \cap V_2$ the coefficients of the Gauss map with respect to the coordinate systems are related by

$$\bar{a}^i_j = \frac{\partial \bar{x}^i}{\partial x^k} \frac{\partial x^l}{\partial \bar{x}^j} a^k_l.$$

This is tantamount to saying that the matrices for the Gauss map relative to different coordinate systems are similar matrices. Hence, if V_1 and V_2 are two overlapping coordinate patches, then the eigenvalue λ is constant over $V_1 \cup V_2$, and then all of $V_1 \cup V_2$ either lies in a plane or on a sphere.

Since S is path-connected, given any two points p and q on S, there exists a path $\alpha : [0,1] \to S$, with $\alpha(0) = p$ and $\alpha(1) = q$. From basic facts in topology, we know that $\alpha([0,1])$ is a compact set and that any open cover of a compact set has a finite subcover. Thus, given a collection of coordinate patches that cover $\alpha([0,1])$, then only a finite subcollection of these coordinate patches is necessary to cover $\alpha([0,1])$. Call this collection $\{V_1, V_2, \ldots, V_r\}$. By the reasoning in the above paragraph, we see that $V_1 \cup V_2 \cup \cdots \cup V_r$ lies either in a plane or on a sphere. Thus all points of S either lie in the same plane or lie on the same sphere. $\qquad\square$

Problems

7.3.1. Show that one can obtain the Codazzi equations from the equation $\vec{N}_{12} = \vec{N}_{21}$ and by using the Gauss equations in Equation (7.29).

7.3.2. Prove that the Riemann symbols defined in Equation (7.46) form the components of a $(1,3)$-tensor.

7.3.3. Prove Equation (7.48) and the claim that supports it.

7.3.4. Suppose that $g_{12} = L_{12} = 0$ (i.e., the coordinate lines are lines of curvature), so that the principal curvatures satisfy

$$\kappa_1 = \frac{L_{11}}{g_{11}} \quad \text{and} \quad \kappa_2 = \frac{L_{22}}{g_{22}}.$$

Prove that the Codazzi equations are equivalent to

$$\frac{\partial \kappa_1}{\partial x^2} = \frac{g_{11,2}}{2g_{11}}(\kappa_2 - \kappa_1) \quad \text{and} \quad \frac{\partial \kappa_2}{\partial x^1} = \frac{g_{22,1}}{2g_{22}}(\kappa_1 - \kappa_2),$$

where by $g_{11,2}$ we mean $\frac{\partial g_{11}}{\partial x^2}$ and similarly for $g_{22,1}$.

7.3.5. Prove that the Riemann symbols R^l_{ijk} defined in Equation (7.46) are antisymmetric in the indices $\{i,j\}$. Conclude that for surfaces, R^l_{ijk} represent at most four distinct functions.

7.3.6. Prove Equation (7.53).

7.3.7. Prove the following formula by Blaschke for the Gaussian curvature:

$$K = -\frac{1}{4\det(g_{ij})^2}\begin{vmatrix} g_{11} & g_{12} & g_{22} \\ g_{11,1} & g_{12,1} & g_{22,1} \\ g_{11,2} & g_{12,2} & g_{22,2} \end{vmatrix}$$

$$-\frac{1}{2\sqrt{\det(g_{ij})}}\left(\frac{\partial}{\partial x^1}\left(\frac{g_{22,1} - g_{12,2}}{\sqrt{\det(g_{ij})}}\right) - \frac{\partial}{\partial x^2}\left(\frac{g_{12,1} - g_{11,2}}{\sqrt{\det(g_{ij})}}\right)\right).$$

7.3.8. Consider the two surfaces parametrized by

$$\vec{X}(u,v) = (v\cos u, v\sin u, \ln v),$$
$$\vec{Y}(u,v) = (v\cos u, v\sin u, u).$$

Prove that these two surfaces have equal Gaussian curvature functions over the same domain but that they do not possess the same metric tensor. (This gives an example of two surfaces with given coordinate systems over which different metric tensor components lead to the same Gaussian curvature function.)

7.3.9. The coordinate curves of a parametrization $\vec{X}(u,v)$ form a *Tchebysheff net* if the lengths of the opposite sides of any quadrilateral formed by them are equal.

(a) Prove that a necessary and sufficient condition for a parametrization to be a Tchebysheff net is

$$\frac{\partial g_{11}}{\partial v} = \frac{\partial g_{22}}{\partial u} = 0.$$

(b) (ODE) Prove that when a parametrization constitutes a Tchebysheff net, there exists a reparametrization of the coordinate neighborhood so that the new components of the metric tensor are

$$g_{11} = 1, \qquad g_{12} = \cos\theta, \qquad g_{22} = 1,$$

where θ is the angle between the coordinate lines at the given point on the surface.

(c) Show that in this case,

$$K = -\frac{\theta_{uv}}{\sin\theta}.$$

7.4   The Fundamental Theorem of Surface Theory

Suppose we consider a regular oriented surface S and coordinate patch V parametrized by $\vec{X} : U \to \mathbb{R}^3$. We have seen that the coefficients (g_{ij}) and (L_{ij}) of the first and second fundamental forms satisfy $\det(g_{ij}) > 0$ and the Gauss-Codazzi equations. That, given these conditions, there exists an essentially unique surface with specified first and second fundamental forms is a profound result, called the Fundamental Theorem of Surface Theory.

Theorem 7.4.1. *If E, F, G and e, f, g are sufficiently differentiable functions of (u, v) that satisfy the Gauss-Codazzi Equations (7.52) and (7.42) and $EG - F^2 > 0$, then there exists a parametrization $\vec{X}$ of a regular orientable surface that admits*

$$g_{11} = E, \quad g_{12} = F, \quad g_{22} = G, \quad L_{11} = e, \quad L_{12} = f, \quad L_{22} = g.$$

Furthermore, this surface is uniquely determined up to its position in space.

The original proof of this theorem was provided by Bonnet in 1855 in [5]. In more recent texts, one can find the proof in the Appendix to Chapter 4 in [10] or in Chapter VI of [29]. We will not provide a complete proof of the Fundamental Theorem of Surface Theory here since it involves solving a system of partial differential equations, but we will sketch the main points behind it.

The setup for the proof is to consider nine functions $\xi_i(u, v)$, $\varphi_i(u, v)$, and $\psi_i(u, v)$ with $1 \le i \le 3$, and think of these functions as the components of the vector functions $\vec{X}_u$, $\vec{X}_v$, and $\vec{N}$ so that $\vec{X}_u = (\xi_1, \xi_2, \xi_3)$, $\vec{X}_v = (\varphi_1, \varphi_2, \varphi_3)$, and $\vec{N} = (\psi_1, \psi_2, \psi_3)$. With this setup, the equations that define Gauss's and Weingarten equations, namely, Equations (7.29) and (6.26), become the following system of 18 partial differential equation: for $i = 1, 2, 3$,

$$\frac{\partial \xi_i}{\partial u} = \Gamma^1_{11}\xi_i + \Gamma^2_{11}\varphi_i + L_{11}\psi_i, \qquad \frac{\partial \xi_i}{\partial v} = \Gamma^1_{12}\xi_i + \Gamma^2_{12}\varphi_i + L_{12}\psi_i,$$

$$\frac{\partial \varphi_i}{\partial u} = \Gamma^1_{21}\xi_i + \Gamma^2_{21}\varphi_i + L_{21}\psi_i, \qquad \frac{\partial \varphi_i}{\partial v} = \Gamma^1_{22}\xi_i + \Gamma^2_{22}\varphi_i + L_{22}\psi_i,$$

$$\frac{\partial \psi_i}{\partial u} = a^1_1\xi_i + a^2_1\varphi_i, \qquad\qquad \frac{\partial \psi_i}{\partial v} = a^1_2\xi_i + a^2_2\varphi_i. \qquad (7.54)$$

In general, when a system of partial differential equations involving n functions $u_i(x_1, \ldots, x_m)$ has $n < m$, the solutions may involve not only constants of integration but also unknown functions that can be any continuous function from $\mathbb{R}$ to $\mathbb{R}$ (on some appropriate interval). However, when $n > m$, i.e., when there are more functions in the system than there are independent variables, the system may be "overdetermined" and may either have less freedom in its solution set or have no solutions at all. In fact, one cannot expect the above system to have solutions if the mixed partial derivatives of $\vec{\xi} = (\xi_1, \xi_2, \xi_3)$, $\vec{\varphi} = (\varphi_1, \varphi_2, \varphi_3)$, and $\vec{\psi} = (\psi_1, \psi_2, \psi_3)$ are not equal. This is usually called the *compatibility condition* for systems of partial differential equations, and, as we see in the above system, this condition imposes relations between the functions $\Gamma^i_{jk}(u, v)$ and $L_{jk}(u, v)$.

The key ingredient behind the Fundamental Theorem of Surface Theory is Theorem V in Appendix B of [29] that, applied to our context, states that if all second derivatives of the Γ^i_{jk} and L_{jk} functions are continuous and if the compatibility condition holds in Equation (7.54), solutions to the system exist and are unique once values for $\vec{\xi}(u_0, v_0)$, $\vec{\varphi}(u_0, v_0)$, and $\vec{\psi}(u_0, v_0)$ are given, where (u_0, v_0) is a point in the common domain of $\Gamma^i_{jk}(u, v)$ and $L_{jk}(u, v)$. The compatibility condition required in this theorem is satisfied if and only if the functions $g_{11} = E$, $g_{12} = F$, $g_{22} = G$, $L_{11} = e$, $L_{12} = f$, and $L_{22} = g$ satisfy the Gauss-Codazzi equations.

Solutions to Equation (7.54) can be chosen in such a way that

$$\vec{\xi}(u_0, v_0) \cdot \vec{\xi}(u_0, v_0) = E(u_0, v_0), \qquad \vec{\varphi}(u_0, v_0) \cdot \vec{\varphi}(u_0, v_0) = G(u_0, v_0),$$

$$\vec{\xi}(u_0, v_0) \cdot \vec{\varphi}(u_0, v_0) = F(u_0, v_0), \qquad \vec{\psi}(u_0, v_0) \cdot \vec{\psi}(u_0, v_0) = 1,$$

$$\vec{\psi}(u_0, v_0) \cdot \vec{\xi}(u_0, v_0) = 0, \qquad \vec{\psi}(u_0, v_0) \cdot \vec{\varphi}(u_0, v_0) = 0,$$

$$\frac{\vec{\xi}(u_0, v_0) \times \vec{\varphi}(u_0, v_0)}{\|\vec{\xi}(u_0, v_0) \times \vec{\varphi}(u_0, v_0)\|} = \vec{\psi}(u_0, v_0). \tag{7.55}$$

The next step of the proof is to show that, given the above initial conditions, the following equations hold for all (u, v) where the solutions are defined:

$$\vec{\xi}(u,v) \cdot \vec{\xi}(u,v) = E(u,v), \qquad \vec{\varphi}(u,v) \cdot \vec{\varphi}(u,v) = G(u,v),$$
$$\vec{\xi}(u,v) \cdot \vec{\varphi}(u,v) = F(u,v), \qquad \vec{\psi}(u,v) \cdot \vec{\psi}(u,v) = 1,$$
$$\vec{\psi}(u,v) \cdot \vec{\xi}(u,v) = 0, \qquad \vec{\psi}(u,v) \cdot \vec{\varphi}(u,v) = 0,$$
$$\frac{\vec{\xi}(u,v) \times \vec{\varphi}(u,v)}{\|\vec{\xi}(u,v) \times \vec{\varphi}(u,v)\|} = \vec{\psi}(u,v).$$

From the solutions for $\vec{\xi}$, $\vec{\varphi}$, and $\vec{\psi}$, we form the new system of differential equations

$$\begin{cases} \vec{X}_u = \vec{\xi}, \\ \vec{X}_v = \vec{\varphi}. \end{cases}$$

One easily obtains a solution for the function $\vec{X}$ over appropriate (u,v) by

$$\vec{X}(u,v) = \int_{u_0}^{u} \vec{\xi}(u,v)\, du + \int_{v_0}^{v} \vec{\varphi}(u_0,v)\, dv. \qquad (7.56)$$

The resulting vector function $\vec{X}$ is defined over an open set $U \subset \mathbb{R}^2$ containing (u_0, v_0), and $\vec{X}$ parametrizes a regular surface S. By construction, the coefficients of the first fundamental form for this surface are

$$g_{11} = E, \qquad g_{12} = F, \qquad g_{22} = G.$$

One then proves that it is also true that the coefficients of the second fundamental form satisfy

$$L_{11} = e, \qquad L_{12} = f, \qquad L_{22} = g.$$

It remains to be shown that this surface is unique up to a rigid motion in $\mathbb{R}^3$. It is not hard to see that the equalities in Equation (7.55) imposed on the initial conditions still allow one the freedom to choose the unit vector $\hat{\xi}(u_0, v_0) = \vec{\xi}(u_0, v_0)/\|\vec{\xi}(u_0, v_0)\|$ and the vector $\vec{\psi}(u_0, v_0)$, which must be perpendicular to $\vec{\xi}(u_0, v_0)$. The vectors

$$\hat{\xi}(u_0, v_0), \quad \vec{\psi}(u_0, v_0), \quad \hat{\xi}(u_0, v_0) \times \vec{\psi}(u_0, v_0)$$

form a positive orthonormal frame, so any two choices allowed by Equation (7.55) differ from each other by a rotation in $\mathbb{R}^3$. Finally, the integration in Equation (7.56) introduces a constant vector of integration. Thus, two solutions to Gauss's and Weingarten equations differ from each other by a rotation and a translation, namely, any rigid motion in $\mathbb{R}^3$.

Problems

7.4.1. (ODE) Consider solutions to Gauss's and Weingarten equations for which the coefficients of the first and second fundamental forms are constant.

    (a) Let E, F, and G be constants such that $EG - F > 0$ and view them as constant functions. Prove that the Gauss-Codazzi equations impose $L_{jk} = 0$.

    (b) Prove that all solutions to the Gauss-Weingarten equations in this situation are planes.

7.4.2. (ODE) Find all regular parametrized surfaces that have

$$g_{11} = 1, \qquad g_{12} = 0, \qquad g_{22} = \cos^2 u$$
$$L_{11} = 1, \qquad L_{12} = 0, \qquad L_{22} = \cos^2 u.$$

7.4.3. Does there exist a surface $\vec{X}(u, v)$ with

$$g_{11} = 1, \qquad g_{12} = 0, \qquad g_{22} = \cos^2 u,$$
$$L_{11} = \cos^2 u, \qquad L_{12} = 0, \qquad L_{22} = 1?$$

CHAPTER 8

Curves on Surfaces

Historians of mathematics often point to Euclid as the inventor, or at least the father in some metaphorical sense, of the synthetic methods of mathematical proofs. Euclid's *Elements*, which is comprised of 13 books, treats a wide variety of topics but focuses heavily on geometry. In his *Elements*, Euclid presents 23 geometric definitions along with five postulates (or axioms), and in Books I–VI and XI–XIII, he proves around 250 propositions about lines, circles, angles, and ratios of quantities in the plane and in space.

Most popular texts about the nature of mathematics agree that Euclid's geometric methods held a foundational importance in the development of mathematics (see, for example, [11, pp. 76–77] or [16, p. 7]). Many such texts also retell the story of the discovery in the nineteenth century of geometries that remain consistent yet do not satisfy the fifth and most debatable of Euclid's postulates ([6], [8, pp. 214–227], [9], [16, pp. 217–223]). One can readily list spherical geometry and hyperbolic geometry as examples of such geometries.

At this point in this book, with the methods from the theory of curves and surfaces now at our disposal, we stand in a position to study geometry on any regular surface, not just on a sphere or on its hyperbolic equivalent, the pseudosphere. As we shall see, on a general regular surface, and already on a sphere, the notion of "straightness" is not intuitive, and one might debate whether such a notion should exist at all. Hence, instead of only trying to consider lines and circles, we first study regular curves on a regular surface in general and only later define notions of shortest distance, straightness, and parallelism.

8.1   Curvatures and Torsion

8.1.1   Natural Frames

Throughout this chapter, we let S be a regular surface of class C^2 and let V be an open set of S parametrized by a vector function $\vec{X} : U \to \mathbb{R}^3$, where U is an open set in $\mathbb{R}^2$. We consider a curve C of class C^2 of the form $\vec{\gamma} = \vec{X} \circ \vec{\alpha}$, where $\vec{\alpha} : I \to U$ is a curve in the domain of $\vec{X}$, with $\vec{\alpha}(t) = (u(t), v(t))$.

Let $p = \vec{\gamma}(t_0)$ be a point on the curve and on the surface. Using the theory of curves in $\mathbb{R}^3$, one typically performs calculations on quantities related to $\vec{\gamma}$ in the Frenet frame $\{\vec{T}, \vec{P}, \vec{B}\}$. This frame is orthonormal and hence has many nice properties. On the other hand, using the local theory of surfaces, one would typically perform calculations in the $\{\vec{X}_u, \vec{X}_v, \vec{N}\}$ reference frame. This frame is not orthonormal but is often the most practical. If a point p on a surface is not an umbilical point, the frame $\{\vec{N}, \vec{e}_1, \vec{e}_2\}$, where the vectors $\vec{e}_i$ are principal directions (well defined up to a change in sign), is a natural orthonormal frame associated to S at p. Though more geometric in nature than $\{\vec{X}_u, \vec{X}_v, \vec{N}\}$, the frame $\{\vec{N}, \vec{e}_1, \vec{e}_2\}$ does not lend itself well to calculations with specific parametrizations. Studying curves on surfaces, one could choose between these three reference frames, but it turns out that a combination will be most helpful.

Borrowing first from surface theory, we use the unit normal vector $\vec{N}$, which is invariant up to a sign under a parametrization of S. We will often consider $\vec{N}(t)$ to be a single-variable vector function over the interval I, by which we mean explicitly $\vec{N}(t) = \vec{N} \circ \vec{\alpha}(t)$. Borrowing from the theory of space curves, we use $\vec{T}$, the unit tangent vector to the curve C at p, which is also invariant up to a sign under reparametrization of the curve. By construction, $\vec{N}$ and $\vec{T}$ are perpendicular to each other, and we just need to choose a third vector to complete the frame. We cannot use $\vec{P}$ as a third vector in the frame because there is no guarantee that the principal normal vector $\vec{P}(t)$ is perpendicular to $\vec{N}$. In fact, Section 6.4 discusses how the second fundamental form measures the relationship between $\vec{N}$ and $\vec{P}$. Consequently, to complete a natural orthonormal frame related to a curve on a surface, we define the new

vector function
$$\vec{U}(t) = \vec{N}(t) \times \vec{T}(t).$$

We remind the reader that in the theory of plane curves, the unit normal $\vec{U}$ to a curve at a point is defined as the counterclockwise rotation of the unit tangent $\vec{T}$. One can rephrase this definition as

$$\vec{U} = \vec{k} \times \vec{T}.$$

Therefore, our definition of a normal vector in the theory of plane curves exactly matches the above definition of $\vec{U}$ for a curve on a surface if one considers the xy-plane a surface in $\mathbb{R}^3$, with $\vec{k}$ as the unit surface normal vector.

From now on, for calculations related to a curve on a surface, we will use the frame $\{\vec{N}, \vec{T}, \vec{U}\}$, which is often called the *Darboux frame*.

8.1.2 Normal Curvature

As in the theory of space curves, we begin the study of curves on surfaces with the derivative $\vec{T}' = s'\kappa\vec{P}$. Since the principal normal vector is perpendicular to $\vec{T}$, this vector, often called the *curvature vector*, is perpendicular to $\vec{T}$, and thus, in the $\{\vec{N}, \vec{T}, \vec{U}\}$ frame, decomposes into an $\vec{N}$ component and a $\vec{U}$ component. In Definition 6.4.1, we already introduced the notion of the normal curvature $\kappa_n(t)$ as the component of $\vec{T}'$ along $\vec{N}$. We now define an additional function $\kappa_g : I \to \mathbb{R}$ such that

$$\vec{T}' = s'(t)\kappa(t)\vec{P} = s'(t)\kappa_n(t)\vec{N} + s'(t)\kappa_g(t)\vec{U} \qquad (8.1)$$

for all $t \in I$. We call κ_g the *geodesic curvature* of C at p on S.

Since $\{\vec{N}, \vec{T}, \vec{U}\}$ is an orthonormal frame, we can already provide one way to calculate these curvatures, namely, using

$$\kappa_n = \kappa\vec{P} \cdot \vec{N} \qquad \text{and} \qquad \kappa_g = \kappa\vec{P} \cdot \vec{U}. \qquad (8.2)$$

If one is given parametric equations for the curve $\vec{\gamma}$ and the surface $\vec{X}$, then Equation (8.2) is often the most direct way of calculating κ_n. If we are given the second fundamental form to the surface S, there is an alternative way to calculate κ_n. We remind the reader of

Figure 8.1. $\vec{N}$, $\vec{T}$, $\vec{U}$, and $\vec{P}$ of a curve on a surface.

Proposition 6.4.2, which relates the second fundamental form of S to the normal curvature by

$$\kappa_n(t_0) = II_p(\vec{T}(t_0)).$$

Therefore, the second fundamental form of a unit vector $\vec{T}$ is the component of a curve's curvature vector $\vec{T}'$ in the normal direction. If $\vec{T}(t_0)$ makes an angle θ with $\vec{e}_1$, the principal direction associated to the maximum principal curvature, then by Euler's curvature formula in Equation (6.32),

$$\kappa_n(t_0) = \kappa_1 \cos^2 \theta + \kappa_2 \sin^2 \theta.$$

We remind the reader that though $\vec{T}$ at a point p on a regular curve C may change sign under a reparametrization, the principal normal $\vec{P}$ is uniquely defined, regardless of the orientation of travel along C. However, for surfaces, the unit normal $\vec{N}$, which is calculated by

$$\vec{N} = \frac{\vec{X}_u \times \vec{X}_v}{\|\vec{X}_u \times \vec{X}_v\|},$$

may change sign under a reparametrization. Thus, in Equation (8.2) one notices that κ_n and κ_g are unique only up to a possible sign change under a reparametrization of the curve or of the surface.

The principal normal vector $\vec{P}$ in the $\{\vec{N}, \vec{T}, \vec{U}\}$ reference frame of a circle on the surface of a sphere is illustrated in Figure 8.1. In this figure, however, since the surface is a sphere, all the points are umbilical, and there does not exist a unique basis $\{\vec{e}_1, \vec{e}_2\}$ of principal directions (even up to sign).

As an application of Euler's formula in the context of curves on surfaces, let us consider any two curves $\vec{\gamma}_1$ and $\vec{\gamma}_2$ on the surface S, each parametrized by arc length in such a way that they intersect orthogonally at the point p at $s = s_0$. If we write

$$\vec{\gamma}_1'(s_0) = (\cos\theta)\vec{e}_1 + (\sin\theta)\vec{e}_2,$$

then

$$\vec{\gamma}_2'(s_0) = \pm((\sin\theta)\vec{e}_1 - (\cos\theta)\vec{e}_2)$$

for either $-$ or $+$ signs as a choice on $\pm$. If we denote $\kappa_n^{(1)}$ and $\kappa_n^{(2)}$ as the normal curvatures of $\vec{\gamma}_1$ and $\vec{\gamma}_2$, respectively, at p, then one obtains the interesting fact that

$$\kappa_n^{(1)} + \kappa_n^{(2)} = \cos^2\theta\kappa_1 + \sin^2\theta\kappa_2 + \sin^2\theta\kappa_1 + \cos^2\theta\kappa_2$$
$$= \kappa_1 + \kappa_2 = 2H.$$

In other words, for any two orthogonal unit tangent directions $\vec{v}_1$ and $\vec{v}_2$ at a point p, the average of the associated normal curvatures is equal to the mean curvature and does not depend on the particular directions of $\vec{v}_1$ and $\vec{v}_2$.

Note that since $\vec{T} \cdot \vec{N} = 0$, it follows that $\vec{N}' \cdot \vec{T} = -\vec{T}' \cdot \vec{N}$, and hence, we deduce that

$$\vec{N}' \cdot \vec{T} = -s'(t)\kappa_n(t).$$

We remind the reader that an asymptotic curve on a surface is a curve that satisfies $\kappa_n(t) = 0$ (see Definition 6.4.11) at all points on the curve. Geometrically, this means that an asymptotic curve is a curve on which $\vec{N}'$ has no $\vec{T}$ component or, vice versa, a curve on which $\vec{T}'$ has no $\vec{N}$ component. Also, from an intuitive perspective, though we introduced κ_n as a measure of how much $\vec{T}$ changes in the normal direction $\vec{N}$, $-\kappa_n$ gives a measure of how much $\vec{N}$ changes in the tangent direction $\vec{T}$.

8.1.3   Geodesic Curvature

The geodesic curvature κ_g of a curve on a surface corresponds to the component of the curvature vector occurring in the tangent plane. Equation (8.2) states this relationship as

$$\kappa_g = \kappa \vec{P} \cdot \vec{U},$$

which is equivalent to

$$s'(t)\kappa_g = \vec{T}' \cdot \vec{U} = \vec{T}' \cdot (\vec{N} \times \vec{T}) = (\vec{T}'\vec{N}\vec{T}) = (\vec{T}\vec{T}'\vec{N}), \qquad (8.3)$$

where $(\vec{u}\vec{v}\vec{w})$ is the triple-vector product in $\mathbb{R}^3$. This already provides a formula to calculate $\kappa_g(t)$ in specific situations. It turns out, however, that the geodesic curvature is an intrinsic quantity, a fact that we show now.

Let us write $\vec{\alpha}(t) = (u(t), v(t))$ and $\vec{\gamma} = \vec{X} \circ \vec{\alpha}$ for the curve on the surface. (To simplify the expression of certain formulas as we did in Chapter 7, we will refer to the coordinates u and v as x^1 and x^2, and we will use the expressions $\dot{x}^1, \dot{x}^2, \ddot{x}^1, \ddot{x}^2$ to refer to the derivatives of u', v', u'', v''.) For the tangent vector, we have

$$s'(t)\vec{T} = \vec{X}_u u'(t) + \vec{X}_v v'(t) = \sum_{i=1}^{2} \dot{x}^i \vec{X}_i. \qquad (8.4)$$

Taking a second derivative of this expression, we get

$$s''\vec{T} + s'\vec{T}' = \vec{X}_{uu}(u')^2 + 2\vec{X}_{uv}u'v' + \vec{X}_{vv}(v')^2 + \vec{X}_u u'' + \vec{X}_v v''$$

$$= \sum_{i=1}^{2}\sum_{j=1}^{2} \dot{x}^i \dot{x}^j \vec{X}_{ij} + \sum_{k=1}^{2} \ddot{x}^k \vec{X}_k. \qquad (8.5)$$

The triple-vector product $(\vec{T}\vec{T}'\vec{N})$ can also be written as $(\vec{T} \times \vec{T}') \cdot \vec{N}$. Therefore, since $\vec{T} \times \vec{T} = \vec{0}$, taking the cross product of the two expressions in Equations (8.4) and (8.5) and then taking the dot product with $\vec{N}$ leads to

$$(s')^2(\vec{T}\vec{T}'\vec{N}) = \Big[(\vec{X}_u \times \vec{X}_{uu})(u')^3 + (2\vec{X}_u \times \vec{X}_{uv} + \vec{X}_v \times \vec{X}_{uu})(u')^2 v'$$

$$+ (\vec{X}_u \times \vec{X}_{vv} + 2\vec{X}_v \times \vec{X}_{uv})u'(v')^2 + (\vec{X}_v \times \vec{X}_{vv})(v')^3\Big] \cdot \vec{N}$$

$$+ (\vec{X}_u \times \vec{X}_v) \cdot \vec{N}(u'v'' - u''v'). \qquad (8.6)$$

It is possible to express all the coefficients of the terms $(u')^3$, $(u')^2v'$, $u'(v')^2$, and $(v')^3$ in terms of the metric tensor coefficients g_{ij} or their derivatives. For example,

$$
\begin{aligned}
(\vec{X}_u \times \vec{X}_{uu}) \cdot \vec{N} &= (\vec{X}_u \times \vec{X}_{uu}) \cdot \frac{\vec{X}_u \times \vec{X}_v}{\sqrt{\det(g)}} \\
&= \frac{(\vec{X}_u \cdot \vec{X}_u)(\vec{X}_{uu} \cdot \vec{X}_v) - (\vec{X}_u \cdot \vec{X}_v)(\vec{X}_{uu} \cdot \vec{X}_u)}{\sqrt{\det(g)}} \\
&= \frac{g_{11}[11,2] - g_{12}[11,1]}{\sqrt{\det(g)}} \\
&= \frac{\det(g)g^{22}[11,2] + \det(g)g^{21}[11,1]}{\sqrt{\det(g)}} \\
&= \Gamma_{11}^2 \sqrt{\det(g)}.
\end{aligned}
$$

Repeating the calculations for all the relevant terms, one gets

$$
\begin{aligned}
(\vec{X}_u \times \vec{X}_{uv}) \cdot \vec{N} &= \Gamma_{12}^2 \sqrt{\det(g)}, & (\vec{X}_u \times \vec{X}_{vv}) \cdot \vec{N} &= \Gamma_{22}^2 \sqrt{\det(g)}, \\
(\vec{X}_v \times \vec{X}_{uu}) \cdot \vec{N} &= -\Gamma_{11}^1 \sqrt{\det(g)}, & (\vec{X}_v \times \vec{X}_{uv}) \cdot \vec{N} &= -\Gamma_{12}^1 \sqrt{\det(g)}, \\
(\vec{X}_v \times \vec{X}_{vv}) \cdot \vec{N} &= -\Gamma_{22}^1 \sqrt{\det(g)}, & (\vec{X}_u \times \vec{X}_v) \cdot \vec{N} &= \sqrt{\det(g)}.
\end{aligned}
$$

Putting these expressions into Equation (8.6) and using Equation (8.3) gives

$$
\begin{aligned}
(s')^3 \kappa_g = &\left(\Gamma_{11}^2 (u')^3 + (2\Gamma_{12}^2 - \Gamma_{11}^1)(u')^2 v' + (\Gamma_{22}^2 - 2\Gamma_{12}^1)u'(v')^2 \right. \\
&\left. -\Gamma_{22}^1 (v')^3 + u'v'' - u''v' \right) \sqrt{g_{11}g_{22} - (g_{12})^2}.
\end{aligned} \tag{8.7}
$$

This shows that, as opposed to the normal curvature, the geodesic curvature is an intrinsic quantity, depending only on the metric tensor and the parametric equations $\vec{\alpha}(t) = (u(t), v(t))$.

Though complete, Equation (8.7) is not written as concisely as it could be using tensor notation. Recall first the following permutation symbols introduced in Section 7.1:

$$
\varepsilon_{h_1 h_2 \cdots h_n} = \begin{cases} +1, & \text{if } h_1 h_2 \ldots h_n \text{ is an even permutation of } 1, 2, \ldots, n, \\ -1, & \text{if } h_1 h_2 \ldots h_n \text{ is an odd permutation of } 1, 2, \ldots, n, \\ 0, & \text{if } h_1 h_2 \ldots h_n \text{ is not a permutation of } 1, 2, \ldots, n. \end{cases}
$$

We shall use this symbol in the simple case with $n = 2$, so with only two indices. Then our indices satisfy $1 \leq i, j \leq 2$ and $\varepsilon_{ii} = 0$, $\varepsilon_{12} = 1$, and $\varepsilon_{21} = -1$. We can now summarize (8.7) as

$$\kappa_g = \frac{\sqrt{\det(g)}}{(s')^3} \left(\sum_{i,j,k,l=1}^{2} \varepsilon_{il} \Gamma_{jk}^l \dot{x}^i \dot{x}^j \dot{x}^k + \sum_{i,j=1}^{2} \varepsilon_{ij} \dot{x}^i \ddot{x}^j \right) \qquad (8.8)$$

or, in other words,

$$\kappa_g = \frac{\sqrt{\det(g)}}{(s')^3} \left(\varepsilon_{il} \Gamma_{jk}^l \dot{x}^i \dot{x}^j \dot{x}^k + \varepsilon_{ij} \dot{x}^i \ddot{x}^j \right), \qquad (8.9)$$

where we use the Einstein summation convention.

Example 8.1.1 (Plane Curves). We modeled the $\{\vec{N}, \vec{T}, \vec{U}\}$ frame off the frame $\{\vec{k}, \vec{T}, \vec{U}\}$, viewing the plane as a surface in $\mathbb{R}^3$ with normal vector $\vec{k}$. Consider then curves in the plane as curves on a surface. Using an orthonormal basis in the plane, i.e., the trivial parametrization $\vec{X}(u, v) = (u, v)$, we get $g_{11} = g_{22} = 1$ and $g_{12} = 0$. In this case, all the Christoffel symbols are 0 and of course $\sqrt{\det(g)} = 1$. Then Equation (8.7) gives

$$\kappa_g = \frac{u'v'' - u''v'}{(s')^3},$$

which is precisely the Equation (1.12) we obtained for the curvature of a plane curve in Section 1.3.

It is an interesting fact, the proof of which we leave as an exercise to the reader (Problem 8.1.6), that the geodesic curvature at a point p on a curve C on a surface S is equal to the geodesic curvature of the plane curve obtained by projecting C orthogonally onto T_pS. From an intuitive perspective, this property indicates why the geodesic curvature should be an intrinsic quantity. Intrinsic properties depend on the metric tensor, which is an inner product on the tangent space T_pS, so any quantity that measures something within the tangent space is usually an intrinsic property.

8.1.4 Geodesic Torsion

Similar to how we viewed the Frenet frame in Chapter 3, we consider the triple $\{\vec{N}, \vec{T}, \vec{U}\}$ as a moving frame based at a point $\vec{\gamma}(t)$. Since $\vec{N}$, $\vec{T}$, and $\vec{U}$ are unit vectors, using the same reasoning as we did to establish Equation (3.4), we can determine that

$$\frac{d}{dt}\begin{pmatrix} \vec{N} & \vec{T} & \vec{U} \end{pmatrix} = \begin{pmatrix} \vec{N} & \vec{T} & \vec{U} \end{pmatrix} A(t),$$

where $A(t)$ is an antisymmetric matrix. We already know most of the coefficient functions in $A(t)$. In Equation (8.1), our definitions of the normal and geodesic curvature gave

$$\vec{T}' = s'(t)\kappa_n(t)\vec{N} + s'(t)\kappa_g(t)\vec{U}. \qquad (8.10)$$

Using the usual dot product relations among the vectors in an orthonormal basis, namely

$$\vec{N} \cdot \vec{N} = 1, \qquad \vec{T} \cdot \vec{T} = 1, \qquad \vec{U} \cdot \vec{U} = 1,$$
$$\vec{N} \cdot \vec{T} = 0, \qquad \vec{N} \cdot \vec{U} = 0, \qquad \vec{T} \cdot \vec{U} = 0,$$

we deduce the following equalities:

$$\vec{N}' \cdot \vec{N} = 0, \qquad \vec{T}' \cdot \vec{T} = 0, \qquad \vec{U}' \cdot \vec{U} = 0,$$
$$\vec{N}' \cdot \vec{T} = -\vec{T}' \cdot \vec{N}, \qquad \vec{N}' \cdot \vec{U} = -\vec{U}' \cdot \vec{N}, \qquad \vec{T}' \cdot \vec{U} = -\vec{U}' \cdot \vec{T}.$$

Consequently, from Equation (8.10),

$$\vec{N}' \cdot \vec{T} = -s'\kappa_n \qquad \text{and} \qquad \vec{U}' \cdot \vec{T} = -s'\kappa_g.$$

Therefore, in order to describe the coefficients of $A(t)$, and thereby express the derivatives $\vec{N}'$, $\vec{T}'$, $\vec{U}'$ in the $\{\vec{N}, \vec{T}, \vec{U}\}$ frame, we need to label one more coefficient function. Define the *geodesic torsion* $\tau_g : I \to \mathbb{R}$ to be the unique function such that

$$\vec{N}' \cdot \vec{U} = s'(t)\tau_g(t) \qquad \text{for all } t \in I.$$

With this function defined, we can imitate Equation (3.4) for the change of the Frenet, and write

$$\frac{d}{dt}\begin{pmatrix} \vec{N} & \vec{T} & \vec{U} \end{pmatrix} = \begin{pmatrix} \vec{N} & \vec{T} & \vec{U} \end{pmatrix} s'(t) \begin{pmatrix} 0 & \kappa_n(t) & -\tau_g(t) \\ -\kappa_n(t) & 0 & -\kappa_g(t) \\ \tau_g(t) & \kappa_g(t) & 0 \end{pmatrix}.$$
$$(8.11)$$

The normal and geodesic curvatures both possess fairly intuitive geometric explanations as to what they measure. However, it is harder to say precisely what the geodesic torsion measures. Using the formula

$$\vec{N}' = s'(-\kappa_n \vec{T} + \tau_g \vec{U}),$$

we could say that the geodesic torsion measures the rate of change of $\vec{N}$ in the $\vec{U}$ direction, but this is not particularly instructive. In the next section, we study geodesic curves of a surface, which in some sense generalize the notion of straight lines on the surface. Problem 8.2.1 shows that the geodesic torsion is the torsion function of the geodesic curves.

Problems

8.1.1. Let $\vec{X} = (R\cos u \sin v, R\sin u \sin v, R\cos v)$, with $(u,v) \in [0, 2\pi] \times [0, \pi]$, be a parametrization for a sphere. Consider the circle on the sphere given by $\vec{\gamma}(t) = \vec{X}(t, \varphi_0)$, where φ_0 is a fixed constant. Calculate the normal curvature, the geodesic curvature, and the geodesic torsion of $\vec{\gamma}$ on $\vec{X}$ (see Figure 8.1 for an illustration with $\varphi_0 = 2\pi/3$).

8.1.2. Consider the cylinder $x^2 + y^2 = 1$, and consider the curve C on the cylinder obtained by intersecting the cylinder with the plane through the x-axis that makes an angle of θ with the xy-plane.

    (a) Show that C is an ellipse.

    (b) Compute the normal curvature, geodesic curvature, and geodesic torsion of C on the cylinder.

8.1.3. Let $\vec{\gamma}(s)$ be a curve parametrized by arc length on a surface parametrized by $\vec{X}(u,v)$. Prove that

$$\kappa_g \sqrt{\det(g)} = \frac{\partial}{\partial u}(\vec{T} \cdot \vec{X}_v) - \frac{\partial}{\partial v}(\vec{T} \cdot \vec{X}_u).$$

8.1.4. Consider the torus parametrized by $\vec{X}(u,v) = ((a + b\sin v)\cos u, (a+b\sin v)\sin u, b\cos v)$, where $a > b$. The curve $\vec{\gamma}(mt, nt)$, where m and n are relatively prime, is called the (m,n)-torus knot. Calculate the geodesic curvature of this (m,n)-torus knot.

8.1.5. Consider the surface that is a function graph $z = f(x,y)$. Calculate the normal curvature, geodesic curvature, and geodesic torsion of a level curve , i.e., a curve of the form $f(x,y) = c$.

8.1.6. Let $\vec{X}$ be the parametrization for a coordinate patch of a regular surface S, and let $\vec{\gamma} = \vec{X} \circ \vec{\alpha}$ be the parametrization for a curve C on S. Consider a point p on S, and let C' be the orthogonal projection of C onto the tangent plane $T_p S$. Prove that the geodesic curvature κ_g at p of C on S is equal to the curvature κ_g of C' at p as a plane curve in $T_p S$.

8.1.7. *Bonnet's Formula.* Suppose that a curve on a surface is given by $\varphi(u, v) = C$, where C is a constant. Prove that the geodesic curvature is given by

$$\kappa_g = \frac{1}{\sqrt{\det(g)}} \left[\frac{\partial}{\partial u} \left(\frac{g_{12}\varphi_v - g_{22}\varphi_u}{g_{11}\varphi_v^2 - 2g_{12}\varphi_u\varphi_v + g_{22}\varphi_u^2} \right) + \frac{\partial}{\partial v} \left(\frac{g_{12}\varphi_u - g_{11}\varphi_v}{g_{11}\varphi_v^2 - 2g_{12}\varphi_u\varphi_v + g_{22}\varphi_u^2} \right) \right].$$

8.1.8. *Liouville's Formula.* Let $\vec{X}(u, v)$ be an orthogonal parametrization of a patch on a regular surface S. Let C be a curve on this patch parametrized by arc length by $\vec{\gamma}(s) = \vec{X}(u(s), v(s))$. Let $\theta(s)$ be a function defined along C that gives the angle between $\vec{T}$ and $\vec{X}_u$. Prove that the geodesic curvature of C is given by

$$\kappa_g = \frac{d\theta}{ds} + \kappa_{(u)} \cos\theta + \kappa_{(v)} \sin\theta,$$

where $\kappa_{(u)}$ is the geodesic curvature along the u-parameter curve (i.e., $v = v_0$) and similarly for $\kappa_{(v)}$.

8.1.9. Let S be a regular surface parametrized by $\vec{X}(u, v)$, and let C be a regular curve on S parametrized by $\vec{X}(t) = \vec{X}(u(t), v(t))$ for $t \in [a, b]$. Consider the normal tube around C parametrized by

$$\vec{Y}_r(t, v) = \vec{X}(t) + r \cos v\, \vec{U}(t) + r \sin v\, \vec{N}(t)$$

for some $r > 0$.

(a) Find the metric tensor associated to $\vec{Y}(t, v)$, and conclude that

$$\det g = (s')^2 r^2 (1 - r\kappa_g \cos v - r\kappa_n \sin v)^2.$$

(b) Show that if r is small enough but still positive, then $\vec{Y}_r$ is a regular parametrization.

(c) Assuming that $\vec{Y}_r$ is regular, calculate the coefficients L_{ij} of the second fundamental form of the normal tube around C. Prove that the Gaussian curvature K satisfies

$$K(t, v) = -\frac{s'(t)}{\sqrt{\det(g)}} (\kappa_n(t) \sin v + \kappa_g(t) \cos v).$$

8.2   Geodesics

Classical geometry in the plane studies in great detail relationships
between points, straight lines, and circles. One could characterize
our theory of surfaces until now as a local theory in that we have
concentrated our attention on the behavior of curves on surfaces at
a point. Consequently, the notion of straightness (a global notion)
and the concept of a circle (also a curve defined by a global property)
do not yet make sense in our theory of curves on surfaces.

Euclid defines a line as a "breadthless width" and a straight
line as "a line which lies evenly with the points on itself." These
definitions do not particularly help us generalize the concept of a
straight line to a general surface. However, it is commonly known
that given any two points P and Q in $\mathbb{R}^n$, a line segment connecting
P and Q provides the path of shortest distance between these two
points.

On a regular surface $S \subset \mathbb{R}^3$ that is not planar, even the notion
of distance *in* S between two points P and Q poses some difficulty
since we cannot assume a straight line in $\mathbb{R}^3$ connecting P and Q
lies in S. However, since it is possible to talk about the arc length
of curves, we can define the *distance on* S between P and Q as

$$\inf\{\text{arc length of } C \,|\, C \text{ is a curve on } S \text{ connecting } P \text{ and } Q\}. \quad (8.12)$$

Therefore, one might wish to take as a first intuitive formulation of
straightness on a regular surface S the following definition: a curve
C on S is "straight" if for all pairs of points on the curve, the arc
length between those two points P and Q is equal to the distance
PQ between them. For general regular surfaces, this proposed definition turns out to be unsatisfactory, but it does lead to a more
sophisticated way of generalizing straightness to surfaces.

Let S be a regular surface, and let P and Q be points of S.
Suppose that P and Q are in a coordinate patch that is parametrized
by $\vec{X} : U \to \mathbb{R}^3$, where $U \subset \mathbb{R}^2$. Consider curves on S parametrized
by $\vec{\gamma}(t) = \vec{X}(u(t), v(t))$ such that $\vec{\gamma}(0) = P$ and $\vec{\gamma}(1) = Q$. According
to Equation (6.1), the arc length of such a curve is

$$s = \int_0^1 \sqrt{g_{11}(u'(t))^2 + 2g_{12}u'(t)v'(t) + g_{22}(v'(t))^2}\, dt, \quad (8.13)$$

where we understand that the g_{ij} coefficients are functions of u and v, which are in turn functions of t. To find a curve that connects P and Q with the shortest arc length, one must find parametric equations $(u(t), v(t))$ that minimize the integral in Equation (8.13). Such problems are studied in calculus of variations, a brief introduction to which is presented in Appendix B of [22].

According to the Euler-Lagrange Theorem in calculus of variations, if we set

$$f = \sqrt{g_{11}(u'(t))^2 + 2g_{12}u'(t)v'(t) + g_{22}(v'(t))^2},$$

then the parametric equations $(u(t), v(t))$ that optimize the arc length s in Equation (8.13) must satisfy

$$\mathcal{L}_u(f) \overset{\text{def}}{=} \frac{\partial f}{\partial u} - \frac{d}{dt}\left(\frac{\partial f}{\partial u'}\right) = 0 \quad \text{and}$$

$$\mathcal{L}_v(f) \overset{\text{def}}{=} \frac{\partial f}{\partial v} - \frac{d}{dt}\left(\frac{\partial f}{\partial v'}\right) = 0.$$

We call $\mathcal{L}_u$ the Euler-Lagrange operator with respect to u and similarly for v. For a function with two intermediate variables u and v, as we have in this instance, let us also define the operator

$$\mathcal{L}(f) \overset{\text{def}}{=} (\mathcal{L}_u(f), \mathcal{L}_v(f)).$$

Proposition 8.2.1. *With* $f = \sqrt{g_{11}(u'(t))^2 + 2g_{12}u'(t)v'(t) + g_{22}(v'(t))^2}$, *the Euler-Langrange operator of* f *satisfies*

$$\mathcal{L}(f) = \sqrt{\det(g)}\kappa_g(t)\left(v'(t), -u'(t)\right). \tag{8.14}$$

Proof: (The proof is left as an exercise for the persistent reader and relies on the careful application of Euler-Lagrange equations and Equation (8.7).) $\qquad\square$

Corollary 8.2.2. *The parametric equations* $(u(t), v(t))$ *optimize the integral in Equation (8.13) if and only if* $\kappa_g(t) = 0$ *or* $(v'(t), u'(t)) = \vec{0}$.

Obviously, $(v'(t), -u'(t)) = \vec{0}$ integrates to $(u, v) = (c_1, c_2)$, where c_1 and c_2 are constants, so the curve degenerates to a point. The other part of Corollary 8.2.2 motivates the following definition, which generalizes to a regular surface the notion of a straight line.

Definition 8.2.3. A *geodesic* is a curve on a surface with geodesic curvature $\kappa_g(t)$ identically 0.

As a first remark, one notices that on a geodesic C, the curvature vector is $\vec{T}' = s'\kappa\vec{P} = s'\kappa_n\vec{N}$, so at each point on the surface, $\kappa = \pm\kappa_n$ and the curve's principal normal vector is equal to the surface unit normal vector up to a possible change of sign. The transition matrix between the Frenet frame and the $\{\vec{N}, \vec{T}, \vec{U}\}$ frame becomes (written in an alternative order)

$$\begin{pmatrix} \vec{T} & \vec{N} & \vec{U} \end{pmatrix} = \begin{pmatrix} \vec{T} & \vec{P} & \vec{B} \end{pmatrix} \begin{pmatrix} 1 & 0 & 0 \\ 0 & \varepsilon & 0 \\ 0 & 0 & \varepsilon \end{pmatrix},$$

where $\varepsilon = \pm 1$. Consequently, along a geodesic, the osculating plane of the curve and the tangent plane to the surface are normal to each other.

From Equation (8.7), a curve $\vec{\gamma} = \vec{X} \circ \vec{\alpha}$, with $\vec{\alpha}(t) = (u(t), v(t))$, is a geodesic if and only if

$$\Gamma^2_{11}(u')^3 + (2\Gamma^2_{12} - \Gamma^1_{11})(u')^2 v' + (\Gamma^2_{22} - 2\Gamma^1_{12})u'(v')^2 - \Gamma^1_{22}(v')^3 + u'v'' - u''v' = 0. \quad (8.15)$$

This formula holds for any parameter t and not just when $\vec{\gamma}$ is parametrized by arc length.

We can approach the task of finding equations for geodesics in an alternative way. Assume now that we consider curves on the surface parametrized by arc length so that $s' = 1$ and $s'' = 0$. Since $\vec{T}'$ is parallel to $\vec{N}$, we conclude that

$$\vec{T}' \cdot \vec{X}_u = 0 \qquad \text{and} \qquad \vec{T}' \cdot \vec{X}_v = 0.$$

Then using Equation (8.5) we deduce that

$$\vec{X}_{uu} \cdot \vec{X}_u(u')^2 + 2\vec{X}_{uv} \cdot \vec{X}_u u'v' + \vec{X}_{vv} \cdot \vec{X}_u(v')^2 + \vec{X}_u \cdot \vec{X}_u u'' + \vec{X}_v \cdot \vec{X}_u v'' = 0,$$
$$\vec{X}_{uu} \cdot \vec{X}_v(u')^2 + 2\vec{X}_{uv} \cdot \vec{X}_v u'v' + \vec{X}_{vv} \cdot \vec{X}_v(v')^2 + \vec{X}_u \cdot \vec{X}_v u'' + \vec{X}_v \cdot \vec{X}_v v'' = 0.$$

Solving algebraically for u'' and v'' in the above two equations, we obtain the following classical equations for a geodesic curve which we will express in tensor notation for simplicity:

$$\frac{d^2 x^i}{ds^2} + \sum_{j,k=1}^{2} \Gamma^i_{jk} \frac{dx^j}{ds} \frac{dx^k}{ds} = 0 \qquad \text{for } i = 1, 2. \quad (8.16)$$

At first sight, one might see a discrepancy between Equation (8.16), which involves two equations and Equation (8.15), which involves only one. However, it is essential to point out that the system in Equation (8.16) holds for geodesics parametrized by arc length. Furthermore, Equation (8.16) is equivalent to the system of equations

$$\kappa_g(t) = 0 \quad \text{and} \quad (s')^2 = g_{11}(u')^2 + 2g_{12}u'v' + g_{22}(v')^2.$$

Finding explicit parametric equations of geodesic curves on a surface is often difficult since it involves solving a system of nonlinear second-order differential equations. However, in specific situations, there sometimes exist simplifications for Equation (8.16) that allow one to find an algebraic or even a differential equation relating u and v that is satisfied by geodesics.

Example 8.2.4 (The Plane: Cartesian Coordinates). As a first example, consider the xy-plane parametrized with the usual Cartesian coordinates. Then the Christoffel symbols are all $\Gamma^i_{jk} = 0$. Using Equation (8.16), the equations for a geodesic in the plane are simply

$$\frac{d^2u}{ds^2} = 0 \quad \text{and} \quad \frac{d^2v}{ds^2} = 0.$$

Integrating both equations twice, one obtains $u(s) = as + c$ and $v(s) = bs + d$. Furthermore, since $(u'(s))^2 + (v'(s))^2 = 1$, these constants must satisfy $a^2 + b^2 = 1$. Therefore, geodesics parametrized by arc length in the plane are given as $\vec{\gamma}(s) = \vec{p} + s\vec{u}$, where $\vec{p}$ is a point and $\vec{u}$ is a unit vector.

Example 8.2.5 (The Plane: Polar Coordinates). In contrast to the previous example, consider the xy-plane parametrized with polar coordinates so that as a surface in $\mathbb{R}^3$, the xy-plane is given by

$$\vec{X}(r, \theta) = (r \cos \theta, r \sin \theta, 0).$$

A short calculation gives

$$\Gamma^2_{12} = \Gamma^2_{21} = \frac{1}{r} \quad \text{and} \quad \Gamma^1_{22} = -r$$

and the remaining five other symbols are 0. Equations (8.16) become

$$\frac{d^2r}{ds^2} - r\left(\frac{d\theta}{ds}\right)^2 = 0, \tag{8.17}$$

$$\frac{d^2\theta}{ds^2} + \frac{2}{r}\frac{dr}{ds}\frac{d\theta}{ds} = 0. \tag{8.18}$$

We transform this system to obtain a differential equation relating r and θ as follows. Note that by repeatedly using the chain rule, we get

$$\frac{dr}{ds} = \frac{dr}{d\theta}\frac{d\theta}{ds} \quad \text{and} \quad \frac{d^2r}{ds^2} = \frac{d^2r}{d\theta^2}\left(\frac{d\theta}{ds}\right)^2 + \frac{dr}{d\theta}\frac{d^2\theta}{ds^2}.$$

Putting these two into Equations (8.17) and (8.18) leads to

$$\left(\frac{d\theta}{ds}\right)^2\left(\frac{d^2r}{d\theta^2} - \frac{2}{r}\left(\frac{dr}{d\theta}\right)^2 - r\right) = 0,$$

which breaks into the pair of equations

$$\frac{d\theta}{ds} = 0 \quad \text{or} \quad \left(\frac{d^2r}{d\theta^2} - \frac{2}{r}\left(\frac{dr}{d\theta}\right)^2 - r\right) = 0. \tag{8.19}$$

One notices that the first equation in Equation (8.19) is solved for $\theta = \text{const}$ which corresponds to a line through the origin. The other equation in Equation (8.19) does not appear particularly tractable but a substitution simplifies it greatly. Using the new variable $u = \frac{1}{r}$, one can check that

$$\frac{d^2r}{d\theta^2} = \frac{2}{u^3}\left(\frac{du}{d\theta}\right)^2 - \frac{1}{u^2}\frac{d^2u}{d\theta^2},$$

and hence the second equation in Equation (8.19) reduces to

$$\frac{d^2u}{d\theta^2} + u = 0. \tag{8.20}$$

Using standard techniques in ordinary differential equations, the general solution to Equation (8.20) can be written as $u(\theta) = C\cos(\theta-\theta_0)$ where C and θ_0 are constants. Therefore, in polar coordinates, equations for geodesics in the plane (i.e., lines) are given by

$$\theta = C \quad \text{or} \quad r(\theta) = C\sec(\theta - \theta_0).$$

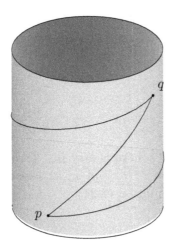

Figure 8.2. Two geodesics on a cylinder.

Example 8.2.6 (Cylinder). Consider a right circular cylinder. A parametrization for this cylinder is $\vec{X}(u,v) = (\cos(u), \sin(u), v)$, with (u,v) in $[0, 2\pi] \times \mathbb{R}$. An easy calculation shows that $\Gamma^i_{jk} = 0$ for all $i, j, k,$ and hence, that geodesics on a cylinder are curves of the form $\gamma = \vec{X} \circ \vec{\alpha}$ where

$$\vec{\alpha}(t) = (at + b, ct + d)$$

for a, b, c, d that are constant. As a result, the geodesics on a cylinder are either straight lines or helices.

Figure 8.2 illustrates how on a surface, two different geodesics may connect two distinct points, a situation that does not occur in the plane. In fact, on a cylinder, there is an infinite number of geodesics that connect any two points. The difference between each geodesic connecting p and q is how many times the geodesic wraps around the cylinder and in which direction it wraps.

Example 8.2.7 (Sphere). Consider the parametrization of the sphere given by

$$\vec{X}(x^1, x^2) = (R \cos x^1 \sin x^2, R \sin x^1 \sin x^2, R \cos x^2),$$

where $x^1 = u$ is the longitude θ in spherical coordinates, and $x^2 = v$ is the angle φ down from the positive z-axis. In Example 7.2.5, we determined the Christoffel symbols for this parametrization. Equations (8.16) for geodesics on the sphere become

$$\frac{d^2x^1}{ds^2} + 2\cot(x^2)\frac{dx^1}{ds}\frac{dx^2}{ds} = 0,$$

$$\frac{d^2x^2}{ds^2} - \sin(x^2)\cos(x^2)\left(\frac{dx^1}{ds}\right)^2 = 0.$$

(8.21)

A geodesic on the sphere is now just a curve of the form $\vec{\gamma}(s) = \vec{X}(x^1(s), x^2(s))$, where $x^1(s)$ and $x^2(s)$ satisfy the system of differential equations in Equation (8.21). Taking a first derivative of $\vec{\gamma}(s)$ gives

$$\vec{\gamma}'(s) = R\left(-\sin x^1 \sin x^2 \frac{dx^1}{ds} + \cos x^1 \cos x^2 \frac{dx^2}{ds},\right.$$

$$\left.\cos x^1 \sin x^2 \frac{dx^1}{ds} + \sin x^1 \cos x^2 \frac{dx^2}{ds}, -\sin x^2 \frac{dx^2}{ds}\right),$$

and the second derivative, after simplification using Equation (8.21), is

$$\frac{d^2\vec{\gamma}}{ds^2} = -\left[\sin^2(x^2)\left(\frac{dx^1}{ds}\right)^2 + \left(\frac{dx^2}{ds}\right)^2\right]\vec{\gamma}(s).$$

However, the term $R^2[\sin^2(x^2)(\frac{dx^1}{ds})^2 + (\frac{dx^2}{ds})^2]$ is the first fundamental form applied to

$$((x^1)'(s), (x^2)'(s)),$$

which is precisely the square of the speed of $\vec{\gamma}(s)$. However, since the geodesic is parametrized by arc length, its speed is identically 1. Thus, Equations (8.21) lead to the differential equation

$$\vec{\gamma}''(s) + \frac{1}{R^2}\vec{\gamma}(s) = 0.$$

Standard techniques with differential equations allow one to show that all solutions to this differential equation are of the form

$$\vec{\gamma}(s) = \vec{a}\cos\left(\frac{s}{R}\right) + \vec{b}\sin\left(\frac{s}{R}\right),$$

where $\vec{a}$ and $\vec{b}$ are constant vectors. Note that $\vec{\gamma}(0) = \vec{a}$ and that $\vec{\gamma}'(0) = \frac{1}{R}\vec{b}$. Furthermore, to satisfy the conditions that $\vec{\gamma}(s)$ lie on the sphere of radius R and be parametrized by arc length, we deduce that $\vec{a}$ and $\vec{b}$ satisfy

$$\|\vec{a}\| = R, \qquad \|\vec{b}\| = R, \qquad \text{and} \qquad \vec{a} \cdot \vec{b} = 0.$$

Therefore, we find that $\vec{\gamma}(s)$ traces out a great arc on the sphere that is the intersection of the sphere and the plane through the center of the sphere spanned by $\vec{\gamma}(0)$ and $\vec{\gamma}'(0)$.

Example 8.2.8 (Surfaces of Revolution). Consider a surface of revolution about the z-axis given by the parametric equations

$$\vec{X}(u, v) = (f(v) \cos u, f(v) \sin u, h(v)),$$

where f and h are functions defined over a common interval I. We assume that over the open interval $(0, 2\pi) \times I$ the function $\vec{X}$ is a regular parametrization, which implies that $f(v) > 0$. A simple calculation reveals that the Christoffel symbols of the second kind are

$$\Gamma^1_{11} = 0, \qquad\qquad \Gamma^1_{12} = \frac{f'}{f}, \qquad \Gamma^1_{22} = 0,$$
$$\Gamma^2_{11} = -\frac{ff'}{(f')^2 + (h')^2}, \qquad \Gamma^2_{12} = 0, \qquad \Gamma^2_{22} = \frac{f'f'' + h'h''}{(f')^2 + (h')^2}. \tag{8.22}$$

Equations (8.16) for geodesics parametrized by arc length on a surface of revolution become

$$\frac{d^2u}{ds^2} + \frac{2f'}{f}\frac{du}{ds}\frac{dv}{ds} = 0$$
$$\frac{d^2v}{ds^2} - \frac{ff'}{(f')^2 + (h')^2}\left(\frac{du}{ds}\right)^2 + \frac{f'f'' + h'h''}{(f')^2 + (h')^2}\left(\frac{dv}{ds}\right)^2 = 0. \tag{8.23}$$

As complicated as Equations (8.23) appear, it is possible to find a "solution" to this system of differential equations for u in terms of v. However, before establishing a general solution, we will determine which meridians ($u = $ const.) and which parallels or latitude lines ($v = $ const.) are geodesics.

Consider first the meridian lines that are defined by $u = C$, where C is a constant. Notice that the first equation in the system in Equation (8.23) is trivially satisfied for meridians and that the second equation becomes

$$\frac{d^2v}{ds^2} + \frac{f'f'' + h'h''}{(f')^2 + (h')^2}\left(\frac{dv}{ds}\right)^2 = 0. \tag{8.24}$$

It is easy to check that the first fundamental form on $(u'(t), v'(t))$ on the surface of revolution is

$$f(v)^2 u'(t)^2 + (f'(v)^2 + h'(v)^2)v'(t)^2. \tag{8.25}$$

However, assuming that we have a meridian that is parametrized by arc length, then

$$(f'(v)^2 + h'(v)^2)v'(s)^2 = 1$$

since the speed function of a curve parametrized by arc length is 1. Consequently,

$$v'(s)^2 = \frac{1}{f'(v)^2 + h'(v)^2},$$

and taking a derivative of this equation with respect to s, one obtains

$$2v'v'' = -\frac{2f'(v)f''(v)v' + 2h'(v)h''(v)v'}{(f'(v)^2 + h'(v)^2)^2} = -2\frac{f'(v)f''(v) + h'(v)h''(v)}{f'(v)^2 + h'(v)^2}(v')^3.$$

Since $v'(s) \neq 0$ on the meridian parametrized by arc length, then

$$\frac{d^2v}{ds^2} = -\frac{f'(v)f''(v) + h'(v)h''(v)}{f'(v)^2 + h'(v)^2}\left(\frac{dv}{ds}\right)^2,$$

which shows that Equation (8.24) is identically satisfied on all meridians.

Now consider the parallel curves on a surface of revolution, which are defined by $v = v_0$, a real constant. In Equation (8.23), the first equation leads to $u'(s) = C$, a constant, and the second equation becomes

$$\frac{ff'}{(f')^2 + (h')^2}\left(\frac{du}{ds}\right)^2 = 0.$$

Since the parallel curves, parametrized by arc length, are regular curves, then $C \neq 0$, and the condition that the surface of revolution

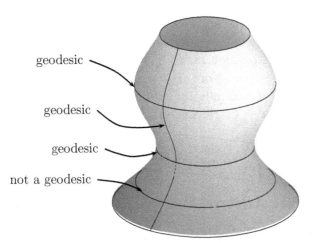

geodesic

geodesic

geodesic

not a geodesic

Figure 8.3. Geodesics on a surface of revolution.

be a regular surface implies that $f(v) \neq 0$. Thus, the second equation is satified for parallels $v = v_0$ if and only if $f'(v_0) = 0$. (See Figure 8.3.)

Even if a geodesic is neither a meridian nor a parallel, the first equation in Equation (8.23) is simple enough that one can nonetheless deduce some interesting consequences. Note that by taking a derivative with respect to the arc length parameter s, we have

$$\frac{d}{ds}(f^2 u') = 2f f' v' u' + f^2 u''.$$

Multiplying the first equation of Equation (8.23) by f^2, we see that it can be written as

$$f(v)^2 u'(s) = C, \qquad (8.26)$$

where C is a constant. Note that when a curve on a surface is parametrized by arc length with coordinate functions $(u'(s), v'(s))$, the angle θ it makes with any given parallel curve satisfies

$$\cos \theta = \frac{I_p((u', v'), (1, 0))}{\sqrt{I_p((u', v'), (u', v'))}\sqrt{I_p((1, 0), (1, 0))}} = u' f.$$

As a geometric interpretation, since f is the radius r of surface of revolution at a given point, Equation (8.26) leads to the relation

$$r \cos \theta = C \qquad (8.27)$$

for all nonparallel geodesics on a surface of revolution, where θ is the angle between the geodesic and the parallels. Equation (8.27), which is equivalent to Equation (8.26) for nonparallel curves, is often called the *Clairaut relation*. Note that a curve satisfying Clairaut's relation is a meridian if and only if $C = 0$.

With the relation $u' = C/f^2$, since the speed is equal to 1, Equation (8.25) leads to

$$\left(\frac{dv}{ds}\right)^2 ((f')^2 + (h')^2) = 1 - \frac{C^2}{f^2}. \tag{8.28}$$

Taking the derivative of this equation with respect to s, one obtains

$$2\frac{dv}{ds}\frac{d^2v}{ds^2}((f')^2 + (h')^2) + \left(\frac{dv}{ds}\right)^3 2(f'f'' + h'h'') = 2\frac{C^2 f'}{f^3}\frac{dv}{ds}$$

which is equivalent to

$$((f')^2+(h')^2)\frac{dv}{ds}\left[\frac{d^2v}{ds^2} - \frac{ff'}{(f')^2+(h')^2}\left(\frac{du}{ds}\right)^2 + \frac{f'f''+h'h''}{(f')^2+(h')^2}\left(\frac{dv}{ds}\right)^2\right] = 0.$$

Therefore, if a geodesic is not a parallel, the first equation of (8.23), which is equivalent to Clairaut's relation, implies the second equation.

Assuming that a geodesic is not a meridian, then from Clairaut's relation, we know that $u'(s)$ is never 0 so one can define an inverse function to $u(s)$, namely $s(u)$, and then v can be given as a function of u by $v = v(s(u))$. Then in Equation (8.28), replacing $\frac{dv}{ds}$ with $\frac{dv}{du}\frac{du}{ds} = \frac{dv}{du}\frac{C}{f^2}$, we obtain

$$\left(\frac{dv}{du}\right)^2 \frac{C^2}{f^4}((f')^2 + (h')^2) = 1 - \frac{C^2}{f^2},$$

and hence,

$$\frac{dv}{du} = \frac{f}{C}\sqrt{\frac{f^2 - C^2}{(f')^2 + (h')^2}}.$$

Over an interval where this derivative is not 0, it is possible to take an inverse function of u with respect to v, and, by integration, this function satisfies

$$u = C \int \frac{1}{f(v)}\sqrt{\frac{f'(v)^2 + h'(v)^2}{f(v)^2 - C^2}}\, dv + D$$

for some constant of integration D. The constants C and D lead to a two-parameter family of solutions that parametrize segments of geodesics.

Theorem 8.2.9. *Let S be a regular surface of class C^3. For every point p on S and every unit vector $\vec{u} \in T_p S$, there exists a unique geodesic on S through p in the direction of $\vec{v}$.*

Proof: Let $\vec{X} : U \to \mathbb{R}^3$ be a regular parametrization of a neighborhood of p on S, and let g_{ij} and Γ^i_{jk} be the components of the metric tensor and Christoffel symbols of the second kind, respectively, in relation to $\vec{X}$. Note that since S is of class C^3, all of the functions Γ^i_{jk} are of class C^1 over their domain.

The proof of this proposition is an application of the existence and uniqueness theorem for first-order systems of differential equations (see [2, Section 31.8]), which states that if $F : (R)^n \to \mathbb{R}^n$ is of class C^1, then there exists a unique function $\vec{x} : I \to \mathbb{R}^n$ that satisfies

$$\vec{x}' = F(\vec{x}) \qquad \text{and} \qquad \vec{x}(t_0) = \vec{C},$$

where I is an open interval containing t_0, and $\vec{C}$ is a constant vector.

Consider now the system of differential equations given in Equation (8.16), where we do not assume that a solution is parametrized by arc length. Setting the dependent variables $v^1 = (x^1)'$ and $v^2 = (x^2)'$, then Equation (8.16) is equivalent to the system of first-order differential equations

$$\begin{cases} (x^1)' &= v^1, \\ (x^2)' &= v^2, \\ (v^1)' &= -\Gamma^1_{11}(v^1)^2 - 2\Gamma^1_{12}v^1 v^2 - \Gamma^1_{22}(v^2)^2, \\ (v^2)' &= -\Gamma^2_{11}(v^1)^2 - 2\Gamma^2_{12}v^1 v^2 - \Gamma^2_{22}(v^2)^2, \end{cases} \qquad (8.29)$$

where the quantities Γ^i_{jk} are functions of x^1 and x^2. Therefore, according to the existence and uniqueness theorem, there exists a unique solution to Equation (8.16) with specific values given for $x^1(s_0)$, $x^2(s_0)$, $(x^1)'(s_0)$, and $(x^2)'(s_0)$. Set $\vec{\alpha}(s) = \vec{X}(x^1(s), x^2(s))$. Now Equation (8.16) is the formula for arc length parametrizations of geodesics. However, it is not hard to prove that if

$$f(s) = g_{11}(x^1(s), x^2(s))(x^1)'(s)^2 + 2g_{12}(x^1(s), x^2(s))(x^1)'(s)(x^2)'(s)$$
$$+ g_{22}(x^1(s), x^2(s))(x^2)'(s)^2,$$

where x^1 and x^2 satisfy Equation(8.16), then $f'(s) = 0$, so $f(s)$ is a constant function over its domain. Consequently, if $(x^1)'(s_0)$ and $(x^2)'(s_0)$ are given so that $\vec{\alpha}'(s_0) = \vec{u}$, then $\|\vec{\alpha}(s)\| = 1$ for all s. If, in addition, $x^1(s_0)$ and $x^2(s_0)$ are chosen so that $\vec{\alpha}(s_0) = p$, then $\vec{\alpha}(s)$ is an arc length parametrization of a geodesic passing through p with direction $\vec{u}$ and is unique.   □

The proof of the above theorem establishes an additional fact concerning the equation for a geodesic curve.

Proposition 8.2.10. *Let $\vec{X}$ be the parametrization of a neighborhood of a regular surface of class C^3. Any solution $(x^1(s), x^2(s))$ to Equation (8.16), where one makes no prior assumptions on the parameter s, is such that the parametric curve $\vec{\gamma}(s) = \vec{X}(x^1(s), x^2(s))$, whose locus is a geodesic, has constant speed.*

Problems

8.2.1. Let S be a regular surface, and let $\vec{\gamma}$ be a geodesic on S. Prove that the geodesic torsion of γ is equal to $\pm\tau$, where τ is the torsion function of $\vec{\gamma}(t)$ as a space curve. [Hint: The possible difference in sign stems from the possible change in sign of κ_n which may come from a reparametrization of the surface. This property of τ_g justifies its name as geodesic torsion.]

8.2.2. (ODE) Consider a right circular cone with opening angle α, where $0 < \alpha < \pi/2$. Consider the coordinate patch parametrized by $\vec{X}(u, v) = (v \sin \alpha \cos u, v \sin \alpha \sin u, v \cos \alpha)$, where we assume $v > 0$. Determine equations for the geodesics on this cone. [Hint: Find a differential equation that expresses $\frac{dv}{du}$ in terms of u and v and then solve this to find an equation for u in terms of v.]

8.2.3. Consider the torus parametrized by $\vec{X}(u, v) = ((a + b \cos v) \cos u, (a + b \cos v) \sin u, b \sin v)$ where $a > b$. Show that the geodesics on a torus satisfy the differential equation

$$\frac{dr}{du} = \frac{1}{Cb} r \sqrt{r^2 - C^2} \sqrt{b^2 - (r - a)^2},$$

where C is a constant and $r = a + b \cos v$.

8.2.4. Find the differential equations that determine geodesics on a function graph $z = f(x, y)$.

8.2.5. If $\vec{X} : U \to \mathbb{R}^3$ is a parametrization of a coordinate patch on a regular surface S such that $g_{11} = E(u)$, $g_{12} = 0$ and $g_{22} = G(u)$ show that

(a) the u-parameter curves (i.e., over which v is a constant) are geodesics;

(b) the v-parameter curve $u = u_0$ is a geodesic if and only if $G_u(u_0) = 0$;

(c) the curve $\vec{x}(u, v(u))$ is a geodesic if and only if

$$v = \pm \int \frac{C\sqrt{E(u)}}{\sqrt{G(u)}\sqrt{G(u) - C^2}} \, du,$$

where C is a constant.

8.2.6. *Pseudosphere.* Consider a surface with a set of coordinates (u, v) defined over the upper half of the uv-plane, i.e., on $H = \{(u, v) \in \mathbb{R}^2 \,|\, v > 0\}$, such that the metric tensor is

$$(g_{ij}) = \begin{pmatrix} 1 & 0 \\ 0 & e^{2v} \end{pmatrix}.$$

(a) Use Gauss's equations to prove that such a surface has constant Gaussian curvature $K = -1$.

(b) Prove that in this coordinate system, all the geodesics appear in H as vertical lines or semicircles with center on the u-axis.

(c) Show that the following parametrization has the desired metric tensor:

$$\vec{X}(u, v) = \left(e^v \cos u, \, e^v \sin u, \, \ln(1 - \sqrt{1 - e^{2v}}) - v + \sqrt{1 - e^{2v}} \right).$$

The image of this parametrization is the pseudosphere.

8.2.7. Fill in the details in the proof of Proposition 8.2.9, namely, prove that if $x^1(s)$ and $x^2(s)$ satisfy Equation (8.16), then the function $f(s)$, which is the square of the speed of $\vec{X}(x^1(s), x^2(s))$, is constant.

8.2.8. *Liouville Surface.* A regular surface is called a *Liouville surface* if it can be covered by coordinate patches in such a way that each patch can be parametrized by $\vec{X}(u, v)$ such that

$$g_{11} = g_{22} = U(u) + V(v) \qquad \text{and} \qquad g_{12} = 0,$$

where U is a function of u alone and V is a function of v alone. (Note that Liouville surfaces generalize surfaces of revolution.) Prove the following facts:

(a) Show that the geodesics on a Liouville surface can be given as solutions to an equation of the form

$$\int \frac{du}{\sqrt{U(u) - c_1}} = \pm \int \frac{dv}{\sqrt{V(v) + c_1}} + c_2,$$

where c_1 and c_2 are constants.

(b) Show that if ω is the angle a geodesic makes with the curve $v = $const., then

$$U \sin^2 \omega - V \cos^2 \omega = C$$

for some constant C.

8.2.9. Let $\vec{\alpha} : I \to \mathbb{R}^3$ be a regular space curve parametrized by arc length with nowhere 0 curvature. Consider the ruled surface parametrized by

$$\vec{X}(s,t) = \vec{\alpha}(s) + t\vec{B}(s),$$

where $\vec{B}$ is the binormal vector of $\vec{\alpha}$. Suppose that $\vec{X}$ is defined over $I \times (-\varepsilon, \varepsilon)$, with $\varepsilon > 0$.

(a) Prove that if ε is small enough, the image S of $\vec{X}$ is a regular surface.

(b) Prove that if S is a regular surface, $\vec{\alpha}(I)$ is a geodesic on S.

(This shows that every regular space curve that has nonzero curvature is the geodesic of some surface.)

8.2.10. Consider the elliptic paraboloid given by $z = x^2 + y^2$. Note that this is a surface of revolution. Consider the geodesics on this surface that are not meridians.

(a) Suppose that the geodesic intersects the parallel at $z = z_0$ with an angle of θ_0. Find the lowest parallel that the geodesic reaches.

(b) Prove that any geodesic that is not a meridian intersects itself an infinite number of times.

8.2.11. Consider the hyperboloid of one sheet given by the equation $x^2 + y^2 - z^2 = 1$ and let p be a point in the upper half-space defined by $z > 0$. Consider now geodesic curves that go through p and make an acute angle of θ_0 with the meridian of the hyperboloid passing through p. Call r_0 the distance from p to the z-axis.

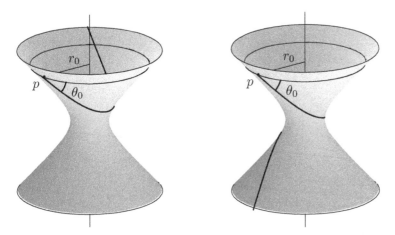

Figure 8.4. Geodesics on the hyperboloid of one sheet.

(a) Prove that if $\cos\theta_0 > 1/r_0$, then the geodesic remains in the upper half-space $z > 0$.

(b) Prove that if $\cos\theta_0 < 1/r_0$, then the geodesic crosses the $z = 0$ plane and descends indefinitely in the negative z-direction.

(c) Prove that if $\cos\theta_0 = 1/r_0$, then the geodesic, as it descends from p, asymptotically approaches the meridian given by $x^2 + y^2 = 1$ at $z = 0$.

(d) Using the parametrization

$$\vec{X}(u, v) = (\cosh v \cos u, \cosh v \sin u, \sinh v),$$

suppose that the initial conditions for the geodesic are $u(0) = 0$ and $v(0) = v_0$, so that $p = \vec{X}(0, v_0)$. Find the initial conditions $u'(0)$ and $v'(0)$ so that the geodesic is parametrized by arc length and satisfies the condition in (c).

(See Figure 8.4 for examples of nonasymptotic behavior.)

8.3 Geodesic Coordinates

8.3.1 General Geodesic Coordinates

Definition 8.3.1. Let S be a regular surface of class C^3. A system of *geodesic coordinates* is an orthogonal regular parametrization $\vec{X}$ of S such that, for one of the coordinates, all the coordinate lines are geodesics.

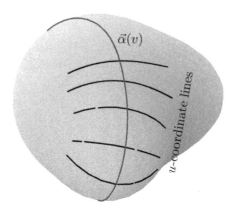

Figure 8.5. Geodesic coordinate system generated by $\vec{\alpha}(v)$.

There are many ways to define geodesic coordinates on a regular surface of class C^3. We introduce one general way first.

Suppose that $\vec{\alpha}(t)$ for $t \in [a, b]$ is a regular curve with image C on S of class C^2. According to Theorem 8.2.9, for each $t_0 \in [a, b]$, there exists a unique geodesic on S through $\vec{\alpha}(t_0)$ perpendicular to C (see Figure 8.5). Furthermore, one can parametrize the geodesics $\vec{\gamma}_{t_0}(s)$ by arc length such that $\vec{\gamma}_{t_0}(0) = \vec{\alpha}(t_0)$ and $t \mapsto \vec{\gamma}'_t(0)$ is continuous in t. Define the function

$$\vec{X}(s, t) = \vec{\gamma}_t(s). \tag{8.30}$$

We wish to show that over an open set containing C, the function $\vec{X}(s, t)$ is a regular parametrization of class C^2. (One desires class C^2 since this is required for the first and second fundamental forms to exist.) However, we must assume that S is of class C^5.

Proposition 8.3.2. *If S is of class C^5, there exists an $\varepsilon > 0$ such that $\vec{X}(s, t)$, as defined in Equation (8.30), is a regular parametrization of class C^2 over $(-\varepsilon, \varepsilon) \times (a, b)$.*

Proof: The proof relies on standard theorems of existence and uniqueness for differential equations as well as on some basic topology.

Let p be a point on C, and let $\vec{X}(x^1, x^2)$ be a regular parametrization of class C^5 of a neighborhood $V_p = \vec{X}(U_p)$ of S containing p. Let $\alpha^1(t)$ and $\alpha^2(t)$ be functions such that $\vec{\alpha}(t) = \vec{X}(\alpha^1(t), \alpha^2(t))$, the given parametrization of C on S in the neighborhood V_p. Call J_p the domain of $\vec{\alpha}$ such that $\vec{\alpha}(J_p) \subset V_p$, and suppose that $p = \vec{\alpha}(t_0)$.

Let $x^1(s,t)$ and $x^2(s,t)$ be functions that map onto the geodesics $\vec{\gamma}_t(s)$ via

$$\vec{X}(x^1(s,t), x^2(s,t)) = \vec{\gamma}_t(s).$$

By Theorem 8.2.9, for all parameters t, over the interval $s \in (-\varepsilon_t, \varepsilon_t)$, there exists a unique solution for $x^1(s,t)$ and for $x^2(s,t)$ to Equation (8.16) given the initial conditions

$$x^1(0,t) = \alpha^1(t), \quad x^2(0,t) = \alpha^2(t), \quad \frac{\partial x^1}{\partial s}(0,t) = u(t), \quad \frac{\partial x^2}{\partial s}(0,t) = v(t),$$

where u and v are any functions of class C^1. Furthermore, these solutions are of class C^2 in the variable s. Now since for each t we want $\vec{\gamma}_t(s)$ to be an arc length parametrization of a geodesic perpendicular to C, we impose the following two conditions on u and v for all t:

(i) $g_{11}u^2 + 2g_{12}uv + g_{22}v^2 = 1,$

(ii) $g_{11}u\dfrac{d\alpha^1}{dt} + g_{12}u\dfrac{d\alpha^2}{dt} + g_{12}v\dfrac{d\alpha^1}{dt} + g_{22}v\dfrac{d\alpha^2}{dt} = 0,$

(8.31)

where (i), for example, means that

$$g_{11}(\alpha^1(t), \alpha^2(t))u(t)^2 + 2g_{12}(\alpha^1(t), \alpha^2(t))u(t)v(t) + g_{22}(\alpha^1(t), \alpha^2(t))v(t)^2 = 1$$

for all t. The requirement to parametrize each $\vec{\gamma}_t(s)$ so that $\vec{\gamma}_t'(0)$ is continuous in t is equivalent to requiring that u and v be continuous. These conditions completely specify $u(t)$ and $v(t)$ for all t. Notice that for (i) and (ii) to both be satisfied, $((\alpha^1)'(t), (\alpha^2)'(t))$ and $(u(t), v(t))$ cannot be linear multiples of each other, and hence,

$$\begin{vmatrix} (\alpha^1)'(t) & u(t) \\ (\alpha^2)'(t) & v(t) \end{vmatrix} \neq 0 \quad \text{for all } t.$$

Now write the equations for geodesics as a first-order system, as in Equation (8.29). Since $\vec{X}$ is of class C^5, then all the functions Γ^i_{jk} are of class C^3. Then according to theorems of dependency of solutions of differential equations on initial conditions (see [2, Theorem 32.4]), solutions to the system from Equation (8.29) are of class C^2 in terms of initial conditions, which implies that

$$\frac{\partial^2 x^1}{\partial s \partial t}, \quad \frac{\partial^2 x^2}{\partial s \partial t}, \quad \frac{\partial^2 x^1}{\partial t^2}, \quad \text{and} \quad \frac{\partial^2 x^2}{\partial t^2}$$

are continuous over a possibly smaller open neighborhood U_p' of $(0, t_0)$. Thus, we conclude that $x^1(s, t)$ and $x^2(s, t)$ are of class C^2 over U_p'.

To prove regularity of $\vec{X}(s, t)$, note that

$$\begin{vmatrix} (\alpha^1)'(t_0) & u(t_0) \\ (\alpha^2)'(t_0) & v(t_0) \end{vmatrix} = \left. \frac{\partial(x^1, x^2)}{\partial(s, t)} \right|_{(0, t_0)},$$

the Jacobian of the change of variables $x^1(s, t)$ and $x^2(s, t)$ at the point $(0, t_0)$. Since this Jacobian is a continuous function and is nonzero at $(0, t_0)$, there is an open neighborhood U_p'' of $(0, t_0)$ such that $\partial(x^1, x^2)/\partial(s, t) \neq 0$. By Proposition 5.4.2,

$$\frac{\partial \vec{X}}{\partial s} \times \frac{\partial \vec{X}}{\partial t} = \frac{\partial(x^1, x^2)}{\partial(s, t)} \left(\frac{\partial \vec{X}}{\partial x^1} \times \frac{\partial \vec{X}}{\partial x^2} \right).$$

so over U_p'' the parametrization $\vec{X}(s, t)$ is regular. Consequently, there exists an $\varepsilon_p > 0$ such that

$$\bar{U}_p \stackrel{\text{def}}{=} (-\varepsilon_p, \varepsilon_p) \times (t_0 - \varepsilon_p, t_0 + \varepsilon_p) \subset U_p' \cap U_p'',$$

and over $\bar{U}_p$, the parametrization $\vec{X}(s, t)$ is regular and of class C^2. Call $\bar{V}_p = \vec{X}(\bar{U}_p)$.

Finally, consider the whole curve C on S. The curve C can be covered by open sets of the form $\bar{V}_p$ for various $p \in C$. Let p and q be two points on C, and let us write $\vec{X}_p : \bar{U}_p \to \bar{V}_p$ and $\vec{X}_q : \bar{U}_q \to \bar{V}_q$ for associated parametrizations of $\bar{V}_p$ and $\bar{V}_q$. If

$$\bar{U}_p = (-\varepsilon_p, \varepsilon_p) \times J_p \quad \text{and} \quad \bar{U}_q = (-\varepsilon_q, \varepsilon_q) \times J_q$$

overlap, since for all $t \in J_p \cap J_q$ we have $\vec{X}_p(s, t) = \vec{X}_q(s, t)$ by the uniqueness of geodesics, then we can extend $\vec{X}_p$ and $\vec{X}_q$ to a function $\vec{X}$ over

$$(-\min(\varepsilon_p, \varepsilon_q), \min(\varepsilon_p, \varepsilon_q)) \times (J_p \cup J_q).$$

Since $[a, b]$ is compact, $[a, b]$ can be covered by a finite number of sets $\bar{U}_p$. Therefore, there exists $\varepsilon > 0$ and a single function $\vec{X}$ defined over $(-\varepsilon, \varepsilon) \times (a, b)$ such that $\vec{X}(0, t) = \vec{\alpha}(t)$, and for $t \in (a, b)$, $\vec{X}(s, t)$ parametrizes a geodesic by arc length. $\qquad\square$

Proposition 8.3.2 leads to the following general theorem.

Theorem 8.3.3. *Let S be a regular surface of class C^5, and let $\vec{\alpha}$: $[a, b] \to S$ be a regular parametrization of a simple curve of class C^2. Then there exists a system of geodesic coordinates $\vec{X}(u, v)$ of class C^2 defined over $-\varepsilon < u < \varepsilon$ and $a < v < b$ such that $\vec{X}(0, v) = \vec{\alpha}(v)$, and the u-coordinate lines parametrize geodesics by arc length.*

Proof: Proposition 8.3.2 constructs the function $\vec{X}(u, v)$ as desired. However, it remains to be proven that $\vec{X}(u, v)$ is an orthogonal parametrization.

Consider the derivative of g_{12} with respect to u

$$g_{12,1} = \frac{\partial g_{12}}{\partial u} = \frac{\partial}{\partial u}(\vec{X}_1 \cdot \vec{X}_2) = \vec{X}_{11} \cdot \vec{X}_2 + \vec{X}_1 \cdot \vec{X}_{12}.$$

Since each u-coordinate line is parametrized by arc length, $\vec{X}_1 \cdot \vec{X}_1 = 1$, so by differentiating with respect to v, we find that $g_{11,2} = 2\vec{X}_1 \cdot \vec{X}_{12} = 0$. By the same reasoning, over the u-coordinate lines, in the $\{\vec{N}, \vec{T}, \vec{U}\}$ frame one has

$$\frac{\partial \vec{X}}{\partial u} = \vec{T} \quad \text{and} \quad \frac{\partial^2 \vec{X}}{\partial u^2} = \kappa \vec{P} = \kappa_g \vec{U} + \kappa_n \vec{N}.$$

But since $\vec{X}(u, v)$ with v fixed is a geodesic, $\kappa_g = 0$. Thus, $\vec{X}_{11} = \kappa_n \vec{N}$ and, in particular, $\vec{X}_{11} \cdot \vec{X}_2 = 0$.

Consequently, $g_{12,1} = 0$, and therefore, g_{12} is a function of v only. We can write $g_{12}(u, v) = g_{12}(0, v)$. However, by construction of $\vec{X}(u, v)$ in Proposition 8.3.2, $g_{12}(0, v) = 0$ for all v. Thus, g_{12} is identically 0, and hence, $\vec{X}(u, v)$ is an orthogonal parametrization.$\square$

The class of geodesic coordinate systems described in Theorem 8.3.3 is of a particular type. Not every geodesic coordinate system needs to be defined in reference to a curve C on S as done above. Let $\vec{X}(u, v)$ be any system of geodesic coordinates where the u-coordinate lines are geodesics. A priori, we know only that $g_{12} = 0$. However, much more can be said.

Proposition 8.3.4. *Let S be a surface of class C^3, and let $\vec{X} : U \to V$ be a regular parametrization of class C^3 of a neighborhood V of S. The parametrization $\vec{X}(u, v)$ is a system of geodesic coordinates in which the u-coordinate lines are geodesics if and only if the metric tensor is of the form*

$$g = \begin{pmatrix} E(u) & 0 \\ 0 & G(u, v) \end{pmatrix}.$$

Proof: First, the parametrization $\vec{X}$ is an orthogonal if and only if $g_{12} = 0$.

By Equation (8.7), along the u-coordinate lines, the geodesic curvature is

$$\kappa_g = \left(\frac{du}{ds} \right)^3 \Gamma_{11}^2 \sqrt{g_{11} g_{22}}.$$

Since the metric tensor must be positive definite everywhere, $g_{11} g_{22}$ is never 0. Since $du/ds \neq 0$, along a u-coordinate line, $\kappa_g = 0$ if and only if $\Gamma_{11}^2 = -\frac{1}{2} g^{22} \frac{\partial g_{11}}{\partial v} = 0$. The result follows. □

Along u-coordinate lines, since $v' = 0$, the speed function $\dfrac{ds}{du} = \sqrt{E(u)}$ is independent of v. Regardless of v, the arc length formula between $u = u_0$ and u is

$$s(u) = \int_{u_0}^{u} \sqrt{E(u)} \, du.$$

Therefore, it is possible to reparametrize u along the u-coordinate lines by arc length, with $u(s) = s^{-1}(u)$. This leads to the following proposition.

Proposition 8.3.5. *Let S be a regular surface of class C^3, and let $\vec{X}(u, v)$ be a system of geodesic coordinates in a neighborhood V of S. Over the same neighborhood V, there exists a system of geodesic coordinates $\vec{X}(s, v)$ such that the s-coordinate lines are geodesics parametrized by arc length. Furthermore, if this is the case, then the coefficients of the metric tensor are of the form*

$$g_{11}(u, v) = 1, \qquad g_{12}(u, v) = 0, \qquad g_{22}(u, v) = G(u, v),$$

and the Gaussian curvature of S is given by

$$K = -\frac{1}{\sqrt{g_{22}}} \frac{\partial^2 \sqrt{g_{22}}}{\partial u^2}. \tag{8.32}$$

Proof: The fact that $g_{11}(u,v) = 1$ follows from the u-coordinate lines being parametrized by arc length and Equation (8.32) is a direct application of Equation (7.53). ☐

8.3.2 Geodesic Polar Coordinates

Let p be a point on a regular surface of class C^3. By Theorem 8.2.9, for every $\vec{v} \in T_pS$, there exists a unique geodesic $\vec{\gamma}_{p,\vec{v}} : (-\varepsilon, \varepsilon) \to S$, with $\vec{\gamma}_{p,\vec{v}}(0) = p$. Furthermore, by Proposition 8.2.10, we know that $\|\vec{\gamma}_{p,\vec{v}}'(t)\| = \|\vec{v}\|$. We remark then that for any constant scalar λ,

$$\vec{\gamma}_{p,\vec{v}}(\lambda t) = \vec{\gamma}_{p,\lambda\vec{v}}(t)$$

for all $t \in (-\varepsilon/\lambda, \varepsilon/\lambda)$.

Definition 8.3.6. Let S be a regular surface of class C^3, and let $p \in S$ be a point. For any tangent $\vec{v} \in T_pS$, we define the *exponential map* at p as

$$\exp_p(\vec{v}) = \begin{cases} \exp_p(\vec{v}) = p, & \text{if } \vec{v} = \vec{0}, \\ \vec{\gamma}_{p,\vec{v}}(1), & \text{if } \vec{v} \neq \vec{0}, \end{cases}$$

whenever $\vec{\gamma}_{p,\vec{v}}(1)$ is well defined.

The map $\exp_p : U \to S$, where U is a neighborhood of $\vec{0}$ in T_pS corresponds to traveling along the geodesic through p with direction $\vec{v}$ over the distance $\|\vec{v}\|$ (see Figure 8.6). We will show that the exponential map can lead to some nice parametrizations of a neighborhood of p on S, but we first need to prove the following two propositions.

Proposition 8.3.7. *For all $p \in S$, the exponential map is defined over an open neighborhood U of $\vec{0}$. Furthermore, if S is of class C^4, then $\exp_p$ is differentiable over U as a function from T_pS into $\mathbb{R}^3$.*

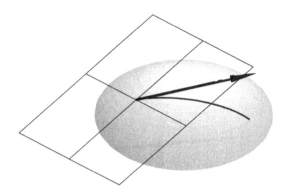

Figure 8.6. Exponential map on an ellipsoid.

Proof: We first show that $\exp_p$ is defined over some open disk centered at $\vec{0}$.

Let $C_1 \subset T_pS$ be the set of all vectors of unit length. Set R to be a large positive real number. For each $\vec{v} \in C_1$, let $\varepsilon(\vec{v})$ be the largest positive real $\varepsilon \leq R$ such that $\vec{\gamma}_{p,\vec{v}} : (-\varepsilon, \varepsilon) \to S$ parametrizes a geodesic, or in other words, solves the system of equations in Equation (8.16). The theorems of existence and uniqueness of solutions to differential equations tell us that the solutions to Equation (8.16) depend continuously on the initial conditions, so $\varepsilon(\vec{v})$ is continuous over C_1. (Setting R as an upper bound ensures that $\varepsilon(\vec{v})$ is defined for all $\vec{v} \in C_1$.)

Since C_1 is a compact set, as a function to $\mathbb{R}$, $\varepsilon(\vec{v})$ attains a minimum ε_0 on C_1. However, since $\varepsilon(\vec{v}) > 0$ for all $\vec{v} \in C_1$, then $\varepsilon_0 > 0$. Consequently, $\exp_p$ is defined over the open ball $B_{\varepsilon_0}(\vec{0})$ and is continuous.

Let $\vec{X} : U' \to S$ be a regular parametrization of a neighborhood of p on S. According to the theorem of dependence of solutions to differential equations on initial conditions, the function

$$\vec{\gamma}_p(\vec{v}, t) : U \times (-\varepsilon, \varepsilon) \longrightarrow \mathbb{R}^3$$

is of class C^1 in the initial conditions $\vec{v}$ if the Christoffel symbol functions are of class C^2, which means that $\vec{X}$ needs to be of class C^4. The result follows. $\square$

Proposition 8.3.8. *Let S be a regular surface of class C^4. There is a neighborhood U of $\vec{0}$ in T_pS such that $\exp_p$ is a homeomorphism onto $V = \exp_p(U)$, which is an open neighborhood of p on S.*

Proof: The tangent space T_pS is isomorphic as a vector space to $\mathbb{R}^2$, and one can therefore view $\exp_p$ as a function from $U' \subset \mathbb{R}^2$ into $\mathbb{R}^3$. Let $\vec{v} \in T_pS$ be a nonzero vector, and let $\vec{\alpha}(t) = t\vec{v}$ be defined for $t \in (-\varepsilon, \varepsilon)$ for some $\varepsilon > 0$. Consider the curve on S defined by

$$\exp_p(\vec{\alpha}(t)) = \exp_p(t\vec{v}) = \vec{\gamma}_{p,t\vec{v}}(1) = \vec{\gamma}_{p,\vec{v}}(t).$$

According to the chain rule,

$$\frac{d}{dt}\left(\exp_p(t\vec{v})\right)\Big|_0 = d(\exp_p)_0 \frac{d\vec{\alpha}}{dt}\Big|_0 = d(\exp_p)_0(\vec{v}),$$

where $\vec{v}$ in this expression is viewed as an element in T_pS, and hence, by isomorphism, in $\mathbb{R}^2$. However, by construction of the geodesics, $\vec{\gamma}'_{p,\vec{v}}(0) = \vec{v}$, where we view $\vec{v}$ as an element of $\mathbb{R}^3$. This result proves, in particular, that $d(\exp_p)_0$ is nonsingular.

Using the Implicit Function Theorem from analysis (see [7, Theorem 8.27]), the fact that $d(\exp_p)_0$ is invertible implies that there exists an open neighborhood U of $\vec{0}$ such that $\exp_p : U \to \exp_p(U)$ is a bijection. Furthermore, setting $V = \exp_p(U)$, by the Implicit Function Theorem, the inverse function $\exp_p^{-1} : V \to U$ is at least of class C^1, and hence, it is continuous.

In conclusion, $\exp_p$ is a homeomorphism between U and $\exp_p(U)$. $\square$

The identification of T_pS with $\mathbb{R}^2$ allows one to define parametrizations of neighborhoods of p with some nice properties.

Definition 8.3.9. Let S be a surface of class C^4. A system of *Riemann normal coordinates* of a neighborhood of p is a parametrization defined by

$$\vec{X}(u,v) = \exp_p(u\vec{w}_1 + v\vec{w}_2),$$

where $\{\vec{w}_1, \vec{w}_2\}$ is an orthonormal basis of T_pS.

Propositions 8.3.7 and 8.3.8 imply that Riemann normal coordinates provide a regular parametrization of a neighborhood V of p.

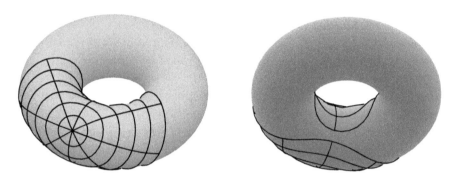

Figure 8.7. Geodesic polar coordinate lines on a torus.

Furthermore, the theorem of the dependence of solutions to differential equations on initial conditions shows that Riemann normal coordinates form a parametrization of class C^r if S is a surface of class C^{r+3}.

Definition 8.3.10. Let S be a surface of class C^4. The *geodesic polar coordinates* of a neighborhood of p give a parametrization defined by

$$\vec{X}(r, \theta) = \exp_p \left((r \cos \theta)\vec{w}_1 + (r \sin \theta)\vec{w}_2\right),$$

where $\{\vec{w}_1, \vec{w}_2\}$ is an orthonormal basis of T_pS. A curve on S that can be parametrized by

$$\vec{\gamma}(t) = \exp_p \left((R \cos t)\vec{w}_1 + (R \sin t)\vec{w}_2\right), \qquad \text{for } t \in [0, 2\pi]$$

is called a *geodesic circle* of center p and radius R.

Figure 8.7 gives an example of coordinate lines of a geodesic polar coordinate system on a torus. It is important to note that a geodesic circle is neither a geodesic curve on the surface nor a circle in $\mathbb{R}^3$.

We present the following propositions about the above coordinate systems but leave the proofs as exercises for the reader.

Theorem 8.3.11. *Let S be a regular surface of class C^5, and let p be a point on S. There exists an open neighborhood U of $(0,0)$ in $\mathbb{R}^2$ such that the Riemann normal coordinates defined in Definition 8.3.9 form a system of geodesic coordinates.*

Proof: This theorem follows immediately from Theorem 8.3.3, where one uses the curve

$$\vec{\alpha}(t) = \exp_p(t\vec{w}_1),$$

where $\vec{w}_1$ is a unit vector in T_pS. □

Proposition 8.3.12. *Let S be a regular surface of class C^5, and let p be a point on S. Let $\vec{X}$ be a parametrization of a Riemann normal coordinate system in a neighborhood of p so that $\vec{X}(0,0) = p$, and let g_{ij} be the coefficients of the metric tensor associated to $\vec{X}$. The coefficients satisfy $g_{ij}(0,0) = \delta_i^j$, and all the first partial derivatives of all the g_{ij} functions vanish at $(0,0)$.*

Proof: (Left as an exercise for the reader. See Problem 8.3.2.) □

The next theorem discusses the existence of geodesic polar coordinate systems in a neighborhood of a point $p \in S$.

Theorem 8.3.13. *Let S be a regular surface of class C^5, and let p be a point on S. There exists an $\varepsilon > 0$ such that the parametrization $\vec{X}(r,\theta)$ in Definition 8.3.10, with $0 < r < \varepsilon$ and $0 < \theta < 2\pi$, is a regular parametrization and defines a geodesic coordinate system in a neighborhood V whose closure contains p in its interior.*

The parametrization $\vec{X}(r,\theta)$ is defined as a function for $-r_0 < r < r_0$ and $\theta \in \mathbb{R}$ for some $r_0 > 0$. However, similar to usual polar coordinates in $\mathbb{R}^2$, one must restrict one's attention to $r > 0$ and $0 < \theta < 2\pi$ to ensure that $\vec{X}$ is a homeomorphism. The image $\vec{X}\left((0,\varepsilon) \times (0,2\pi)\right)$ is then a "geodesic disk" on S centered at p with a "geodesic radius" removed. By definition, the parametrization $\vec{X}(r,\theta)$ is such that all the coordinate lines for one of the variables are geodesics, but Theorem 8.3.13 asserts that, within a small enough radius, the parametrization $\vec{X}(r,\theta)$ is regular and orthogonal.

The existence of geodesic polar coordinates at any point p on a surface S of high enough class leads to interesting characterizations of the Gaussian curvature K of S at p. First, we remind the reader that if $\vec{X}(r,\theta)$ is a system of geodesic polar coordinates, then the coefficients of the associated metric tensor are of the form

$$g_{11}(r,\theta) = 1, \quad g_{12}(r,\theta) = 0, \quad g_{22}(r,\theta) = G(r,\theta) \qquad (8.33)$$

for some function $G(r,\theta)$ defined over the domain of $\vec{X}$.

Proposition 8.3.14. *Let S be a surface of class C^5, and let $\vec{X}(r, \theta)$ be a system of geodesic polar coordinates at p on S. Then the function $G(r, \theta)$ in Equation (8.33) satisfies*

$$\sqrt{G(r, \theta)} = r - \frac{1}{6}K(p)r^3 + R(r, \theta),$$

where $K(p)$ is the Gaussian curvature of S at p and $R(r, \theta)$ is a function that satisfies

$$\lim_{r \to 0} \frac{R(r, \theta)}{r^3} = 0.$$

Proof: (Left as an exercise for the reader. See Problem 8.3.4.)   □

Proposition 8.3.14 shows that the perimeter of a geodesic circle at p of radius R is

$$C = \int_0^{2\pi} \sqrt{G(R, \theta)}\, d\theta = 2\pi R - \frac{1}{3}K(p)\pi R^3 + F(R) \qquad (8.34)$$

where $F(r)$ is a function of the radial variable r and satisfies

$$\lim_{r \to 0} \frac{F(r)}{r^3} = 0.$$

This, and a similar consideration of the area of geodesic disks around p, leads to the following geometric characterization of the Gaussian curvature at p.

Theorem 8.3.15. *Let S be a surface of class C^5, and let $p \in S$ be a point. Define $C(r)$ (respectively $A(r)$) as the perimeter (respectively the area) of the geodesic circle (respectively disk) centered at p and of radius r. The Gaussian curvature $K(p)$ of S at p satisfies*

$$K(p) = \lim_{r \to 0} \frac{3}{\pi}\left(\frac{2\pi r - C(r)}{r^3} \right) \quad and \quad K(p) = \lim_{r \to 0} \frac{12}{\pi}\left(\frac{\pi r^2 - A(r)}{r^4} \right).$$

Proof: (Left as an exercise for the reader. See Problem 8.3.5.)   □

Problems

8.3.1. Prove that if $\vec{X} : U \to V$ is a parametrization in which both families of coordinate lines are families of geodesics, then the Gaussian curvature satisfies $K(u, v) = 0$.

8.3.2. Prove Proposition 8.3.12. [Hint: First use Equation (8.16) to prove that all Γ^i_{jk} vanish at $(0,0)$, i.e., at p.]

8.3.3. (*) Prove Theorem 8.3.13.

8.3.4. Prove Proposition 8.3.14. [Hint: Use a Taylor series expansion of $G(r, \theta)$ and Equation (8.32).]

8.3.5. Prove Theorem 8.3.15.

8.3.6. Consider the usual parametrization of the sphere S of radius R

$$\vec{X}(u, v) = (R \cos u \sin v, R \sin u \sin v, R \cos v).$$

Define the new parametrization $Y(r, \theta) = \vec{X}(\theta, r/R)$.

   (a) Prove that $Y(r, \theta)$ is a geodesic polar coordinate system of S at $p = (R, 0, 0)$.

   (b) Prove the result of Proposition 8.3.14 directly and determine the corresponding remainder function $R(r, \theta)$.

8.4 Gauss-Bonnet Theorem and Applications

No course on classical differential geometry is complete without the Gauss-Bonnet Theorem. Arguably the most profound theorem in the differential geometry study of surfaces, the Gauss-Bonnet Theorem simultaneously encompasses a total curvature theorem for surfaces, the total geodesic curvature formula for plane curves, and other famous results, such as the sum of angles formula for a triangle in plane, spherical, or hyperbolic geometry. Other applications of the theorem extend much further and lead to deep connections between topological invariants and differential geometric quantities, such as the Gaussian curvature.

In his landmark paper [14], Gauss proved an initial version of what is now called the Gauss-Bonnet Theorem. The form in which we present it was first published by Bonnet in 1848 [4]. Various alternative proofs exist (e.g., [31]) and a search of the literature turns

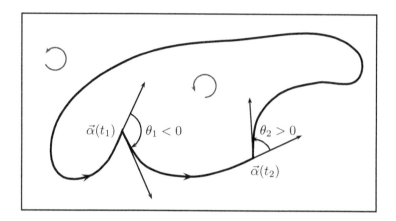

Figure 8.8. A regular piecewise curve.

up a large variety of generalizations (e.g., to polyhedral surfaces and to higher-dimensional manifolds). Despite the far-reaching consequences of the theorem, the difficulty of the proof resides in one essentially topological property of curves on surfaces for which we provide a reference. The theorem then follows easily from Green's Theorem for simple closed curves in the plane.

In order to present the Gauss-Bonnet Theorem, we need to first establish a few definitions about the geometry of curves on surfaces.

We call a set of points $\{t_i\}_{i \in \mathcal{K}}$ in $\mathbb{R}$ *discrete* if

$$\inf\{|t_i - t_j| \,|\, i, j \in \mathcal{K} \text{ and } i \neq j\} > 0,$$

i.e., if any two distinct points are separated by at least some fixed, positive real number. The indexing set $\mathcal{K}$ may be finite, say $\mathcal{K} = \{1, \dots, k\}$, or, if it is infinite, it may be taken as either the set of nonnegative integers $\mathbb{N}$ or the set of integers $\mathbb{Z}$.

We say that a curve $\vec{\alpha} : I \to \mathbb{R}^n$ is *piecewise regular* if $\vec{\alpha}$ is continuous over I and if there exists a discrete set of points $\{t_i\}_{i \in \mathcal{K}}$ such that $\vec{\alpha}$ is regular over $[t_i, t_{i+1}]$ for all $i \in \mathcal{K}$ (such that $i + 1$ is also in $\mathcal{K}$). The points $\vec{\alpha}(t_i)$ on the curve are called *vertices* (or *corners*) of the curve. If $\vec{\alpha}$ is regular over an interval I', then the trace $\vec{\alpha}(I')$ is called a *regular arc* of the curve.

Let $\vec{\alpha}(t_i)$ be a vertex of a piecewise regular curve. For some $\varepsilon > 0$, $\vec{\alpha}$ is regular on $(t_i - \varepsilon, t_i]$ and on $[t_i, t_i + \varepsilon)$. Therefore, we can take the limit of $\vec{\alpha}'(t)$ as t approaches t_i from the right and from the left. Furthermore, both one-sided limits will be nonzero by the regularity condition. Call

$$\vec{\alpha}'(t_i^-) = \lim_{t \to t_i^-} \vec{\alpha}'(t) \qquad \text{and} \qquad \vec{\alpha}'(t_i^+) = \lim_{t \to t_i^+} \vec{\alpha}'(t).$$

We call a vertex $\vec{\alpha}(t_i)$ a *cusp* if $\vec{\alpha}'(t_i^-)$ and $\vec{\alpha}'(t_i^+)$ are collinear. We assume for the moment that none of the corners are cusps.

We point out that our definitions for piecewise regular curves apply to any curve in $\mathbb{R}^m$. However we now also assume that $\vec{\alpha}$ has the properties that will interest us for the Gauss-Bonnet Theorem, namely that $\vec{\alpha}$ is a simple, closed, piecewise regular curve on a regular oriented surface S with orientation n. In this case, I is a closed and bounded interval, and there can be at most a finite number of vertices. Furthermore, we impose the criterion that the curve $\vec{\alpha}$ traces out the image $\vec{\alpha}(I)$ in the same direction as the orientation of S.

For all vertices $\vec{\alpha}(t_i)$ (that are not cusps), we define the *external angle* θ_i of the vertex as the angle $-\pi < \theta_i < \pi$ swept out from $\vec{\alpha}'(t_i^-)$ to $\vec{\alpha}'(t_i^+)$ in the plane through the point $\vec{\alpha}(t_i)$ and spanned by $\vec{\alpha}'(t_i^-)$ and $\vec{\alpha}'(t_i^+)$ (see Figure 8.8).

8.4.1   The Local Gauss-Bonnet Theorem

We are now in a position to approach the Gauss-Bonnet Theorem. First, however, we motivate the Gauss-Bonnet Theorem with the following intuitive example.

Example 8.4.1 (The Moldy Potato Chip). We particularly encourage the reader to consult the demo applet for this example.

Consider a region $\mathcal{R}$ on a regular surface S such that the boundary curve $\partial\mathcal{R}$ is a regular curve that is simple and convex on the surface S. We now create the "moldy potato chip" as the surface that consists of taking the region $\mathcal{R}$ and spreading it out over every possible normal direction by a distance of r, where r is a fixed real number. As the demo shows, the surface of the moldy potato chip (MPC) consists of three pieces: two pieces that are the parallel surfaces of S "above" and "below" $\mathcal{R}$, and one region that consists of a half-tube around $\partial\mathcal{R}$. (In the applet, these portions are colored by magenta and orange, respectively.)

The total Gaussian curvature $\iint_{\text{MPC}} K\,dS$ of the MPC is the same as the area of the portion of the unit sphere covered by the Gauss map of the MPC. Now, if $\mathcal{R}$ were small, i.e. almost a point, then the MPC would be almost just a sphere, and hence, the image of the Gauss map would be exactly the unit sphere, which has area 4π. As one stretches $\mathcal{R}$ out in a continuous manner, the area of the image of the Gauss map must also change continuously but it must always be an integral multiple of 4π since the Gauss map of a surface without boundary covers the unit sphere a fixed number of times. Hence, the total curvature of the moldy potato chip is always 4π.

However, Problem 6.5.9 showed that if $r > 0$ is small enough, the total Gaussian curvature for each of the two parallel surfaces above and below $\mathcal{R}$ is $\iint_{\mathcal{R}} K\,dS$. Furthermore (and we leave the full details of this until Example 8.1.9), the total Gaussian curvature of the half-tube around $\partial\mathcal{R}$ is $2\int_{\partial\mathcal{R}} \kappa_g\,ds$. Hence, we conclude that

$$\iint_{\text{MPC}} K\,dS = 4\pi \iff 2\iint_{\mathcal{R}} K\,dS + 2\int_{\partial\mathcal{R}} \kappa_g\,ds = 4\pi$$

$$\iff \iint_{\mathcal{R}} K\,dS + \int_{\partial\mathcal{R}} \kappa_g\,ds = 2\pi, \qquad (8.35)$$

which establishes a global result from local properties (the geodesic curvature of $\partial\mathcal{R}$ and the Gaussian curvature of S over $\mathcal{R}$).

In what follows, we work to establish Equation (8.35) rigorously and expand the hypotheses under which it holds, the end result being the celebrated global Gauss-Bonnet Theorem. Furthermore, we provide an intrinsic proof of the Gauss-Bonnet Theorem, which establishes it as long as we have a metric tensor and without the assumption that the surface is in an ambient Euclidean three-space.

Certain types of regions on surfaces play an important role in what follows. We call a region $\mathcal{R}$ on a regular surface S simple if it is homeomorphic to a disk.

Let S be a surface of class C^3, and let $\vec{X} : U \to \mathbb{R}^3$ be an orthogonal parametrization of a neighborhood $V = \vec{X}(U)$ of S. Consider a regular curve $\vec{\gamma}(s)$ parametrized by arc length whose image lies in V.

Liouville's Formula (see Problem 8.1.8) states that the geodesic curvature of $\vec{\gamma}(s)$ satisfies

$$\kappa_g(s) = \frac{d\varphi}{ds} + \kappa_{(u)} \cos\varphi + \kappa_{(v)} \sin\varphi,$$

where $\kappa_{(u)}$ is the geodesic curvature along the u-parameter curve (i.e., $v = v_0$) and similarly for $\kappa_{(v)}$ and $\varphi(s)$ is the angle $\vec{\gamma}'(s)$ makes with $\vec{X}_u$. Using Equation (8.7) to calculate $\kappa_{(u)}$ and $\kappa_{(v)}$ and writing $\cos\varphi$ and $\sin\varphi$ in terms of the metric tensor, one can rewrite Liouville's Formula as

$$\kappa_g(s) = \frac{d\varphi}{ds} + \frac{1}{2\sqrt{g_{11}g_{22}}} \frac{\partial g_{22}}{\partial u} \frac{dv}{ds} - \frac{1}{2\sqrt{g_{11}g_{22}}} \frac{\partial g_{11}}{\partial v} \frac{du}{ds}. \qquad (8.36)$$

Now consider a simple, closed, piecewise regular curve $\vec{\alpha} : I \to \mathbb{R}^3$ on S in the neighborhood V. Suppose that $\vec{\alpha}$ is parametrized by arc length so that $I = [0, l]$ and has vertices at $s_1 < s_2 < \cdots < s_k$. Set $s_0 = 0$ and $s_{k+1} = l$, and call C the image $\vec{\alpha}(I)$. Also call C_i the image of $\vec{\alpha}([s_{i-1}, s_i])$ for $1 \leq i \leq k+1$. Suppose, additionally, that $\vec{\alpha}(s) = \vec{X}(u(s), v(s))$, where the mapping

$$s \longmapsto (u(s), v(s))$$

traces out a simple, closed, regular curve C' in the domain U. Call U' the region inside of C'. Then integrating the geodesic curvature around the curve C, we get

$$\sum_{i=0}^{k} \int_{C_i} \kappa_g \, ds = \sum_{i=0}^{k} \int_{s_i}^{s_{i+1}} \kappa_g(s) \, ds$$

$$= \sum_{i=0}^{k} \int_{s_i}^{s_{i+1}} \left(\frac{1}{2\sqrt{g_{11}g_{22}}} \frac{\partial g_{22}}{\partial u} \frac{dv}{ds} - \frac{1}{2\sqrt{g_{11}g_{22}}} \frac{\partial g_{11}}{\partial v} \frac{du}{ds} \right) ds + \sum_{i=0}^{k} \int_{s_i}^{s_{i+1}} \frac{d\varphi}{ds} \, ds.$$

Applying Green's Theorem (see Theorem 2.1.3, which generalizes to simple closed piecewise regular curves) to the first term, we get

$$\sum_{i=0}^{k} \int_{C_i} \kappa_g \, ds = \iint_{U'} \frac{1}{2} \left(\frac{\partial}{\partial u} \left(\frac{1}{\sqrt{g_{11}g_{22}}} \frac{\partial g_{22}}{\partial u} \right) + \frac{\partial}{\partial v} \left(\frac{1}{\sqrt{g_{11}g_{22}}} \frac{\partial g_{11}}{\partial v} \right) \right) du \, dv + \sum_{i=0}^{k} \int_{s_i}^{s_{i+1}} \frac{d\varphi}{ds} \, ds.$$

By Equation (7.53), this becomes

$$\sum_{i=0}^{k} \int_{C_i} \kappa_g \, ds = -\iint_{U'} K \sqrt{g_{11} g_{22}} \, du \, dv + \sum_{i=0}^{k} \int_{s_i}^{s_{i+1}} \frac{d\varphi}{ds} \, ds,$$

which leads to

$$\sum_{i=0}^{k} \int_{C_i} \kappa_g \, ds = -\iint_{\vec{X}(U')} K \, dS + \sum_{i=0}^{k} (\varphi(s_{i+1}) - \varphi(s_i)). \qquad (8.37)$$

It remains for us to interpret in some way the last term in Equation (8.37). However, this is the content of the Theorem of Turning Tangents, proved by H. Hopf in [17], which we state without proving.

Lemma 8.4.2 (Theorem of Turning Tangents). *Let $\vec{\alpha}$ be a simple, closed, piecewise regular curve on a regular surface S of class C^3. Let $\vec{\alpha}(s_i)$ be the vertices with external angles θ_i, and let $\varphi(s)$ be as defined above. Then*

$$\sum_{i=0}^{k} (\varphi(s_{i+1}) - \varphi(s_i)) = \pm 2\pi - \sum_{i=1}^{k} \theta_i,$$

where the sign of 2π depends on the orientation of $\vec{\alpha}$.

Putting together Equation (8.37) and Lemma 8.4.2 establishes a local version of the Gauss-Bonnet Theorem.

Theorem 8.4.3 (Local Gauss-Bonnet Theorem). *Let $\vec{X} : U \to \mathbb{R}^3$ be an orthogonal parametrization of a neighborhood $V = \vec{X}(U)$ of an oriented surface S of class C^3. Let $\mathcal{R} \subset V$ be a simple region of S, and suppose that the boundary is $\partial \mathcal{R} = \vec{\alpha}([0,l])$ for some simple, closed, piecewise regular, positively oriented curve $\vec{\alpha} : [0,l] \to S$ of class C^2. Let $\vec{\alpha}(s_i)$, with $1 \leq i \leq k$, be the vertices of $\partial \mathcal{R}$, and let θ_i be their external angles. Call C_i the regular arcs of $\partial \mathcal{R}$. Then*

$$\sum_{i=0}^{k} \int_{C_i} \kappa_g \, ds + \iint_{\mathcal{R}} K \, dS + \sum_{i=1}^{k} \theta_i = 2\pi.$$

(Note that in this statement of the theorem, there are k vertices and $k+1$ regular arcs because the formulation assumes that $\vec{\alpha}(0)$ is not a vertex. This technicality is removed in the statement of the global version of the theorem.)

We have called the above formulation of the Gauss-Bonnet Theorem a local version since, as stated, it requires that the region $\mathcal{R}$ of S be inside a coordinate neighborhood of S that admits an orthogonal parametrization. By Problem 6.1.12, we know that if S is a regular surface of class C^2, then at every point $p \in S$ there exists a neighborhood of p that can be parametrized by a regular orthogonal parametrization. To obtain a global version of the theorem, we will "piece together" local instances of the above theorem.

The above proof of the local Gauss-Bonnet Theorem is an intrinsic proof in that it relies exclusively on intrinsic quantities: the metric tensor, the geodesic curvature, and angles on the surface. The main component of the proof is Green's Theorem. The following example gives an extrinsic proof of the local Gauss-Bonnet Theorem in that it consider a normal variation over the surface.

Example 8.4.4 (The Moldy Patch). The motivating example, Example 8.4.1 with the moldy potato chip, falls just shy of giving an extrinsic (assumes we have/know a normal vector to the surface) proof of the local Gauss-Bonnet Theorem. We provide the details here.

Consider a simply-connected (no holes) region $\mathcal{R}$ on a regular surface S as described in Theorem 8.4.3. Suppose that $\mathcal{R}$ is parametrized by $\vec{X} : U \to \mathbb{R}^3$ for some $U \subset \mathbb{R}^2$, and call $\vec{N}$ the associated normal vector.

Define the surface $\mathcal{T}_r$ as the tubular neighborhood of $\mathcal{R}$ with radius r. This means that $\mathcal{T}_r$ consists of

1. two pieces for the normal variation to $\mathcal{R}$ parametrized, respectively, by $\vec{X} + r\vec{N}$ and $\vec{X} - r\vec{N}$ over U (which we call respectively $U_{(+r)}$ and $U_{(-r)}$);

2. k half-tubes of radius r around the smooth pieces of the boundary $\partial\mathcal{R}$ pointing "away" from the region $\mathcal{R}$;

3. k lunes of spheres (of radius r) at the k vertices of $\partial\mathcal{R}$.

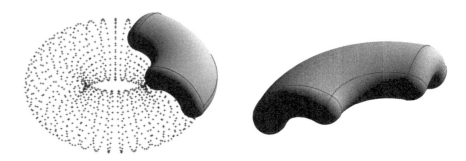

Figure 8.9. A "moldy patch."

Figure 8.9 shows the pieces of $\mathcal{T}_r$ for a patch on a torus. We assume from now on that r is small enough so that each of the pieces of $\mathcal{T}_r$ is a regular surface.

Let K be the Gaussian curvature of S over $\mathcal{R}$, and call $K_{\mathcal{T}}$ the Gaussian curvature of the moldy patch $\mathcal{T}_r$. By Proposition 6.5.2, the quantity $K_{\mathcal{T}}\,dS_{\mathcal{T}}$ is a signed area element of the image on the unit sphere of $\mathcal{T}_r$ under its Gauss map. So in calculating the total curvature of the moldy patch, one is adding or subtracting area of the sphere depending on the sign of $K_{\mathcal{T}}$. We now reason why

$$\iint_{\mathcal{T}_r} K_{\mathcal{T}}\,dS_{\mathcal{T}} = 4\pi. \tag{8.38}$$

Since $\mathcal{T}_r$ has no boundary, then the Gauss map of $\mathcal{T}_r$ covers the unit sphere an integer number of times, where we add area for positive curvature and subtract for negative curvature. Thus, the integral in Equation (8.38) must be equal to $4\pi h$, where $h \in \mathbb{Z}$. A region $\mathcal{R}$ that is simply-connected is also *contractible*, which means it can be shrunk continuously to a point. Now if $\mathcal{R}$ is a point, then $\mathcal{T}_r$ is a sphere of radius r. In this case, the Gauss map for $\mathcal{T}_r$ is a bijective map onto the unit sphere, and hence, the integral in Equation(8.38) gives precisely the surface of the unit sphere, so is equal to 4π. However, as one "uncontracts" from a point to $\mathcal{R}$, the integral in Equation (8.38) must vary continuously. However, a continuous function to a discrete set, namely, $4\pi\mathbb{Z}$, is constant. Though we have not spelled out the whole topological background, this reasoning justifies Equation (8.38).

We now break down Equation (8.38) according to various pieces of $\mathcal{T}_r$: the normal variation patches, the half-tubes, and the lunes of spheres.

By Problem 6.5.9, if we call $K_{(+r)}$ and $K_{(-r)}$ the respective Gaussian curvatures of $\vec{X} + r\vec{N}$ and $\vec{X} - r\vec{N}$ over U, then

$$2 \iint_U K\, dS = \iint_{U_{(+r)}} K_{(+r)}\, dS + \iint_{U_{(-r)}} K_{(-r)}\, dS.$$

In Problem 8.1.9, one calculates the Gaussian curvature of the normal tube around a curve on the surface. Consider a regular arc C of $\partial\mathcal{R}$, and assume it is parametrized by $t \in [a, b]$. As a consequence of Problem 8.1.9, over the normal half-tube pointing away from $\vec{U}$ (i.e., away from the inside of $\mathcal{R}$), the total Gaussian curvature is

$$\iint_{\text{half-tube around } C} K\, dS = \int_a^b \int_{\pi/2}^{3\pi/2} -s'(t)\left((\sin v)\kappa_n(t) + (\cos v)\kappa_g(t)\right)\, dt$$

$$= 2\int_a^b s'(t)\kappa_g(t)\, dt = 2\int_C \kappa_g\, ds.$$

Finally, we consider the lunes of $\mathcal{T}_r$ that are around the vertices of $\partial\mathcal{R}$. Under the image of the Gauss map, the lunes map to the same corresponding lune on the unit sphere. Thus, in Equation (8.38), a lune around a vertex with exterior angle α_i contributes $(\alpha_i/2\pi)4\pi = 2\alpha_i$ to the surface of the unit sphere.

Combining each of these results, we find that

$$\iint_{\mathcal{T}_r} K_{\mathcal{T}}\, dS_{\mathcal{T}} = 2\iint_U K\, dS + 2\sum_{i=0}^{k} \int_{C_i} \kappa_g\, ds + 2\sum_{i=1}^{k} \theta_i,$$

and the local Gauss-Bonnet Theorem follows immediately from Equation (8.38).

In our presentation of the local Gauss-Bonnet Theorem, we did not allow corners to be cusps. The main difficulty lies in deciding whether to assign a value of π or $-\pi$ to the angle θ_i of any given cusp in order to retain the validity of the local Gauss-Bonnet Theorem. We can allow for cusps on the boundary $\partial\mathcal{R}$ if we employ the following sign convention for the angles of cusps: if the cusp

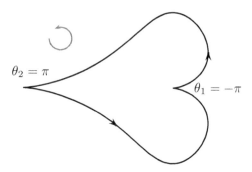

Figure 8.10. External angles at cusps.

$\vec{\alpha}(t_i)$ points into the interior of the closed curve (i.e., $\vec{\alpha}(t_i^-)$ points into the interior) then $\theta_i = -\pi$; and if the cusp $\vec{\alpha}(t_i)$ points away from the interior of the closed curve (i.e., $\vec{\alpha}(t_i^-)$ points away from the interior) then $\theta_i = \pi$. (See Figure 8.10. The orientation of the surface matters since it determines the direction of travel around the boundary curve $\partial \mathcal{R}$.)

8.4.2   The Global Gauss-Bonnet Theorem

In order to extend our methods to a global presentation, we need to briefly discuss triangulations of surfaces, the classification of orientable surfaces, and the Euler characteristic of regions of surfaces in $\mathbb{R}^3$. These topics are typically considered in the area of topology, but we summarize the results that we need in order to give a full treatment to the global Gauss-Bonnet Theorem without insisting that the reader have mastery of the supporting topology. (The authors include technical details behind these concepts in Appendices A.5 and A.6 of [22]. Otherwise, the interested reader could consult Chapters 6 and 7 of [1].)

   In intuitive terms, a *triangulation* of a surface consists of a network of a finite number of regular curve segments on the surface such that any point on the surface either lies on one of the curves or lies in a region that is bounded by precisely three curve segments. The first picture in Figure 8.11 depicts a triangulation on a torus. As an additional technical requirement, one should be able to continuously deform the surface with its triangulation so that each "triangle"

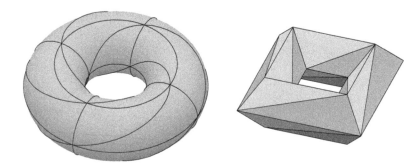

Figure 8.11. Torus triangulation.

becomes a true triangle without changing the topological nature of the surface. (Compare the two pictures in Figure 8.11.)

In a triangulation, a *vertex* is an endpoint of one of the curve segments on the surface. We call the curve segments *edges*, and the regions enclosed by edges (the "triangles") we call *faces*. An interesting and useful result, first proved by Rado in 1925, is that every regular compact surface admits a triangulation.

A result of basic topology is that given a compact regular surface S, the quantity

$$\#(\text{vertices}) - \#(\text{edges}) + \#(\text{faces}) \tag{8.39}$$

is the same regardless of any triangulation of S. This number is called the *Euler characteristic* of S and is denoted by $\chi(S)$. Furthermore, from the definition of triangulation, one can deduce that the Euler characteristic does not change if the surface is deformed continuously (no cutting or pinching). One often restates this last property by saying that the Euler characteristic is a *topological invariant*.

For example, the torus triangulation in Figure 8.11 has 16 vertices, 48 edges, and 32 faces. Thus, the Euler characteristic of the torus is 0. As another example consider the tetrahedron, which is homeomorphic to the sphere. A tetrahedron has four vertices, six edges and four faces, so its Euler characteristic, and therefore the Euler characteristic of the sphere, is $\chi = 4 - 6 + 4 = 2$.

A fundamental theorem in topology, the Classification Theorem of Surfaces, states that every orientable surface without boundary

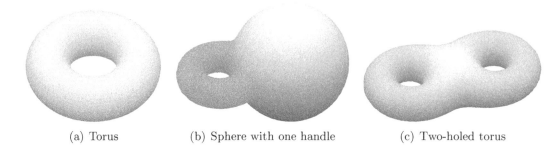

(a) Torus   (b) Sphere with one handle   (c) Two-holed torus

Figure 8.12. Tori.

is homeomorphic to a sphere or to a sphere with a finite number of "handles" added to it. Figure 8.12(a) shows a torus while Figure 8.12(b) shows a sphere with one handle added to it. These two surfaces are in fact the same under a continuous deformation, i.e., they are homeomorphic. Figure 8.12(c) depicts a two-holed torus that, in the language of the Classification Theorem of Surfaces, is called a sphere with two handles.

It is not hard to show that the Euler characteristic of a sphere with g handles added is

$$\chi(S) = 2 - 2g. \tag{8.40}$$

The notion of the Euler characteristic applies equally well to a surface with boundary as long as the boundary is completely covered by edges and vertices of the triangulation. For example, we encourage the reader to verify that a sphere with a small disk removed has Euler characteristic of 1.

We must now discuss orientations on a triangulation. When considering adjacent triangles, we can think of the orientation of a triangle as a direction of travel around the edges. Two adjacent triangles have a compatible orientation if the orientation of the first leads one to travel along the common edge in the opposite direction of the orientation on the second triangle (see Figure 8.13). It turns out that if a surface is orientable, then it is possible to choose an orientation of each triangle in the triangulation such that adjacent triangles have compatible orientations.

A compact regular surface S is covered by a finite number of coordinate neighborhoods given by regular parametrizations. In general,

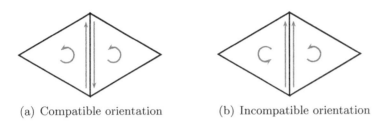

(a) Compatible orientation   (b) Incompatible orientation

Figure 8.13. Adjacent oriented triangles.

if $\mathcal{R}$ is a regular region of S, it may not lie entirely in one coordinate patch. However, it is possible to show that not only does a regular region $\mathcal{R}$ admit a triangulation, but every regular region $\mathcal{R}$ admits a triangulation such that each triangle is contained in a coordinate neighborhood. This comment and two lemmas show that it makes sense to talk about the surface integral over the whole region $\mathcal{R} \subset S$.

Lemma 8.4.5. *Suppose that $\vec{X}_1 : U_1 \to \mathbb{R}^3$ and $\vec{X}_2 : U_2 \to \mathbb{R}^3$ are two systems of coordinates of a regular surface S. Call (u_i, v_i) the coordinates of $\vec{X}_i$, and call $g^{(i)}$ the corresponding metric tensor. Suppose that $T \subset \vec{X}_1(U_1) \cap \vec{X}_2(U_2)$. Then for any function $f : T \to \mathbb{R}$, we have*

$$\iint_{\vec{X}_1^{-1}(T)} f(u_1, v_1)\sqrt{\det(g^{(1)})}\, du_1\, dv_1 = \iint_{\vec{X}_2^{-1}(T)} f(u_2, v_2)\sqrt{\det(g^{(2)})}\, du_2\, dv_2.$$

Proof: By Equation (6.7), one deduces that

$$\det(g^{(1)}) = \left(\frac{\partial(u_2, v_2)}{\partial(u_1, v_1)}\right)^2 \det(g^{(2)}).$$

However, $\frac{\partial(u_2, v_2)}{\partial(u_1, v_1)}$ is the Jacobian of the coordinate transformation $F : \vec{X}_1^{-1}(T) \to \vec{X}_2^{-1}(T)$ defined by $\vec{X}_2^{-1} \circ \vec{X}_1$ restricted to $\vec{X}_1^{-1}(T)$. The result follows as an application of the change of variables formula in double integrals. $\qquad\square$

Lemma 8.4.6. *Let S be a regular oriented surface, and let $\mathcal{R}$ be a regular compact region of S, possibly with a boundary. Given a collection $\{\vec{X}_i\}_{i \in I}$ of coordinate neighborhoods that cover S, $\{T_j\}_{j \in J}$ triangles* *of a triangulation of $\mathcal{R}$, and $i : J \to I$ such that T_j is in the image of $\vec{X}_{i(j)}$, define the sum*

$$\sum_{j \in J} \iint_{\vec{X}_{i(j)}^{-1}(T_j)} f(u_{i(j)}, v_{i(j)}) \sqrt{\det g^{(i(j))}} \, du_{i(j)} \, dv_{i(j)}.$$

This sum is indepenedent of the choice of triangulation of $\mathcal{R}$, collection of coordinate patches, and function i.

Proof: That the sum is independent of the collection of coordinate neighborhoods follows from Lemma 8.4.5. That the sum does not depend on the choice of triangulation is a little tedious to prove and is left as an exercise for the reader. ☐

This leads to a definition of a surface integral over any region on a regular surface.

Definition 8.4.7. Let S be a regular oriented surface, and let $\mathcal{R}$ be a regular compact region of S, possibly with a boundary. We call the common sum described in Lemma 8.4.6 the surface integral of f over $\mathcal{R}$ and denote it by

$$\iint_{\mathcal{R}} f \, dS.$$

We point out that if $\vec{X} : U \to \mathbb{R}^3$ is a regular parametrization of a region of S such that $\vec{X}(U)$ is dense in $\mathcal{R}$, then

$$\iint_{\mathcal{R}} f \, dS = \iint_U f(\vec{X}(u, v)) \| \vec{X}_u \times \vec{X}_v \| \, dA,$$

where the right-hand side is the usual double integral.

We can now state the main theorem of this section.

Theorem 8.4.8 (Global Gauss-Bonnet Theorem). *Let S be a regular oriented surface of class C^3, and let $\mathcal{R}$ be a compact region of S with boundary $\partial\mathcal{R}$. Suppose that $\partial\mathcal{R}$ is a simple, closed, piecewise regular, positively oriented curve. Suppose that $\partial\mathcal{R}$ has k regular arcs C_i of class C^2, and let θ_i be the external angles of the vertices of $\partial\mathcal{R}$. Then*

$$\sum_{i=1}^{k} \int_{C_i} \kappa_g \, ds + \iint_{\mathcal{R}} K \, dS + \sum_{i=1}^{k} \theta_i = 2\pi\chi(\mathcal{R}),$$

where $\chi(\mathcal{R})$ is the Euler characteristic of $\mathcal{R}$.

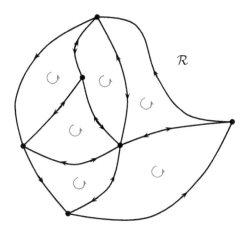

Figure 8.14. A triangulation of a region $\mathcal{R}$.

Proof: Suppose that $\mathcal{R}$ is covered by a collection of coordinate patches. Let $\{T_j\}_{j\in J}$ be the triangles of a triangulation of $\mathcal{R}$ in which all the triangles T_j on S are subsets of some coordinate patch. Suppose also that the set of triangles $\{T_j\}$ are equipped with an orientation that is compatible with the orientation of S.

For each triangle T_j, for $1 \le l \le 3$, call E_{jl} the edges of T_j (as curves on S), call V_{jl} the vertices of T_j, and let β_{jl} be the interior angle of T_j at V_{jl}. For this triangulation, call a_0 the number of vertices, a_1 the number of edges, and a_2 the number of triangles. By construction, the local Gauss-Bonnet Theorem applies to each triangle T_j on S, so on each T_j, we have

$$\sum_{l=1}^{3} \int_{E_{jl}} \kappa_g\, ds + \iint_{T_j} K\, dS = 2\pi - \sum_{l=1}^{3}(\pi - \beta_{jl}) = -\pi + \sum_{l=1}^{3} \beta_{jl}. \quad (8.41)$$

Now consider the sum of Equation (8.41) over all the triangles T_j. Since each triangle has an orientation compatible with the orientation of S, then whenever two triangles share an edge, the edge is traversed in opposite orientations on the adjacent triangles (see Figure 8.14). Consequently, in the sum of Equation (8.41), each integral $\int_{E_{jl}} \kappa_g\, ds$ cancels out another similar integral along any edge that is not a part of the boundary of $\mathcal{R}$. Therefore, applying Lemma 8.4.6, the left-hand side of the sum of Equation (8.41) is precisely

$$\sum_{i=1}^{k} \int_{C_i} \kappa_g \, ds + \iint_{\mathcal{R}} K \, dS. \tag{8.42}$$

The right-hand side of the sum of Equation (8.41) is

$$-\pi a_2 + \sum_{j} \sum_{i=1}^{3} \beta_{jl}.$$

In the double sum $\sum\sum \beta_{jl}$, the sum of interior angles associated to a vertex on the interior of $\mathcal{R}$ contributes 2π, and the sum of angles associated to a vertex V on the boundary $\partial\mathcal{R}$ contributes $\pi - $ (exterior angle of V). Thus,

$$\sum_{j} \left(-\pi + \sum_{l=1}^{3} \beta_{jl} \right) = -\sum_{i=1}^{k} \theta_i - \pi a_2 + 2\pi(\#\text{interior vertices}) + \pi(\#\text{exterior vertices}).$$

Since there are as many vertices on the boundary as there are edges, we rewrite this as

$$\sum_{j} \left(-\pi + \sum_{l=1}^{3} \beta_{jl} \right) = -\sum_{i=1}^{k} \theta_i - \pi a_2 + 2\pi a_0 - \pi(\#\text{exterior edges}). \tag{8.43}$$

However, since each triangle has three edges,

$$-a_2 = 2a_2 - 3a_2 = 2a_2 - 2(\#\text{interior edges}) - (\#\text{exterior edges}). \tag{8.44}$$

Consequently, taking the sum of Equation (8.41) and combining Equations (8.42), (8.43) and (8.44) one obtains

$$\sum_{i} \int_{C_i} \kappa_g \, ds + \iint_{\mathcal{R}} K \, dS = -\sum_{i=1}^{k} \theta_i + 2\pi(a_2 - a_1 + a_0).$$

By definition of the Euler characteristic, $\chi(\mathcal{R}) = a_2 - a_1 + a_0$. The theorem follows. $\square$

The global version of the Gauss-Bonnet Theorem directly generalizes the local version so when one refers to *the* Gauss-Bonnet Theorem, one means the global version. Even at a first glance, the

Gauss-Bonnet Theorem is profound because it connects local properties of curves on a surface (the geodesic curvature κ_g, the Gaussian curvature K, and angles associated to vertices) with global properties (the Euler characteristic of a region of a surface). In fact, since the Euler characteristic is a topological invariant, the Gauss-Bonnet Theorem connects local geometric properties to a topological property.

Let S be a compact regular surface (without boundary). An interesting particular case of the global Gauss-Bonnet Theorem occurs when we consider $\mathcal{R} = S$, which implies that $\partial\mathcal{R}$ is empty. This situation leads to the following strikingly simple and profound corollary.

Corollary 8.4.9. *Let S be an orientable, compact, regular surface of class C^3. Then*

$$\iint_S K \, dS = 2\pi\chi(S).$$

Example 8.4.10. Consider the torus T parametrized by $\vec{X} : [0, 2\pi]^2 \to \mathbb{R}^3$, with

$$\vec{X}(u, v) = ((a + b\cos v)\cos u, (a + b\cos v)\sin u, b\sin v),$$

where $a > b$. We note that $\vec{X}((0, 2\pi)^2)$ covers all of T except for two curves on the torus, so $\vec{X}((0, 2\pi)^2)$ is dense in T. By the comment after Definition 8.4.7, we can use the usual surface integral over this one coordinate neighborhood to calculate $\iint_S K \, dS$ directly.

It is not hard to calculate that over the coordinate neighborhood described above, we have

$$K(u, v) = \frac{\cos v}{b(a + b\cos v)} \quad \text{and} \quad \|\vec{X}_u \times \vec{X}_v\| = b(a + b\cos v).$$

Thus,

$$\iint_S K \, dS = \int_0^{2\pi} \int_0^{2\pi} \frac{\cos v}{b(a + b\cos v)} \cdot b(a + b\cos v) \, du \, dv = \int_0^{2\pi} \int_0^{2\pi} \cos v \, du \, dv = 0,$$

which proves that $\chi(T) = 0$. This agrees with the calculation provided by the triangulation in Figure 8.11.

We shall now present a number of applications of the Gauss-Bonnet Theorem, as well as leave a few as exercises for the reader.

Proposition 8.4.11. *A compact, regular, orientable surface of class C^3 with positive curvature everywhere is homeomorphic to a sphere.*

Proof: Since $K > 0$ over the whole surface S, then for the total Gaussian curvature we have

$$\iint_S K\, ds > 0.$$

By Corollary 8.4.9, this integral is $2\pi\chi(S)$. However, by the Classification Theorem of Surfaces and Equation (8.40), since $\chi(S) > 0$, we have $\chi(S) = 2$, and therefore S is homeomorphic to a sphere. $\square$

Proposition 8.4.12. *Let S be a compact, connected, orientable, regular surface of positive Gaussian curvature. If there exist two simple, closed geodesics γ_1 and γ_2 on S then they intersect.*

Proof: By Proposition 8.4.11, S is homeomorphic to a sphere. Suppose that γ_1 and γ_2 do not intersect. Then they form the boundary of region $\mathcal{R}$ that is homeomorphic to a cylinder with boundary. It is not hard to verify by supplying $\mathcal{R}$ with a triangulation that $\chi(\mathcal{R}) = 0$. However, applying the Gauss-Bonnet Theorem to this situation, we obtain

$$\iint_{\mathcal{R}} K\, dS = 0,$$

which is a contradiction since $K > 0$. $\square$

We now discuss applications of the Gauss-Bonnet Theorem to plane, spherical, and hyperbolic geometry and show how the theorem simultaneously generalizes many well-known theorems in these areas.

8.4.3   Plane, Spherical, and Hyperbolic Geometry

As mentioned in the introduction to this chapter, in his *Elements*, Euclid proves all his propositions through from 23 definitions and five postulates. The fifth postulate reads as follows:

> If a straight line crossing two straight lines makes the interior angles on the same side less than two right angles, the two straight lines, if extended indefinitely, meet on that side on which are the angles less than the two right angles.

Much wordier than the other four, many mathematicians over the centuries attempted to prove the fifth postulate from the others but never succeeded. Studying Euclid's *Elements*, one notices that Euclid himself uses the fifth postulate sparingly. One could speculate that Euclid did this so that, were someone to prove the fifth postulate as a consequence of the others, then the proofs of theorems would still be given in a minimal form. Indeed, many commonly known properties about plane geometry hold without reference to the fifth postulate. However, examples of two commonly known theorems that appear to rely on it are the following:

1. Given a line L and a point P not on L, through P passes exactly one line parallel to L.

2. The sum of the interior angles of a triangle is π.

In the nineteenth century, mathematicians Lobachevsky and Bolyai constructed examples of what we now call non-Euclidean geometries, in which one replaces Euclid's fifth postulate with one of two options. Given a line L and a point P not on L, through P passes (S) no parallel lines to L or (H) more than one parallel line to L. These geometries are called, respectively, *spherical geometry* and *hyperbolic geometry*. Using elementary techniques, one can study these geometries and establish many theorems, including theorems in spherical trigonometry. (See [9] and [27] for comprehensive treatment of non-Euclidean geometry using elementary techniques.) One well-known result in these geometries is that the sum of the interior angles of a triangle is greater than π in the case of spherical geometry and less than π in the case of hyperbolic geometry. Figure 8.15 gives an example of a triangle on a sphere in which each vertex has a right angle, and hence, the sum of the interior angles is $3\pi/2$.

We now apply the Gauss-Bonnet Theorem to plane, spherical, and hyperbolic geometries, beginning with plane geometry.

In plane geometry, we consider the plane to be a surface with Gaussian curvature identically 0 and curves in the plane we consider to be curves on a surface. Consider now a region $\mathcal{R}$ in the plane such that the boundary $\partial\mathcal{R}$ is a piecewise regular, simple, closed curve.

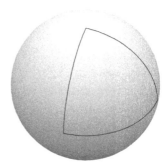

Figure 8.15. Triangle on a sphere.

It is easy to calculate that the Euler characteristic of a region that is homeomorphic to a disk is $\chi(\mathcal{R}) = 1$.

As a first case, let $\partial\mathcal{R}$ be a polygon. Since the regular arcs are straight lines, the geodesic curvature of each regular arc of $\partial\mathcal{R}$ is identically 0. Gauss-Bonnet's formula then reduces to the well-known fact that the sum of the exterior angles $\{\theta_1, \theta_2, \ldots, \theta_k\}$ around a polygon is 2π. Also, since the exterior angle θ_i at any corner is $\pi - \alpha_i$, where α_i is the interior angle, we deduce that the sum of the interior angles of any polygon is

$$\sum_{i=1}^{n} \alpha_i = (n-2)\pi.$$

This statement is in fact equivalent to a theorem that occurs in every high school geometry curriculum, namely, that the sum of the interior angles of a triangle is π radians.

As a second case of the Gauss-Bonnet Theorem applied to plane geometry, suppose that $\partial\mathcal{R}$ is a regular curve. In this case, $\partial\mathcal{R}$ has no vertices and hence no exterior angles. Then the Gauss-Bonnet Theorem reduces to the formula

$$\oint_{\partial\mathcal{R}} \kappa_g \, ds = 2\pi,$$

and hence, the Gauss-Bonnet Theorem generalizes the proposition that the rotation index of a simple, closed curve is 1 (see Propositions 2.2.1 and 2.2.8).

In order to apply the results of this chapter to non-Euclidean geometry, one must clarify the use of the terms "line" and "parallel." In the geometry on a surface of class C^3, the word "line" (or straight line) means a geodesic. The word "parallel" is a little more problematic. In Problem 1.3.12, we defined a parallel curve to a given curve $\vec{\gamma}$ as a curve whose locus is always a fixed orthogonal distance r from the locus of $\vec{\gamma}$. This is true of parallel lines in the plane. However, in the sense of classical geometry, two lines are called parallel if they do not intersect.

Example 8.2.7 showed that geodesics on a sphere are great arcs of the sphere, i.e., an arc on the circle of intersection between the sphere and a plane through the center of the sphere. Therefore, any two geodesic lines intersect at at least two points: the common intersection of the sphere and the two planes supporting the geodesics. This remark justifies the definition of spherical geometry as a geometry in which there exist no parallel lines.

Consider now a triangle $\mathcal{R}$ on a sphere of radius R. By definition of a geometric triangle, $\partial \mathcal{R}$ has three vertices, and its regular arcs are geodesic curves. The Gaussian curvature of a sphere of radius R is $\frac{1}{R^2}$. In this situation, the Gauss-Bonnet Theorem gives

$$(\theta_1 + \theta_2 + \theta_3) + \frac{A}{R^2} = 2\pi, \tag{8.45}$$

where A is the surface area of $\mathcal{R}$, and θ_i are the exterior angles of the vertices. If we call $\alpha_i = \pi - \theta_i$ the interior angles of the triangle, then Equation (8.45) reduces to

$$\alpha_1 + \alpha_2 + \alpha_3 = \pi + \frac{A}{R^2}. \tag{8.46}$$

This shows that in a spherical triangle, the excess sum of the angles, i.e., $\alpha_1 + \alpha_2 + \alpha_3 - \pi$, is 4π times the ratio of the surface area of the triangle to the surface area of the whole sphere. As an example, the spherical triangle in Figure 8.15 has an excess of $\pi/2$ and covers $1/8$ of the sphere.

We propose to study (geodesic) circles on a sphere of radius R. Recall that a circle on a sphere is the set of points that are at a fixed distance r from a point, where distance here means geodesic

Figure 8.16. Hyperbolic triangle.

distance. Call $\mathcal{D}$ the closed disk inside this circle. Using the parametric equations for the sphere,

$$\vec{X}(u,v) = (R\cos u \sin v, R\sin u \sin v, R\cos v),$$

a circle of radius r around the north pole $(0,0,R)$ is given by $\vec{\gamma}(t) = \vec{X}(t,v_0)$, where $r = Rv_0$. It is not hard to tell by symmetry that the geodesic curvature κ_g of a circle on a sphere is a constant. The Gauss-Bonnet Theorem then gives

$$\kappa_g(\text{perimeter of } \partial\mathcal{D}) + \frac{1}{R^2}(\text{surface area of } \mathcal{D}) = 2\pi. \qquad (8.47)$$

Using the surface area formula for a surface of revolution from single-variable calculus, it is possible to prove that the area of $\mathcal{D}$ is

$$\text{surface area of } \mathcal{D} = 2\pi R^2(1 - \cos v_0) = 2\pi R^2 \left(1 - \cos\left(\frac{r}{R}\right)\right).$$

It is not hard to show that the perimeter of $\mathcal{D}$ is

$$\text{perimeter of } \mathcal{D} = 2\pi R \sin v_0 = 2\pi R \sin\left(\frac{r}{R}\right).$$

Consequently, the Gauss-Bonnet Theorem gives the curvature of a circle on a sphere as

$$\kappa_g = \frac{1}{R}\cot\left(\frac{r}{R}\right).$$

In contrast to spherical geometry, hyperbolic geometry is the geometry of a surface with constant negative Gaussian curvature. Example 6.5.5 presented the pseudosphere as a surface with $K = -1$

everywhere. In this particular case, using the Gauss-Bonnet Theorem, it is easy to calculate that the sum of the interior angles of a triangle is

$$\alpha_1 + \alpha_2 + \alpha_3 = \pi - A,$$

where A is the area of the triangle. (See Figure 8.16 for an example. Note that the regular arcs are indeed geodesic segments.) Interestingly enough, this formula for the sum of the interior angles of a hyperbolic triangle puts an upper bound of π for the possible area of a triangle on the pseudosphere with $K = -1$.

Problems

8.4.1. Provide all the details that establish Equation (8.36).

8.4.2. Provide the details for the proof of Lemma 8.4.6.

8.4.3. Calculate the surface area of a disk on a sphere using the surface area formula for a surface of revolution. Calculate directly the geodesic curvature function of a circle on a sphere. [Hint: See the paragraph around Equation (8.47).]

8.4.4. Verify directly the Gauss-Bonnet Theorem for the rectangular region $\mathcal{R}$ on a torus specified by $u_1 \leq u \leq u_2$ and $v_1 \leq v \leq v_2$, where the u-coordinate lines are the parallels and the v-coordinate lines are the meridians of the torus, as a surface of revolution.

8.4.5. Let S be a regular, orientable, compact surface with positive Gaussian curvature. Prove that the surface area of S is less than $4\pi/K_{\min}$, where $K_{\min} > 0$ is the minimum Gaussian curvature.

8.4.6. Let S be a regular, orientable surface of class C^3 in $\mathbb{R}^3$ that is homeomorphic to the sphere. Let γ be a simple closed geodesic in S. The curve γ separates S into two regions A and B that share γ as their boundary. Let $n : S \to \mathbb{S}^2$ be the Gauss map induced from a given orientation of S. Prove that $n(A)$ and $n(B)$ have the same area.

8.4.7. Let S be an orientable surface with Gaussian curvature $K \geq 0$, and let $p \in S$.

   (a) Let γ_1 and γ_2 be two geodesics that intersect at p. Prove that γ_1 and γ_2 do not intersect at another point q in such a way that γ_1 and γ_2 form the boundary of a simple region $\mathcal{R}$.

   (b) Prove also that a geodesic on S cannot intersect itself in such a way as to enclose a simple region.

8.4.8. *Jacobi's Theorem.* Let $\vec{\alpha} : I \to \mathbb{R}^3$ be a closed, regular, parametrized curve. Suppose also that $\vec{\alpha}(t)$ has a curvature function $\kappa(t)$ that is never 0. Suppose also that the principal normal indicatrix, i.e., the curve $\vec{P} : I \to \mathbb{S}^2$, is simple, that is, that it cuts the sphere into only two regions. By viewing $\vec{P}(I)$ as a curve on the sphere $\mathbb{S}^2$, use the Gauss-Bonnet Theorem to prove that $\vec{P}(I)$ separates the sphere into two regions of equal area.

8.5   Intrinsic Geometry

We remind the reader that an intrinsic property of a regular surface or a regular curve on a surface is a property that depends only on the coefficients of the metric tensor and the coordinates of a point or coordinate functions of a curve. Intrinsic geometry is the subset of differential geometry where one restricts one's attention to intrinsic properties.

In Section 6.1, we saw that the arc length of a curve on a surface, the area of a region on the surface, and the angle between two intersecting curves are all intrinsic properties of surfaces. As a consequence, a surface integral and an integral along a path (parametrized by arc length) of functions given in terms of the coordinates are intrinsic quantities. In contrast, the normal vector to a surface at a point and even the second fundamental form are not intrinsic. On the other hand, in Section 7.2 we saw that the Christoffel symbols Γ_{jk}^i are intrinsic quantities, and the Theorema Egregium (Theorem 7.3.2) shows that the Gaussian curvature is also an intrinsic property of the surface.

As far as curves on surfaces, Equation (8.7) shows the geodesic curvature to be an intrinsic property. Furthermore, since the differential equations for geodesics involve only the Christoffel symbols and the coordinate functions of the curve, geodesics are objects in intrinsic geometry. However, the functional dependence of the normal curvature and the geodesic torsion on the unit normal vector, as given in Equation (8.11), indicates that these quantities are not intrinsic. Since the Gauss-Bonnet formula on one side only involves κ_g and K, one may say it is a theorem in intrinsic geometry.

The value of narrowing one's focus to intrinsic geometry is that, in this context, one does not need a specific parametrization (coordinate patch) of a surface to calculate the quantities or determine curves of interest. In Chapters 6, 7, and 8, we have almost exclusively considered regular surfaces in $\mathbb{R}^3$. However, if $\vec{X} : U \to \mathbb{R}^n$, where U is an open subset of $\mathbb{R}^2$, though it is impossible to define *the* unit normal vector (even up to sign), it is still possible to calculate

$$g_{ij} = \vec{X}_i \cdot \vec{X}_j.$$

Therefore, given the metric tensor, the dimension of the ambient Euclidean space of the surface or even how the surface sits in the ambient space is irrelevant. One simply refers to the coordinates in a given coordinate patch and the associated metric tensor for all calculations.

The concept of a differentiable manifold, the next step in the generalization of regular surfaces, avoids any reference to an ambient space. However, without an ambient space, one cannot really do geometry on a differentiable manifold (measure lengths, angles, volumes, curvatures, etc.) without a metric tensor. A manifold equipped with a metric tensor is called a *Riemannian manifold*. Chapters 3 through 5 in [22] study differentiable and Riemannian manifolds.

Example 8.5.1 (Flat Torus). The topological definition of a torus is a topological space that is homeomorphic to $\mathbb{S}^1 \times \mathbb{S}^1$, where $\mathbb{S}^1$ is a circle.

Throughout this book, we have used the following parametrization $\vec{X} : [0, 2\pi]^2 \to \mathbb{R}^3$ of the torus as a surface in $\mathbb{R}^3$:

$$\vec{X}(u, v) = ((R + r \cos v) \cos u, (R + r \cos v) \sin u, r \sin v).$$

The locus of this parametrization is the usual doughnut shape that is the surface of revolution of circle of radius r with center a distance R from the axis of revolution. As long as $0 < r < R$, the image of this parametrization is indeed a topological torus. In the context of differential geometry it is this shape that one calls a torus. As the exercises in Chapter 6 showed, this parametrization has a nontrivial metric tensor.

Problem 6.1.14 introduced the parametrization $\vec{X} : [0, 2\pi]^2 \to \mathbb{R}^4$ given by the functions

$$\vec{X}(u, v) = (\cos u, \sin u, \cos v, \sin v).$$

This parametrization explicitly traces out $\mathbb{S}^1 \times \mathbb{S}^1$. It is easy to see that the metric tensor (g_{ij}) associated to this parametrization is the identity matrix everywhere. This justifies the name of flat torus for this shape. In some intuitive sense, the reason why it is possible for a flat torus to exist is that, like the right circular cylinder in $\mathbb{R}^3$ that has $g_{ij} = \delta^i_j$, there is "enough room" in $\mathbb{R}^4$ to have a torus that has a metric tensor that is everywhere the identity matrix.

Example 8.5.2 (Poincaré Half-Plane). We consider the upper half-plane $H = \{(x, y) \in \mathbb{R}^2 \mid y > 0\}$ with the metric tensor

$$(g_{ij}) = \begin{pmatrix} y^{-2} & 0 \\ 0 & y^{-2} \end{pmatrix}.$$

Using the physics-style line element notation, the metric is written as $ds^2 = \frac{1}{y^2}(dx^2 + dy^2)$. The area element in H with this metric is

$$dS = \sqrt{\det(g)}\, dA = \frac{1}{y^2}\, dx\, dy.$$

So, for example, the surface integral of a function $f : H \to \mathbb{R}$ over a region $\mathcal{R}$ is

$$\iint_{\mathcal{R}} f\, dS = \iint_{\mathcal{R}} \frac{f(x, y)}{y^2}\, dx\, dy.$$

The area of a rectangle $\mathcal{R} = \{(x, y) \in H \mid x_1 \leq x \leq x_2,\ y_1 \leq y \leq y_2\}$ is

$$A = \int_{y_1}^{y_2} \int_{x_1}^{x_2} \frac{1}{y^2}\, dx\, dy = (x_2 - x_1)\left(\frac{1}{y_1} - \frac{1}{y_2}\right).$$

The Poincaré half-plane has many interesting properties that are explored in the exercises. One remarkable property is that the geodesic lines in H with the Poincaré metric are either vertical lines or half-circles with center on the x-axis.

Problems

8.5.1. Consider the upper half-plane $H = \{(x, y) \in \mathbb{R}^2 \mid y > 0\}$ with the metric tensor

$$(g_{ij}) = \begin{pmatrix} e^{-2y} & 0 \\ 0 & 1 \end{pmatrix}.$$

(a) Prove that the Gaussian curvature of such a surface has $K = -1$ everywhere.

(b) Show that the parametrization $\vec{X} : H \to \mathbb{R}^3$ defined by

$$\vec{X}(x, y) = \left(e^{-y} \cos x, e^{-y} \sin x, \ln(1 + \sqrt{1 - e^{-2y}}) + y - \sqrt{1 - e^{-2y}} \right)$$

has the above metric coefficients.

(c) Prove that $\vec{X}(x, y)$ is a regular reparametrization of the pseudosphere as described in Example 6.5.5.

(d) (ODE,*) Find parametric equations $(u(t), v(t))$ for the geodesic curves on H with this given metric.

8.5.2. Consider the Poincaré half-plane presented in Example 8.5.2. Calculate the Gaussian curvature function over H.

8.5.3. Consider the Poincaré half-plane presented in Example 8.5.2. Prove that the geodesic lines are either vertical lines $x = x_0$ or are half-circles with center on the x-axis, i.e., satisfying $(x - x_0)^2 + y^2 = C^2$.

8.5.4. Consider the Poincaré half-plane presented in Example 8.5.2. Let P and Q be two points on H. We determine the distance $d(P, Q)$ between these two points in the Poincaré metric on H (see Figure 8.17).

(a) If P and Q lie on a geodesic that is a half-circle, prove that the distance between them is

$$d(P, Q) = \left| \ln \frac{|PA|/|PB|}{|QA|/|QB|} \right|,$$

where A and B are points as shown in Figure 8.17 and $|PA|$ is the usual Euclidean distance between P and A and similarly for all the others.

(b) If P and Q lie on a geodesic that is a vertical line, prove that the distance between them is

$$d(P, Q) = \left| \ln \frac{|P_2 A_2|}{|Q_2 A_2|} \right|,$$

where A_2 is again as shown in Figure 8.17.

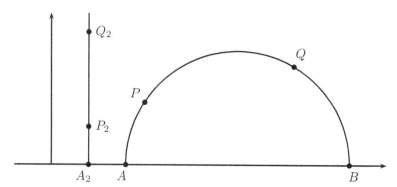

Figure 8.17. Geodesics in the Poincaré half-plane.

8.5.5. Consider the Poincaré half-plane presented in Example 8.5.2. We can consider $\mathbb{R}^2$ as the set of complex numbers $\mathbb{C}$. In this context, $H = \{z \in \mathbb{C} \,|\, \text{Im}(z) > 0\}$. Consider a fractional linear transformation of $\mathbb{C}$ of the form

$$w = \frac{az+b}{cz+d},$$

where $a, b, c, d \in \mathbb{R}$ and $ad - bc = 1$. Write $z = x + iy$ and $w = u + iv$.

(a) Prove that the function $w = f(z)$ sends H into H bijectively.

(b) Prove that the metric tensor for H is unchanged under this transformation, more precisely, that the metric coefficients of H in the w-coordinate are

$$(\bar{g}_{kl}) = \begin{pmatrix} v^{-2} & 0 \\ 0 & v^{-2} \end{pmatrix}.$$

We say that the Poincaré metric on the upper half-plane is invariant under the action of $\text{SL}(2, \mathbb{R})$.

Bibliography

[1] M. A. Armstrong. *Basic Topology*. Undergraduate Texts in Mathematics. New York: Springer-Verlag, 1983.

[2] Vladimir I. Arnold. *Ordinary Differential Equations*. Cambridge, MA: MIT Press, 1973.

[3] Josu Arroyo, Oscar J. Garay, and Jose J. Mencía. "When Is a Periodic Function the Curvature of a Closed Plane Curve." *American Mathematical Monthly* 115:5 (2008), 405–414.

[4] Pierre Ossian Bonnet. "La théorie générale des surfaces." *J. École Polytechnique* 19 (1848), 1–146.

[5] Pierre Ossian Bonnet. "Sur quelques propriétés des lignes géodésiques." *C. R. Acad. Sci. Paris* 40 (1855), 1311–1313.

[6] Roberto Bonola. *Non-Euclidean Geometry: A Critical and Historical Study of its Developments*. New York: Dover Publications, 1955.

[7] Andrew Browder. *Mathematical Analysis: An Introduction*. Undergraduate Texts in Mathematics. New York: Springer-Verlag, 1996.

[8] Richard Courant, Herbert Robbins, and Ian Stewart. *What Is Mathematics?*, Second edition. Oxford, UK: Oxford University Press, 1996.

[9] H. S. M. Coxeter. *Non-Euclidean Geometry*, Sixth edition. Washington, DC: Mathematical Association of America, 1998.

[10] Manfredo do Carmo. *Differential Geometry of Curves and Surfaces*. Upper Saddle River, NJ: Prentice Hall, 1976.

[11] William Dunham. *The Mathematical Universe*. New York: John Wiley and Sons, Inc., 1994.

[12] Grant R. Fowles. *Analytical Mechanics*, Fourth edition. Philadelphia: Saunders College Publishing, 1986.

[13] Joseph A. Gallian. *Contemporary Abstract Algebra*, Sixth edition. New York: Houghton Mifflin Company, 2006.

[14] Karl Friedrich Gauss. "Disquisitiones generales circa superficies curvas." *Comm. Soc. Göttingen* 6 (1823), 99–146.

[15] Michael C. Gemignani. *Elementary Topology*, Second edition. New York: Dover Publications, 1972.

[16] Reuben Hersh and Phillip J. Davis. *The Mathematical Experience.* Boston: Birkhäuser, 1981.

[17] Heinz Hopf. "Über die Drehung der Tangenten und Sehnen ebener Kurven." *Compositio Mathematica* 2 (1935), 50–62.

[18] R. A. Horn. "On Fenchel's Theorem." *American Mathematical Monthly* 78:4 (1971), 381–382.

[19] Serge Lang. *Algebra*, Third edition. Reading, MA: Addison-Wesley, 1993.

[20] H. Blaine Lawson. *Lectures on Minimal Submanifolds.* Rio de Janeiro: Monografias de Matemática, IMPA, 1973.

[21] W. B. Raymond Lickorish. *An Introduction to Knots*, Graduate Texts in Mathematics, 175. New York: Springer-Verlag, 1997.

[22] Stephen T. Lovett. *Differential Geometry of Manifolds with Applications to Physics.* Wellesley, MA: A K Peters, Ltd., 2010.

[23] Ib H. Masden and Jürgen Tornehave. *From Calculus to Cohomology.* Cambridge, UK: Cambridge University Press, 1997.

[24] James Munkres. *Topology.* Englewood Cliffs, NJ: Prentice Hall, 1975.

[25] Johannes C. C. Nitsche. *Lectures on Minimal Surfaces, Volume 1.* Cambridge, UK: Cambridge University Press, 1989.

[26] Robert Osserman. *A Survey of Minimal Surfaces.* New York: Dover Publications, 1969.

[27] Patrick J. Ryan. *Euclidean and Non-Euclidean Geometry.* Cambridge, UK: Cambridge University Press, 1986.

[28] James Stewart. *Calculus*, Sixth edition. Belmont, CA: Thomson Brooks/Cole, 2003.

[29] J. J. Stoker. *Differential Geometry.* New York: Wiley-Interscience, 1969.

[30] Dirk J. Struik. *Lectures on Classical Differential Geometry*, Second edition. New York: Dover Publications, 1961.

[31] E. R. van Kampen. "The theorems of Gauss-Bonnet and Stokes." *Am. Jour. Math.* 60:1 (1938), 129–138.

Index

acceleration
    centripetal, 22
    tangential, 22
algebraic geometry, viii
angle
    at a corner, 16
    on a regular surface, 149
angular velocity, 3, 22
arc length, 13, 148
area formula, 151
asymptote
    of Dupin indicatrix, 185
asymptotic curve, 185, 263
asymptotic direction, 185

bending invariants, 248
Bertrand mate, 77
bijection, 40
bitangent, 58
Bolyai, 315
Bonnet's Formula, 269

cardioid, 8, 26, 42
catenary, 34
catenoid, 188, 207
center of curvature, 31
central projection, 130
Christoffel symbols
    of the first kind, 240
    of the second kind, 240
circle, 2, 5, 62
Clairaut relation, 280
class C^r, 12

Classification Theorem of Surfaces, 307
Codazzi equations, 246
compact, 151, 311
compatibility condition
    of a system of PDEs, 256
concave set, 53
cone, 115, 248, 282
conic surface, 108
conics, vii
conjugate directions, 188, 243
contact, order of, 27, 28, 79
continuity
    of a vector function, 4
contraction, 220
contravariant
    vector, 217
convex set, 53
coordinate functions, 5, 62
coordinate lines, 112, 183
coordinate neighborhood, 122
coordinate patch, 111
coordinate system
    on a regular surface, 122
coordinates
    geodesic, 285
    geodesic polar, 291, 294
    normal, 293
    polar, 273
corner, 16, 298
covariant
    vector, 217